Study Guide

for Stewart's
CALCULUS
Concepts AND Contexts

MULTIVARIABLE

Robert Burton

Dennis Garity

Oregon State University

BROOKS/COLE PUBLISHING COMPANY

I(T)P® An International Thomson Publishing Company

Pacific Grove • Albany • Belmont • Boston • Cincinnati • Johannesburg • London
Madrid • Melbourne • Mexico City • New York • Scottsdale • Singapore • Tokyo • Toronto

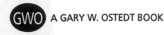 **A GARY W. OSTEDT BOOK**

Project Development Editor: *Beth Wilbur*
Marketing Team: *Caroline Croley and Christine Davis*
Editorial Associate: *Carol Benedict*
Production Editor: *Mary Vezilich*

Cover Design: *Vernon T. Boes*
Cover Photo: *Ian Sabell, Linear Photographs Ltd.*
   Violin created by David Bruce Johnson.
Printing and Binding: *Webcom Ltd.*

COPYRIGHT© 1999 by Brooks/Cole Publishing Company
A division of International Thomson Publishing Inc.
I(T)P The ITP logo is a registered trademark used herein under license.

For more information, contact:

BROOKS/COLE PUBLISHING COMPANY
511 Forest Lodge Road
Pacific Grove, CA 93950
USA

International Thomson Editores
Seneca 53
Col. Polanco
11560 México, D. F., México

International Thomson Publishing Europe
Berkshire House 168-173
High Holborn
London WC1V 7AA
England

International Thomson Publishing Japan
Hirakawacho Kyowa Building, 3F
2-2-1 Hirakawacho
Chiyoda-ku, Tokyo 102
Japan

Thomas Nelson Australia
102 Dodds Street
South Melbourne, 3205
Victoria, Australia

International Thomson Publishing Asia
60 Albert Street
#15-01 Albert Complex
Singapore 189969

Nelson Canada
1120 Birchmount Road
Scarborough, Ontario
Canada M1K 5G4

International Thomson Publishing GmbH
Königswinterer Strasse 418
53227 Bonn
Germany

All rights reserved. No part of this work may be reproduced, stored in a retrieval system, or transcribed, in any form or by any means—electronic, mechanical, photocopying, recording, or otherwise—without the prior written permission of the publisher, Brooks/Cole Publishing Company, Pacific Grove, California 93950.

Printed in Canada

10   9   8   7   6   5   4   3   2   1

ISBN 0-534-34440-2

For my mother and father

For Vicki

                                        RMB

For my mother and father – for their love and encouragement,

and for Marie, Diana, and Dylan – for their love, patience, and understanding

                                        DKG

Preface

Note to Students — How to use this Study Guide

This *Study Guide* is designed to be used in a number of different ways:

- Before you attend a class that covers material from the text, read the first few paragraphs in the corresponding part of the *Study Guide* to become familiar with the key concepts and skills that you will need to master. Watch for where these concepts and skills appear during class.

- After you attend a class that covers material from the text, read the corresponding part of the *Study Guide* to get more explanation of concepts and skills in that section.

- If you get stuck while working on a problem in the text, find the corresponding SkillMaster that is needed for the problem. Read the explanation and try the problems in the *Study Guide* corresponding to that SkillMaster.

- Work through the Worked Examples to get extra practice on skills that you feel you need to spend more time on. Try each of the Worked Examples on your own before reading the solutions.

- When you are preparing for a test, use the master list of SkillMasters at the end of each chapter to find skills that you need to review. Use this list to get more practice in skills that need reinforcing.

Organization of Study Guide:

There is a section in the *Study Guide* corresponding to each section in the text. Each section contains most of the following:

- A brief introduction to the ideas in the section,
- A short list of key concepts,
- A short list of skills to master, called SkillMasters,
- A more detailed explanation of the concepts and skills, and
- Worked Examples for each of the SkillMasters.

The Worked Examples are a key feature of the *Study Guide*. They are presented in a three column format. The problems are listed in the left column. Hints are often given in the middle column. The right column provides details of the solution. You should always try to work the problems on your own before looking at the hints or solutions.

At the end of each chapter, a complete list of the SkillMasters for that chapter is provided.

Along with the material in each section of the *Study Guide* are links in the margin to earlier and later material in the text. The words that a link corresponds to are italicized, and the link is provided in the margin with the following graphic.

Acknowledgments

The authors are grateful to Chris Bryant, Marie Franzosa, and Kathy Smith for proofreading, for checking mathematical correctness, and for helpful suggestions of examples. The authors also thank James Stewart for informal advice and much useful discussion. We are especially grateful for the editorial assistance, support and patience of Beth Wilbur, Carol Benedict and Gary Ostedt. Finally, the support and encouragement of the Oregon State University Mathematics Department and its Chair, John Lee is gratefully acknowledged.

Comments

Please send any comments on this *Study Guide* by e-mail to the following address:

studygd@math.orst.edu

Contents

Chapter 10 - Vector Functions   527

Chapter 11 - Partial Derivatives   567

Chapter 12 - Multiple Integrals   662

Chapter 13 - Vector Calculus   742

Chapter 8 - Infinite Sequences and Series

$$\sum_{n=1}^{\infty} a \cdot r^{n-1} = \frac{a}{1-r}$$

Section 8.1 – Sequences

Key Concepts:

- The limit of a sequence
- Limit Laws
- The Monotonic Sequence Theorem

Skills to Master:

- Find a defining equation for a sequence.
- Use the laws of limits together with known examples to determine if a sequence is divergent or convergent, and if convergent, to find the limit.
- Determine if a sequence is monotonic and use the Monotonic Convergence Theorem to show some sequences are convergent.

Discussion:

page 101.

This section introduces the concepts of sequences and limits of sequences. You should review the material on the *limit of a function* from earlier in the text. Sequences arise in many situations. For example, if you let P_n be the number bacteria in a culture dish n hours after the start of an experiment, then the numbers $\{P_1, P_2, P_3, \ldots\}$ form a sequence.

Key Concept: The limit of a sequence

A sequence $\{a_n\}$ has limit L, or *converges*, if you can make the terms a_n as close to L as you like by taking n sufficiently large. If the sequence $\{a_n\}$ has limit L you write

$$\lim_{n\to\infty} a_n = L \qquad \text{or} \qquad a_n \to L \text{ as } n \to \infty.$$

414

page 562.

The sequence *diverges* if it does not have limit. The meaning of diverging to ∞ or to $-\infty$ is similar to the corresponding concepts for functions. Pay careful attention to *Theorem 2* in this section relating the limit of a sequence to the limit of a function. Note that the values that n takes on when $\{a_n\}$ is a sequence are positive integer values.

Key Concept: Limit Laws

page 111.

The limit laws in this section should look familiar to you. Compare them with the limit laws for functions in Section 2.3. Just as before,

The limit of a $\left\{\begin{array}{l}\text{sum} \\ \text{difference} \\ \text{constant multiple} \\ \text{product} \\ \text{quotient}\end{array}\right\}$ is the $\left\{\begin{array}{l}\text{sum} \\ \text{difference} \\ \text{constant multiple} \\ \text{product} \\ \text{quotient}\end{array}\right\}$ of the limit(s).

Again, for these laws to hold, the individual limits must exist and for the last law to hold, the limit of the denominator must not be 0. There is also a Squeeze Theorem for sequences. Another important result is Theorem 4:

$$\text{If } \lim_{n\to\infty} |a_n| = 0 \text{ then } \lim_{n\to\infty} a_n = 0$$

Key Concept: The Monotonic Sequence Theorem

You should make sure that you understand the meaning of the terms relating to the Monotonic Sequence Theorem. The terms increasing, decreasing, and monotonic have essentially the same meaning for sequences that they had for functions. The meaning of bounded above, bounded below and bounded as they apply to sequences should be clear. The Monotonic Sequence theorem states that every bounded monotonic sequence converges.

SkillMaster 8.1: Find a defining equation for a sequence.

To find a defining equation for the sequence, examine the terms that you are given to see if you can find a pattern. If the terms involve fractions, try to find a pattern for the numerator and a pattern for the denominator. If the terms alternately are positive and negative, include a factor of $(-1)^n$ or $(-1)^{n+1}$ in your equation.

SkillMaster 8.2: Use the laws of limits together with known examples to determine if a sequence is divergent or convergent, and if convergent, to find the limit.

If $r > 0$, then $\lim\limits_{n \to \infty} \dfrac{1}{r^n} = 0$. The sequence $\{r^n\}$ converges to 0 if $-1 < r < 1$, converges to 1 if $r = 1$, and diverges for all other values of r. These basic sequences, together with the laws of limits, should allow you to compute many limits.

SkillMaster 8.3: Determine if a sequence is monotone and use the Monotone Convergence Theorem to show some sequences are convergent.

pages 565-566.

page 88.

Make sure that you understand the techniques used in *Examples 9, 10* and *11* in this section. They show you how to determine if a sequence is monotonic and bounded. Review the technique of mathematical induction if you need to.

Worked Examples

For each of the following examples, first try to find the solution without looking at the middle or right columns. Cover the middle and right columns with a piece of paper. If you need a hint, uncover the middle column. If you need to see the worked solution, uncover the right column.

Example	Tip	Solution
SkillMaster 8.1.		
• Find a formula for the general term a_n of the sequence below, assuming the pattern of the first few terms continues. $$\left\{\frac{1}{1}, \frac{4}{2}, \frac{9}{4}, \frac{16}{8}, \cdots\right\}$$	The pattern of the numerators is that of squares of positive integers, the pattern of the denominators is that of powers of 2.	$$a_n = \frac{n^2}{2^{n-1}}$$
• Write out the first few terms of the sequence $\{a_n\}$ where $$a_n = \frac{(-1)^n \ln(n^2 + 1)}{3^n}$$	Substitute 1 for n in the definition of a_n to find the first term, then substitute 2 for n in the definition of a_n to find the second term, and so on.	$$\left\{\frac{-\ln(2)}{3}, \frac{\ln(5)}{9}, \frac{-\ln(10)}{27}, \frac{\ln(17)}{81}, \cdots\right\}$$
SkillMaster 8.2.		
• Determine if the sequence diverges or converges. If the sequence converges, find the limit. $$a_n = \ln(\sqrt{n} + 1)$$	Recall properties of the logarithm function.	The sequence is unbounded and diverges toward infinity.

- Determine if the sequence diverges or converges. If the sequence converges, find the limit.

$$a_n = \frac{(-1)^n n^2}{n^3 + 1}$$

Divide the numerator and denominator by n^3 and apply Theorem 4.

$$a_n = \frac{(-1)^n n^2}{n^3 + 1}$$

$$|a_n| = \frac{n^2}{n^3 + 1}$$

$$= \frac{1/n}{1 + 1/n^3}$$

$$\lim_{n \to \infty} |a_n| = \lim_{n \to \infty} \frac{1/n}{1 + 1/n^3}$$

$$= \frac{0}{1 + 0}$$

$$= 0$$

$$\lim_{n \to \infty} |a_n| = 0, \text{ so}$$

$$\lim_{n \to \infty} a_n = 0.$$

- Determine if the sequence diverges or converges. If the sequence converges, find the limit.

$$a_n = \frac{\ln(e^n + 1)}{n}$$

The answer is not obvious. Both the numerator and denominator diverge to ∞. This suggests we consider the limit of the function below and use L'Hospital's Rule.

$$\frac{\ln(e^x + 1)}{x}$$

$$\lim_{x \to \infty} \frac{\ln(e^x + 1)}{x}$$

$$= \lim_{x \to \infty} \frac{\frac{d}{dx}(\ln(e^x + 1))}{\frac{d}{dx}(x)}$$

$$= \lim_{x \to \infty} \frac{e^x/(e^x + 1)}{1}$$

$$= \lim_{x \to \infty} \frac{e^x}{(e^x + 1)(1)}$$

$$= \lim_{x \to \infty} \frac{1}{1 + e^{-x}}$$

$$= \frac{1}{1 + 0}$$

$$= 1$$

Thus

$$\lim_{n \to \infty} a_n$$

$$= \lim_{n \to \infty} \frac{\ln(e^n + 1)}{n}$$

$$= 1$$

SkillMaster 8.3.

• Determine whether the sequence is monotonic.

$$a_n = \frac{2n}{n+1}$$

The first few terms are
$$\{1, 4/3, 6/4, 8/5, ...\}$$
$$=$$
$$\{1, 1.3\bar{3}, 1.5, 1.6, ...\}$$
so if the sequence is monotonic it must be increasing. Check whether this is true or not.

To show the sequence is increasing we must show
$$a_{n+1} > a_n$$
or

$$\frac{2(n+1)}{(n+1)+1} > \frac{2n}{n+1}$$
$$\Longleftrightarrow \quad \frac{2n+2}{n+2} > \frac{2n}{n+1}$$
$$\Longleftrightarrow \quad (2n+2)(n+1) > (2n)(n+2)$$
$$\Longleftrightarrow \quad 2n^2 + 4n + 2 > 2n^2 + 4n$$
$$\Longleftrightarrow \quad 2 > 0.$$

The last inequality is true so the preceding inequalities must also be true and the sequence is increasing, and hence monotonic.

• Show the following sequence is increasing and bounded by 3. Conclude that the sequence converges and find its limit.

$$a_1 = \sqrt[3]{6}$$
$$a_{n+1} = \sqrt[3]{6 + a_n}$$

Notice that
$$a_{n+1} = \sqrt[3]{6 + \sqrt[3]{6 + \ldots + \sqrt[3]{6}}}$$
and
$$a_n = \sqrt[3]{6 + \sqrt[3]{6 + \ldots + \sqrt[3]{0}}}$$
where there are $n + 1$ nested radicals. Since each of the radicals are increasing functions it is clear that

$$a_{n+1} > a_n$$
so the sequence is increasing.

The sequence is increasing by the comments in the tip. As the sequence is increasing, it is bounded below by the first term. To show that it is bounded above by 3, use induction.

$$a_1 = \sqrt[3]{6} < 3 \text{ since}$$
$$6 < 3^3 = 27.$$

Suppose, by induction

$$a_n < 3.$$

Then we must show

$$a_{n+1} = \sqrt[3]{6 + a_n} < 3, \text{ or}$$
$$6 + a_n < 3^3,$$
$$\text{i.e. } a_n < 21.$$

This is true since we know
$$a_n < 3 < 21.$$

The sequence is bounded so the Monotonic Sequence Theorem applies and there is a limit L.

$$L = \lim_{n \to \infty} a_{n+1}$$
$$= \lim_{n \to \infty} \sqrt[3]{6 + a_n}$$
$$= \sqrt[3]{6 + L}$$

So $L = \sqrt[3]{6 + L}$
$$L^3 = 6 + L$$
$$0 = L^3 - L - 6.$$

Inspection shows that $L = 2$ is a root of this equation and that this must be the limit. (The derivative of $L^3 - L - 6$ is positive for $L > 1$ so $L^3 - L - 6$ can't have more than one root ≥ 1. Recall the limit is trapped between 1 and 3 because the sequence is.)

Section 8.2 – Series

Key Concepts:

- The sum of a series
- The sum of a geometric series
- The laws for series

Skills to Master:

- Determine if a geometric series is convergent, and if convergent, find the sum of the series.

- Use the laws of series together with known examples to determine if a series is divergent or convergent, and if convergent, find the sum.

Discussion:

This section introduces infinite series. You can think of a series as an infinite sum, or as the result of adding the terms of a sequence. To give a precise meaning to the sum of a series, the concept of partial sums is introduced.

Key Concept: The sum of a series

An infinite series is an expression of the form

$$a_1 + a_2 + a_3 + \cdots \text{ or } \sum_{n=1}^{\infty} a_n \, .$$

The nth partial sum of a series is the finite sum

$$s_n = a_1 + a_2 + a_3 + \cdots + a_n \, .$$

If the sequence of partial sums $\{s_n\}$ of a series converges to a number s we say

that the series converges and write

$$\sum_{n=1}^{\infty} a_n = s.$$

Make sure that you don't confuse the sequence of partial sums of a series with the sequence of terms in the series.

Key Concept: The sum of a geometric series

A geometric series is a series of the form

$$\sum_{n=1}^{\infty} ar^{n-1} \quad a \neq 0.$$

Note that each term in the series is obtained from the previous one by multiplying by r. Such a series is convergent if $|r| < 1$ in which case it converges to

$$\frac{a}{1-r}.$$

If $|r| \geq 1$, the geometric series diverges.

Key Concept: Test for Divergence and the laws for series

The Test for Divergence states that

if $\lim_{n \to \infty} a_n$ does not exist, or if $\lim_{n \to \infty} a_n \neq 0$, then $\sum_{n=1}^{\infty} a_n$ is divergent.

page 575.

The laws for series are given in *Theorem 8*.

SkillMaster 8.4: Determine if a geometric series is
convergent, and if convergent, find the sum of the series.

To determine whether a given series is geometric, divide the $(n + 1)^{st}$ term by the nth and see if you get a constant r, no matter what n is. If you always get the same constant, the series is geometric and r is the ratio associated with the series. You may need to do some algebraic manipulation to get the series in the form

$$\sum_{n=1}^{\infty} ar^{n-1}.$$

Once you have the series in this form, the limit is

$$\frac{a}{1 - r}$$

if $|r| < 1$, and the series diverges if $|r| \geq 1$.

SkillMaster 8.5: Use the laws for series together with known examples to determine if a series is divergent or convergent, and if convergent, find the sum.

If a series is geometric, it is easy to determine convergence or divergence. If the terms of a series do not approach 0, then the series diverges. These facts, together with the laws for series allow you to determine convergence or divergence of other kinds of series. More techniques for determining convergence or divergence will be developed in the next few sections.

Worked Examples

For each of the following examples, first try to find the solution without looking at the middle or right columns. Cover the middle and right columns with a piece of paper. If you need a hint, uncover the middle column. If you need to see the worked solution, uncover the right column.

Example	Tip	Solution
SkillMaster 8.4.		
• Determine if the following geometric series converges and if so find the sum. $$\sum_{n=1}^{\infty} \frac{(-1)^n 2^{2n}}{3^n}$$	Here $a_n = (-1)^n 2^{2n}/3^n$. Rewrite 2^{2n} as 4^n.	$$a_n = (-1)^n 2^{2n}/3^n$$ $$= (-4/3)^n$$ Thus $r = -4/3$. Since $|r| > 1$ the series is divergent.
• Determine if the following geometric series converges and if so, find the sum. $$\sum_{n=1}^{\infty} \frac{2^{n+2}}{\pi^{n-2}}$$	First factor the terms so that the general term appears with a power of $n-1$.	$$\sum_{n=1}^{\infty} \frac{2^{n+2}}{\pi^{n-2}}$$ $$= \frac{2^3}{\pi^{-1}} \sum_{n=1}^{\infty} \frac{2^{n-1}}{\pi^{n-1}}$$ $$= 8\pi \sum_{n=1}^{\infty} \left(\frac{2}{\pi}\right)^{n-1}$$ So $a = 8\pi$ and $r = 2/\pi$. The series is convergent because $|r| < 1$ and the sum is $$\frac{a}{1-r}$$ $$= \frac{8\pi}{1 - 2/\pi}$$ $$= \frac{8\pi^2}{\pi - 2}$$

• Express the number

$$0.135135\ldots$$

as a ratio of integers.

Express the number as a geometric series and determine its sum.

$$0.135135\ldots$$
$$= \sum_{n=1}^{\infty} \frac{135}{(1000)^n}$$

$$0.135135\ldots$$
$$= \frac{135}{1000} \sum_{n=1}^{\infty} \frac{1}{(1000)^{n-1}}$$
$$= \frac{135}{1000} \left(\frac{1}{1 - 1/1000} \right)$$
$$= \frac{135(1000)}{1000(999)}$$
$$= \frac{135}{999}$$
$$= \frac{5}{37}$$

SkillMaster 8.5.

• Determine whether the series is convergent or divergent. If it is convergent, find its sum.

$$\sum_{n=1}^{\infty} [4(0.2)^{n+1} - (0.1)^n]$$

Use the laws for series to express this series as the difference of two geometric series.

$$\sum_{n=1}^{\infty} [4(0.2)^{n+1} - (0.1)^n]$$

$$= 4(0.2)^2 \left[\sum_{n=1}^{\infty} (0.2)^{n-1} \right]$$

$$- (0.1) \left[\sum_{n=1}^{\infty} (0.1)^{n-1} \right]$$

$$= \frac{4(0.2)^2}{1 - 0.2} - \frac{0.1}{1 - 0.1}$$

$$= \frac{0.16}{0.8} - \frac{0.1}{0.9}$$

$$= \frac{1}{5} - \frac{1}{9}$$

$$= 4/45$$

• Determine whether the series is convergent or divergent. If it is convergent, find its sum.

$$\sum_{n=1}^{\infty} \frac{2n}{n+1}$$

Check to see if $a_n = (2n)/(n+1)$ converges to 0. If this limit is not 0 then the series diverges.

$$\lim_{n \to \infty} a_n$$
$$= \lim_{n \to \infty} \frac{2n}{n+1}$$
$$= \lim_{n \to \infty} \frac{2}{1 + 1/n}$$
$$= 2 \neq 0$$

The series diverges.

- Determine whether the series is convergent or divergent. If it is convergent find its sum.

$$\sum_{n=1}^{\infty} [\tan(1/n) - \tan(1/(n+1))]$$

Notice that this series has a "telescoping" appearance as in Example 6 page 573 of the text. Use the definition of the sum of a series to determine if the series has a sum.

$$\sum_{n=1}^{\infty} [\tan(1/n) - \tan(1/(n+1))]$$

$$= \lim_{N \to \infty} \sum_{n=1}^{N} [\tan(1/n) - \tan(1/(n+1))]$$

$$= \lim_{N \to \infty} [\tan(1/1) - \tan(1/N)]$$

$$= \tan(1) - \lim_{N \to \infty} \tan(1/N)$$

$$= \tan(1) - \tan(0)$$

$$= \tan(1)$$

$$\approx 1.5574$$

Section 8.3 – The Integral and Comparison Tests; Estimating Sums.

Key Concepts:

- Integral Test
- Comparison Test
- Series that are p-series
- Remainder Estimate for Integral Test

Skills to Master:

- Use the Integral Test to determine whether a series with positive decreasing terms is convergent.
- Use the Comparison Test to determine whether a series is convergent or divergent.
- Estimate the error in approximating the sum of a series.

Discussion:

This section continues the development of series and gives some new methods for determining the convergence or divergence of a series. For certain kinds of series, a technique to estimate the sum of the series is introduced.

Key Concept: Integral Test

The Integral Test has a number of conditions that need to be satisfied by a series before it can be applied. Make sure that you check all the conditions. The Integral Test is stated as follows.

427

Integral Test: Suppose that f is continuous, positive and decreasing on $[1, \infty)$ and that $a_n = f(n)$.
Then the series

$$\sum_{n=1}^{\infty} a_n$$

is convergent if and only if the improper integral

$$\int_1^{\infty} f(x) \, dx$$

is convergent.

page 427.

Review the section on *improper integrals* if you need to.

Key Concept: Comparison Test

The Comparison Test can only be applied to series with positive terms. This test is stated as follows.

Comparison Test:
Suppose that $\sum a_n$ and $\sum b_n$ are series with positive terms.
(a) If $\sum b_n$ is convergent and $a_n \leq b_n$ for all n, then $\sum a_n$ is also convergent.
(b)If $\sum b_n$ is divergent and $a_n \geq b_n$ for all n, then $\sum a_n$ is also divergent.

Note that you can only conclude convergence using this test by comparing with a series with larger terms that converges, and you can only conclude divergence with this test by comparing with a series with smaller terms that diverges.

Key Concept: Series that are p-series

A series of the form

$$\sum_{n=1}^{\infty} \frac{1}{n^p}$$

is called a *p*-series. Such a series converges if $p > 1$ and diverges if $p \leq 1$.

Key Concept: Remainder Estimate for Integral Test

If a series

$$\sum_{n=1}^{\infty} a_n$$

converges to s by the Integral Test and if $R_n = s - s_n$, then

$$\int_{n+1}^{\infty} f(x)\, dx \leq R_n \leq \int_{n}^{\infty} f(x)\, dx.$$

page 585.

Figures 3 and *4* in this section give the geometric reasoning behind this error estimate. Remember to check that the Integral Test applies before using this error estimate.

SkillMaster 8.6: Use the Integral Test to determine if a series with positive decreasing terms is convergent.

Before applying the integral test to a series $\sum_{n=1}^{\infty} a_n$, make sure that you check that the terms of the series you are considering are positive and decreasing. Next, find a function $f(x)$ so that $f(n) = a_n$. Finally, determine whether the improper integral

$$\int_{1}^{\infty} f(x)\, dx$$

converges or diverges. This tells you whether the original series converges or diverges.

SkillMaster 8.7: Use the Comparison Test to determine whether a series is convergent or divergent.

Given a series $\sum_{n=1}^{\infty} a_n$, with positive terms, you should first determine whether it looks like a series that you already know either converges or diverges. If you think the series converges, try to compare it with a series with larger positive terms that converges. If you think that the series diverges, try to compare it with a series with smaller terms that diverges. You may need to multiply or divide by positive constants to get the inequalities to work out so that the Comparison Test can be correctly applied.

SkillMaster 8.8: Estimate the error in approximating the sum of a series.

If a series converges by the Integral Test, you will often want to estimate the sum of the series. For example, if you want to estimate the sum s of the series $\sum_{n=1}^{\infty} a_n$ to within .0001, you want to find an m so that $R_m < .0001$. You know that

$$R_m \leq \int_m^{\infty} f(x) \, dx,$$

so if you can find an m so that

$$\int_m^{\infty} f(x) \, dx < .0001,$$

you will have found an m that works. To obtain the actual estimate, compute

$$\sum_{n=1}^{m} a_n.$$

Worked Examples

For each of the following examples, first try to find the solution without looking at the middle or right columns. Cover the middle and right columns with a piece of paper. If you need a hint, uncover the middle column. If you need to see the worked solution, uncover the right column.

Example	Tip	Solution

SkillMaster 8.6.

Example	Tip	Solution
• Determine if the series is convergent or divergent. $$\sum_{n=1}^{\infty} ne^{-n}$$	Try the Integral Test. The terms are positive. Use the first derivative test to show that they are decreasing and then the Integral Test to determine convergence.	Let $f(x) = xe^{-x}$; note that f is continuous and positive on $[1, \infty)$. We wish to know that f is decreasing for $x > 1$. $$\begin{aligned} f'(x) &= e^{-x} - xe^{-x} \\ &= (1-x)e^{-x}. \end{aligned}$$ The exponential $e^{-x} > 0$ for all x. Also $(1-x) < 0$ for $x > 1$ so $$f'(x) < 0 \text{ for } x > 1.$$ Thus f is decreasing for $x > 1$ and the sequence is decreasing. The Integral Test applies. The series diverges or converges together with the divergence or convergence of the integral $\int_1^{\infty} xe^{-x}dx$. Use integration by parts. $$\begin{aligned} u &= x & dv &= e^{-x} \\ du &= dx & v &= -e^{-x} \end{aligned}$$ $$\int_1^{\infty} xe^{-x}dx$$ $$= \lim_{N \to \infty} -xe^{-x}\big]_1^N + \int_1^{\infty} e^{-x}dx$$ $$= 0 + e^{-1} - \lim_{N \to \infty} e^{-x}\big]_1^N$$ $$= 2e^{-1} < \infty$$ The integral converges so the series is also convergent.

• Determine if the series is convergent or divergent.

$$\sum_{n=2}^{\infty} \frac{1}{n(\ln(n))^2}$$

The terms are positive and decreasing. Try the Integral Test.

The series diverges or converges according to the divergence or convergence of the integral.

$$\int_2^{\infty} \frac{1}{x(\ln(x))^2} dx$$

Use a substitution.

$$
\begin{aligned}
u &= \ln(x) \\
du &= (1/x)dx \\
x &= 2, u = \ln(2) \\
x &= \infty, u = \infty
\end{aligned}
$$

$$
\int_2^{\infty} \frac{1}{x(\ln(x))^2} dx
$$
$$
= \int_{\ln(2)}^{\infty} \frac{1}{u^2} du
$$
$$
\lim_{N \to \infty} \left. \frac{-1}{u} \right]_{\ln(2)}^{N}
$$
$$
= 1/\ln(2)
$$

Since the integral is convergent, the series is convergent also.

SkillMaster 8.7.

• Determine if the series is convergent or divergent.

$$\sum_{n=1}^{\infty} \frac{2}{3n + \pi}$$

This series looks very much like the harmonic series $\sum(1/n)$. Use the comparison test by comparing the series with a multiple of the harmonic series. Start by observing that
$$\frac{2}{3n + \pi} > \frac{2}{3n + \pi n}.$$

$$\frac{2}{3n + \pi} > \frac{2}{3n + \pi n}$$
$$= 2\frac{1}{(3 + \pi)n}$$
$$= \left(\frac{2}{3 + \pi}\right)\frac{1}{n}.$$

The comparison test shows that the series diverges because it is larger, term by term, than the divergent series

$$\sum_{n=1}^{\infty} \left(\frac{2}{3 + \pi}\right)(1/n).$$

432

• Determine if the series is convergent or divergent.

$$\sum_{n=1}^{\infty} \frac{\cos^2(n)}{n^{1.1}}$$

Use the Comparison Test. Notice that
$$0 \leq \cos^2(n) \leq 1.$$

$$\sum_{n=1}^{\infty} \frac{\cos^2(n)}{n^{1.1}}$$

$$\leq \sum_{n=1}^{\infty} \frac{1}{n^{1.1}}$$

The last series converges since it is p−series with $p > 1$. The original series must also converge.

SkillMaster 8.8.

• We want to approximate the sum

$$\sum_{n=1}^{\infty} \frac{1}{n^5}$$

by summing the first several terms but we need to have an error less than 0.001. How many terms must we take and what is the approximation?

First note that the series is a p−series with $p = 5 > 1$ so the series converges. Compute the remainder term R_n and determine the value of n that will be make this less than 0.001.

If we sum the first n terms the error (the remainder term R_n) is

$$R_n \leq \int_n^{\infty} \frac{1}{x^5} dx$$

$$= \lim_{N\to\infty} \left. (\frac{-1}{4}) x^{-4} \right]_n^N$$

$$= 0 - (-1/4)n^{-4}$$

$$= n^{-4}/4.$$

We need this to be less than 0.001.

$$n^{-4}/4 < 0.001$$
$$\Longleftrightarrow \quad n^{-4} < 0.004$$
$$\Longleftrightarrow \quad 1 < 0.004n^4$$
$$\Longleftrightarrow \quad 250 < n^4$$
$$\Longleftrightarrow \quad \sqrt[4]{250} < n$$
$$\Longleftrightarrow \quad n > 3.98$$

It is sufficient to take $n = 4$ terms.

$$\sum_{n=1}^{4} \frac{1}{n^5}$$

$$= \frac{1}{1} + \frac{1}{2^5} + \frac{1}{3^5} + \frac{1}{4^5}$$
$$\approx 1.0363.$$

Section 8.4 – Other Convergence Tests.

Key Concepts:

- Alternating Series Test
- Alternating Series Estimation Theorem
- Absolute Convergence
- Ratio Test

Skills to Master:

- Use the Alternating Series Test to determine if an alternating series is convergent.
- Use the Alternating Series Estimation Theorem to estimate the error in approximating the sum of an alternating series by a finite sum.
- Determine whether a series is absolutely convergent.
- Use the ratio test to determine whether a series is absolutely convergent.

Discussion:

This section gives a few additional tests for deciding the convergence or divergence of certain series. For alternating series, an estimate is developed for the error in approximating the actual sum of the series by the first n terms.

Key Concept: Alternating Series Test

A series is alternating if the terms are alternately positive and negative. The Alternating Series Test is stated as follows.

Alternating Series Test:
If the alternating series

$$\sum_{n=1}^{\infty}(-1)^{n-1}b_n \;=\; b_1 - b_2 + b_3 - b_4 + \cdots$$

$$\text{where } b_n \;>\; 0$$

satisfies

(a) $b_{n+1} \le b_n$ for all n, and

(b) $\displaystyle\lim_{n\to\infty} b_n = 0$,

then the series is convergent.

Note that you just need to check that the terms alternate, decrease in absolute value, and go to 0.

Key Concept: Alternating Series Estimation Theorem

If $s = \sum_{n=1}^{\infty}(-1)^{n-1}b_n$ is the sum of an alternating series that satisfies the conditions of the Alternating Series Test, and if $R_n - s - s_n$ is the error made in stopping after n terms, then

$$|R_n| = |s - s_n| \le b_{n+1}.$$

Note that this says that the error is at most the magnitude of the first term not used.

Key Concept: Absolute Convergence

A series $\sum_{n=1}^{\infty} a_n$ converges absolutely if $\sum_{n=1}^{\infty} |a_n|$ converges. If a series is absolutely convergent, it converges. Note that a series may converge without converging absolutely. For example, the series

$$\sum_{n=1}^{\infty}(-1)^{n-1}\frac{1}{n}$$

converges by the Alternating Series Test, but the series

$$\sum_{n=1}^{\infty} \left| (-1)^{n-1} \frac{1}{n} \right| = \sum_{n=1}^{\infty} \frac{1}{n}$$

diverges by the Integral Test.

Key Concept: Ratio Test

The Ratio Test is a strong test for determining absolute convergence. It is stated as follows.

Ratio Test:
(a) If $\lim\limits_{n\to\infty} \left| \frac{a_{n+1}}{a_n} \right| = L < 1$, then the series $\sum_{n=1}^{\infty} a_n$
converges absolutely.

(b) If $\lim\limits_{n\to\infty} \left| \frac{a_{n+1}}{a_n} \right| = L > 1$, or $\lim\limits_{n\to\infty} \left| \frac{a_{n+1}}{a_n} \right| = \infty$ then the series $\sum_{n=1}^{\infty} a_n$
is divergent.

Make sure that you understand the Ratio Test gives no information if the limit involved is equal to 1.

SkillMaster 8.9: Use the Alternating Series Test to determine if an alternating series is convergent.

In using the Alternating Series Test, make sure that you check all the needed conditions. The terms must alternate, decrease in magnitude, and approach 0.

SkillMaster 8.10: Use the Alternating Series Estimation
Theorem to estimate the error in approximating the sum of an alternating series by a finite sum.

Once you have determined that a series converges by using the Alternating Series Test, it is easy to approximate the sum and to estimate the error involved in the approximation. The error is no bigger than the magnitude of the first term not used.

SkillMaster 8.11: Determine whether a series is absolutely convergent.

To determine whether a series is absolutely convergent, form a new series by taking the absolute value of each term. Then use a test that you already know to determine whether the new series converges. If it does, the original series is absolutely convergent.

SkillMaster 8.12: Use the ratio test to determine whether a series is absolutely convergent.

The Ratio Test is a good way to determine whether or not a series is absolutely convergent. The test depends on taking the ratio of the absolute values of successive terms and seeing if these ratios converge to a number less than 1.

Worked Examples

For each of the following examples, first try to find the solution without looking at the middle or right columns. Cover the middle and right columns with a piece of paper. If you need a hint, uncover the middle column. If you need to see the worked solution, uncover the right column.

Example	Tip	Solution

SkillMaster 8.9.

• Determine if the series is convergent or divergent. $$\sum_{n=1}^{\infty} \frac{(-1)^n n}{\sqrt{n^3+1}}$$	The series is alternating so try the Alternating Series Test. This test will apply if the absolute values of the terms are decreasing.	We need to check if the absolute values of the terms are decreasing.

$$\frac{n+1}{\sqrt{(n+1)^3+1}} < \frac{n}{\sqrt{n^3+1}}$$

$$\Longleftrightarrow \quad \frac{(n+1)^2}{(n+1)^3+1} < \frac{n^2}{n^3+1}$$

$$\Longleftrightarrow \quad \frac{(n+1)^3+1}{(n+1)^2} > \frac{n^3+1}{n^2}$$

$$\Longleftrightarrow \quad n+1+\frac{1}{(n+1)^2} > n+\frac{1}{n^2}$$

$$\Longleftrightarrow \quad 1 > \frac{1}{n^2} - \frac{1}{(n+1)^2}$$

The last inequality must be true because 1 is always larger than the difference of two unequal numbers between 0 and 1. Now we must check that the terms are converging to 0.

$$0 \le \lim_{n \to \infty} \frac{n}{\sqrt{n^3+1}}$$

$$\le \quad \lim_{n \to \infty} \frac{n}{\sqrt{n^3}} = \lim_{n \to \infty} \frac{1}{\sqrt{n}} = 0$$

The terms approach 0 by the Squeeze Theorem. The Alternating Series Test applies and the series is convergent.

SkillMaster 8.10.

• Show that the Alternating Series Test applies and show the following series is convergent. How many terms must we take for the partial sum to approximate the sum of the series with an error less than 0.001?

$$\sum_{n=1}^{\infty} \frac{(-1)^n}{\ln(n + 1)}$$

To check that the conditions of the Alternative Series Test are satisfies we must check that the terms alternate, that the absolute value of the terms are decreasing and that the terms converge to 0.

The terms clearly alternate. The absolute value of the nth term is $1/\ln(n + 1)$ which is decreasing because the reciprocals, $\ln(n + 1)$ are increasing. The terms converge to 0 because the absolute values of the reciprocals, $\ln(n + 1)$ are diverging toward infinity. The Alternating Series Test applies and the series is convergent.

The error in estimating the nth partial series is less than the absolute value of the $(n + 1)^{st}$ term.

$$|R_n| \leq \frac{1}{\ln((n + 1) + 1)}$$

We need this estimate to be less than 0.001.

$$\frac{1}{\ln(n + 2)} < 0.001$$
$$\iff 1/(0.001) < \ln(n + 2)$$
$$\iff 1000 < \ln(n + 2)$$
$$\iff e^{1000} - 2 < n$$

This number is approximately 2×10^{434} i.e. 2 followed by 434 0's. This is not a practical way to estimate the sum.

SkillMaster 8.11.

• Determine whether the series is absolutely convergent.

$$\sum_{n=1}^{\infty} \frac{(-1)^n n}{(n+1)^{3/2}}$$

The series is absolutely convergent if and only if the series

$$\sum_{n=1}^{\infty} \left| \frac{(-1)^n n}{(n+1)^{3/2}} \right|$$

$$= \sum_{n=1}^{\infty} \frac{n}{(n+1)^{3/2}}$$

is convergent. This looks like a p−series so use the Comparison Test.

$$\sum_{n=1}^{\infty} \left| \frac{(-1)^n n}{(n+1)^{3/2}} \right|$$

$$= \sum_{n=1}^{\infty} \frac{n}{(n+1)^{3/2}}$$

$$> \sum_{n=1}^{\infty} \frac{n}{(2n)^{3/2}} = \frac{1}{2^{3/2}} \sum_{n=1}^{\infty} \frac{n}{(n)^{3/2}}$$

$$= \frac{1}{2^{3/2}} \sum_{n=1}^{\infty} n^{-0.5}$$

which is divergent because it is a p−series with $p = 0.5 < 1$.
The original series is not absolutely convergent.

SkillMaster 8.12.

• Use the Ratio Test to determine' if the series is absolutely convergent.

$$\sum_{n=1}^{\infty} \frac{(-1)^n n^{10}}{n!}$$

Here

$$a_n = \frac{(-1)^n n^{10}}{n!}.$$

If

$$\lim_{n \to \infty} |a_{n+1}/a_n| < 1$$

then the Ratio Test allows us to conclude that the series is absolutely convergent.

$$|a_{n+1}/a_n|$$

$$= \left(\frac{(n+1)^{10}}{(n+1)!} \right) \left(\frac{n!}{n^{10}} \right)$$

$$= \left(\frac{n+1}{n} \right)^{10} \frac{n!}{(n+1)!}$$

$$= \left(1 + \frac{1}{n} \right)^{10} \frac{1}{n+1}$$

$$\lim_{n \to \infty} |a_{n+1}/a_n|$$

$$= \lim_{n \to \infty} \left(1 + \frac{1}{n} \right)^{10} \frac{1}{n+1}$$

$$= (1 + 0)^{10} \lim_{n \to \infty} \frac{1}{n+1}$$

$$= 0 < 1$$

The Ratio Test applies and the series is absolutely convergent.

Section 8.5 – Power Series.

Key Concepts:

- Power series
- Radius and interval of convergence of a power series

Skills to Master:

- Find the radius of convergence and the interval of convergence of a power series.

Discussion:

This section and the two sections that follow deal with power series. A power series is a series of the form

$$\sum_{n=0}^{\infty} c_n x^n = c_0 + c_1 x + c_2 x^2 + \cdots$$

where the c_n are constants and x is a variable. A question of considerable interest is: when can a function be represented as a power series? If it can be, many computations involving the function are made easier.

Key Concept: Power series

A power series in $(x - a)$ is a series of the form

$$\sum_{n=0}^{\infty} c_n (x - a)^n = c_0 + c_1 (x - a) + c_2 (x - a)^2 + \cdots .$$

This series may or may not converge depending on the particular value of x. A power series can be viewed as a function with domain the set of x values for which it converges. Substituting a specific x in the domain into the power series

441

produces a function value.

Key Concept: Radius and interval of convergence of a power series

---

For a given power series,

$$\sum_{n=0}^{\infty} c_n(x - a)^n,$$

there are only three possibilities:

(i) The series converges only when $x = a$.

(ii) The series converges for all x.

(iii) There is a positive number R such that the series converges if $|x - a| < R$ and diverges if $|x - a| > R$.

In case (iii), the number R is called the *radius of convergence* of the series. In case (i) the radius of convergence is 0 and in case (ii) the radius of convergence is ∞. The *interval of convergence* consists of all values of x for which the series converges. In case (iii), it is one of

$$(a - R, a + R),\ (a - R, a + R],\ [a - R, a + R),\ \text{or } [a - R, a + R].$$

SkillMaster 8.13: Find the radius of convergence and the interval of convergence of a power series.

To find the radius of convergence of a power series

$$\sum_{n=0}^{\infty} c_n(x - a)^n,$$

apply one of the techniques for determining convergence of a series to this series. This should tell you for which values of x the series converges. Study *Examples 4 and 5* in this section to see how to do this.

pages 600-601.

Worked Examples

For each of the following examples, first try to find the solution without looking at the middle or right columns. Cover the middle and right columns with a piece of paper. If you need a hint, uncover the middle column. If you need to see the worked solution, uncover the right column.

Example	Tip	Solution

SkillMaster 8.13.

- Find the radius of convergence and the interval of convergence of the following power series.

$$\sum_{n=0}^{\infty} (-1)^{n+1} n^6 x^n$$

Tip: Use the ratio test. If a_n is the nth term in the power series then x is in the interval of convergence if $\lim_{n \to \infty} |a_{n+1}/a_n| < 1$.

Solution:

$$a_n = (-1)^{n+1} n^6 x^n$$
$$|a_n| = n^6 |x|^n$$
$$\lim_{n \to \infty} |a_{n+1}/a_n| =$$
$$\lim_{n \to \infty} \left((n+1)^6 |x|^{n+1} \right) / (n^6 |x|^n) =$$
$$\lim_{n \to \infty} \left(\frac{n+1}{n} \right)^6 |x| = |x|$$

Thus
$$\left| \frac{a_{n+1}}{a_n} \right| < 1$$

if and only if
$$|x| < 1.$$

So, $(-1, 1)$ is the interval of convergence and the radius of convergence is 1. Note that the endpoints ± 1 are not in the interval of convergence because the terms do not approach 0.

- Find the radius of convergence and the interval of convergence of the following power series.

$$\sum_{n=0}^{\infty} \frac{n^n}{n!} x^n$$

(Note: we adopt the convention that $0^0 = 1$ so the first term is 1.)

$|a_n| = \dfrac{n^n}{n!} |x|^n$

Use the Ratio Test.

$$\lim_{n \to \infty} |a_{n+1}/a_n|$$

$$= \lim_{n \to \infty} \left| \frac{(n+1)^{(n+1)} x^{n+1}}{(n+1)!} \frac{n!}{n^n x^n} \right|$$

$$= \lim_{n \to \infty} \left| \frac{(n+1)^n (n+1) n! x^{n+1}}{n^n (n+1)! x^n} \right|$$

$$= \lim_{n \to \infty} \left| \left(\frac{n+1}{n} \right)^n \frac{(n+1)!}{(n+1)!} x \right|$$

$$= \lim_{n \to \infty} |(1+1/n)^n x|$$

$$= e|x|$$

Note that $e|x| < 1$ if and only if $|x| < 1/e$, so the radius of convergence is $1/e$ and the interval of convergence is $(-1/e, 1/e)$.

- Find the radius of convergence and the interval of convergence of the following power series.

$$\sum_{n=0}^{\infty} \frac{4n+1}{\sqrt{n+1}} (x-2)^n$$

$a_n =$

$$\frac{4n+1}{\sqrt{n+1}} (x-2)^n$$

$$\lim_{n \to \infty} |a_{n+1}/a_n| =$$

$$\lim_{n \to \infty} \left| \frac{(4(n+1)+1)(x-2)^{n+1}}{\sqrt{(n+1)+1}} \cdot \right.$$

$$\left. \frac{\sqrt{n+1}}{(4n+1)(x-2)^n} \right| =$$

$$\lim_{n \to \infty} \left| \frac{4n+5}{4n+1} \sqrt{\frac{n+1}{n+2}} (x-2) \right| =$$

$$\lim_{n \to \infty} \left| \frac{4+5/n}{4+1/n} \sqrt{\frac{1+1/n}{1+2/n}} (x-2) \right| =$$

$$\left| \frac{4+0}{4+0} \sqrt{\frac{1+0}{1+0}} (x-2) \right| =$$

$$|x-2|$$

Note that $|x-2| < 1$ if and only if $1 < x < 3$, which is the interval of convergence. The radius of convergence is 1.

Section 8.6 – Representation of Functions as Power Series.

Key Concepts:

- The power series for $\dfrac{1}{1-x}$
- Differentiation and integration of power series

Skills to Master:

- Represent a function as a power series using the known example $\dfrac{1}{1-x}$.

- Represent a function as a power series by differentiating and integrating known examples.

Discussion:

Many functions can be represented as power series. If a function can be represented as a power series, it can be differentiated and integrated by differentiating or integrating the power series term by term.

Key Concept: The power series for $\dfrac{1}{1-x}$

From the formula for the sum of the geometric series

$$\sum_{n=0}^{\infty} x^n = 1 + x + x^2 + x^3 \cdots$$

you can obtain the representation

$$\sum_{n=0}^{\infty} x^n = \frac{1}{1-x} \text{ when } |x| < 1.$$

445

By replacing x by other expressions, you can obtain power series representations for many other functions.

Key Concept: Differentiation and integration of power series

page 604.

Make sure that you understand the statement of *Theorem 2* in this section. This theorem tells you that if a function is represented by a power series with radius of convergence R, then the function is differentiable and can be differentiated or integrated by differentiating or integrating the power series term by term. This is an extremely important result. Note that the radius of convergence of the series that you obtain by differentiating or integrating the original series is the same as the radius of convergence of the original series.

SkillMaster 8.14: Represent a function as a power series using the known example for $\dfrac{1}{1-x}$.

Using the representation $\sum_{n=0}^{\infty} x^n = \frac{1}{1-x}$ when $|x| < 1$, you can obtain representation for many functions that are similar in form to $\frac{1}{1-x}$. By using Theorem 2 on differentiating and integrating power series, you can obtain power series representations for still more functions.

pages 605-606.

Study *Examples 4 through 8* in this section to see how this works.

SkillMaster 8.15: Represent a function as a power series by differentiating and integrating known examples.

Once you have a power series representation for a function, you also have a power series representation for the derivative and integral of the function. This leads to many useful power series representations.

446

Worked Examples

For each of the following examples, first try to find the solution without looking at the middle or right columns. Cover the middle and right columns with a piece of paper. If you need a hint, uncover the middle column. If you need to see the worked solution, uncover the right column.

Example	Tip	Solution

SkillMaster 8.14.

• Represent the function as a power series and give the radius of convergence.

$$f(x) = \frac{x^2}{1 - x}$$

We know

$$\frac{1}{1 - x} =$$

$$1 + x + x^2 + \ldots$$

$$= \sum_{n=0}^{\infty} x^n$$

and that this has a radius of convergence 1. Simply multiply this series by x^2.

$$\begin{aligned} f(x) &= \frac{x^2}{1 - x} \\ &= x^2 \left(\frac{1}{1 - x} \right) \\ &= x^2 \left(\sum_{n=0}^{\infty} x^n \right) \\ &= \sum_{n=0}^{\infty} x^{n+2} \end{aligned}$$

The radius of convergence is the same as the series for $1/(1 - x)$, namely 1.

• Represent the function as a power series and give the radius of convergence.

$$f(x) = \frac{1 + x^2}{1 - x^2}$$

Rewrite this as

$$\frac{1}{1 - x^2} + x^2 \frac{1}{1 - x^2}.$$

The power series for

$$\frac{1}{1 - x^2}$$

may be obtained by substituting x^2 for x in the power series for

$$\frac{1}{1 - x}.$$

$$
\begin{aligned}
f(x) &= \frac{1 + x^2}{1 - x^2} \\
&= \frac{1}{1 - x^2} + x^2 \frac{1}{1 - x^2} \\
&= \left(\sum_{n=0}^{\infty} (x^2)^n \right) + x^2 \left(\sum_{n=0}^{\infty} (x^2)^n \right) \\
&= \left(\sum_{n=0}^{\infty} x^{2n} \right) + x^2 \left(\sum_{n=0}^{\infty} x^{2n} \right) \\
&= \left(\sum_{n=0}^{\infty} x^{2n} \right) + \left(\sum_{n=0}^{\infty} x^{2n+2} \right) \\
&= \left(\sum_{n=0}^{\infty} x^{2n} \right) + \left(\sum_{n=1}^{\infty} x^{2n} \right) \\
&= 1 + \sum_{n=1}^{\infty} 2x^{2n} \\
&= 1 + 2x^2 + 2x^4 + 2x^6 + \ldots
\end{aligned}
$$

The series converges for all $x^2 < 1$ or $-1 < x < 1$ so the radius of convergence is 1.

• Represent the function as a power series and give the radius of convergence.

$$f(x) = \frac{1}{4 + x}$$

Rewrite the function algebraically so that it looks like a function of the form

$$1/(1 - x).$$

$$
\begin{aligned}
f(x) &= \frac{1}{4 + x} \\
&= \frac{1}{4} \left(\frac{1}{1 - (-x/4)} \right)
\end{aligned}
$$

Substitute $(-x/4)$ for x into the power series for $\dfrac{1}{1 - x}$.

$$
\begin{aligned}
f(x) &= \frac{1}{4} \sum_{n=0}^{\infty} (-x/4)^n \\
&= \sum_{n=0}^{\infty} \frac{(-1)^n x^n}{4^{n+1}}
\end{aligned}
$$

The series converges for $|-x/4| < 1$ or $|x| < 4$ or $-4 < x < 4$. The radius of convergence is 4.

SkillMaster 8.15.

- Represent the function as a power series.

$$f(x) = x\ln(1 - x^2)$$

Recall

$$\frac{1}{1-x} =$$

$$1 + x + x^2 + \ldots$$

In Example 6 on page 605 of the text the following series was derived.

$$\ln(1 - x) =$$

$$-x - \frac{x^2}{2} - \frac{x^3}{3} - \ldots$$

$$= \sum_{n-1}^{\infty} -\frac{x^n}{n}.$$

Substitute x^2 for x in this series and multiply by x.

$$f(x) = x\ln(1 - x^2)$$

$$= x\sum_{n=1}^{\infty} -\frac{(x^2)^n}{n}$$

$$= -\sum_{n=1}^{\infty} \frac{x^{2n+1}}{n}$$

- Represent the function as a power series.
$$f(x) = \sqrt{x}\tan^{-1}(\sqrt{x})$$

In Example 7 on page 606 of the text, the series

$$\tan^{-1}(x) =$$

$$\sum_{n=0}^{\infty} \frac{(-1)^n x^{2n+1}}{2n + 1}$$

was derived. Use this to find the required power series.

$$f(x) = \sqrt{x}\tan^{-1}(\sqrt{x})$$

$$= x^{1/2}\left(\sum_{n=0}^{\infty} \frac{(-1)^n (x^{1/2})^{2n+1}}{2n + 1}\right)$$

$$= x^{1/2}\sum_{n=0}^{\infty} \frac{(-1)^n x^{n+1/2}}{2n + 1}$$

$$= \sum_{n=0}^{\infty} \frac{(-1)^n x^{n+1}}{2n + 1}$$

Notice that the original function was defined only for $x \geq 0$. The power series converges for all $-1 < x < 1$, so the power series represents the function f for all $0 \leq x < 1$.

449

• Evaluate the indefinite integral as a power series.

$$\int \frac{x}{1 - x^4}\, dx$$

Express the integrand as a power series and integrate term by term.

$$\int \frac{x}{1 - x^4}\, dx$$

$$= \int x\frac{1}{1 - x^4}\, dx$$

$$= \int x\sum_{n=0}^{\infty} (x^4)^n\, dx$$

$$= \int \sum_{n=0}^{\infty} x^{4n+1}\, dx$$

$$= C + \sum_{n=0}^{\infty} \frac{x^{4n+2}}{4n + 2}$$

Section 8.7 – Taylor and Maclaurin Series.

Key Concepts:

- The Taylor and Maclaurin series of a function
- Taylor's Inequality
- Maclaurin series for familiar functions

Skills to Master:

- Find the Taylor and Maclaurin series for a function using the definition.

Discussion:

This section discusses the important concept of the Taylor series of a function. If a function can be represented by a power series, then the power series representation is the same as the Taylor series for the function. Many familiar functions are equal to their Taylor series. These series can be used to approximate the values of the functions.

Key Concept: The Taylor and Maclaurin series of a function.

Suppose f has derivatives of all orders at a. The Taylor series of a function f centered at a is

$$\sum_{n=0}^{\infty} \frac{f^{(n)}(a)}{n!}(x-a)^n = f(a) + \frac{f'(a)}{1!}(x-a) + \frac{f''(a)}{2!}(x-a)^2 + \frac{f'''(a)}{3!}(x-a)^3 + \cdots .$$

If $a = 0$, the Taylor series is called the Maclaurin series. If f has a power series representation at a, then the power series representation is the Taylor series. Be careful to note that even if f has derivatives of all orders at a, and thus has a Taylor series, f may not be equal to a power series.

451

Key Concept: Taylor's Inequality

A key question is whether a given function f is equal to the sum of its Taylor series at a. If
$$f(x) = T_n(x) + R_n(x)$$
where $T_n(x)$ is the nth degree Taylor polynomial of f at a and if
$$\lim_{n \to \infty} R_n(x) = 0 \text{ for } |x - a| < R,$$
then f is equal to the sum of its Taylor series for all x where $|x - a| < R$. Taylor's Inequality can help you determine if $\lim_{n \to \infty} R_n(x) = 0$ in the above setting.

Taylor's Inequality states that if
$$\left| f^{(n+1)}(x) \right| \le M \text{ for } |x - a| < R,$$
then the remainder $R_n(x)$ of the Taylor series satisfies the inequality
$$R_n(x) \le \frac{M}{(n+1)!} |x - a|^{n+1}.$$
A key fact to remember in applying this to specific situations is that
$$\lim_{n \to \infty} \frac{x^n}{n!} = 0 \text{ for every real number } x.$$

Key Concept: Maclaurin series for familiar functions

page 616.

The *table* in this section gives Maclaurin series for some familiar functions and also gives the intervals of convergence for these series. Make sure that you learn these series and understand how they were obtained.

SkillMaster 8.16: Find the Taylor and Maclaurin series for a function using the definition.

To find Taylor and Maclaurin series for specific functions, use the formula

$$\sum_{n=0}^{\infty} \frac{f^{(n)}(a)}{n!}(x-a)^n = f(a) + \frac{f'(a)}{1!}(x-a) + \frac{f''(a)}{2!}(x-a)^2 + \frac{f'''(a)}{3!}(x-a)^3 + \cdots .$$

The coefficient of $(x-a)^n$ is $\dfrac{f^{(n)}(a)}{n!}$, so to find the series, you need to be able to compute the derivatives of f at a and find a pattern, if possible. This will allow you to determine the general term in the Taylor or Maclaurin series.

Worked Examples

For each of the following examples, first try to find the solution without looking at the middle or right columns. Cover the middle and right columns with a piece of paper. If you need a hint, uncover the middle column. If you need to see the worked solution, uncover the right column.

Example	Tip	Solution

SkillMaster 8.16.

• Find the Maclaurin series for the function using the definition.

$$f(x) = \frac{1}{x+2}$$

Tip: Compute $f'(x), f''(x), \ldots$ until you see the pattern for $f^{(n)}(x)$.

Solution:

$$f(x) = \frac{1}{x+2} = (x+2)^{-1}$$
$$f'(x) = -(x+2)^{-2}$$
$$f''(x) = 2(x+2)^{-3}$$
$$f^{(3)}(x) = -2 \cdot 3(x+2)^{-4}$$
$$f^{(n)}(x) = (-1)^n n!(x+2)^{-(n+1)}$$
$$f^{(n)}(0) = (-1)^n n!/2^{n+1}$$

$$f(x) = \sum_{n=0}^{\infty} \frac{f^{(n)}(0)}{n!} x^n$$
$$= \sum_{n=0}^{\infty} \frac{(-1)^n n!}{n! 2^{n+1}} x^n$$
$$= \sum_{n=0}^{\infty} \frac{(-1)^n}{2^{n+1}} x^n$$

- Find the Taylor series for the function $f(x)$ at the given point a.

$$f(x) = \frac{1}{x+2}, a = 2$$

The nth derivative of f was computed in the previous worked out example.

$$f^{(n)}(x) =$$

$$(-1)^n n!(x+2)^{-(n+1)}.$$

$$f^{(n)}(2) =$$

$$(-1)^n n!(4)^{-n+1}$$

$$
\begin{aligned}
f(x) &= \sum_{n=0}^{\infty} \frac{f^{(n)}(a)}{n!}(x-a)^n \\
&= \sum_{n=0}^{\infty} \frac{(-1)^n n!}{n! 4^{n+1}}(x-2)^n \\
&= \sum_{n=0}^{\infty} \frac{(-1)^n}{4^{n+1}}(x-2)^n
\end{aligned}
$$

- Find the Taylor series for the function $f(x)$ at the given point a.

$$f(x) = \frac{1}{\sqrt{x}}, a = 1$$

Find a pattern for the nth derivative of f.

$$
\begin{aligned}
f(x) &= \frac{1}{\sqrt{x}} = x^{-1/2} \\
f'(x) &= (-1/2)x^{-3/2} \\
f''(x) &= \frac{1 \cdot 3}{2 \cdot 2}x^{-5/2} \\
f^{(3)}(x) &= -\frac{1 \cdot 3 \cdot 5}{2 \cdot 2 \cdot 2}x^{-7/2} \\
f^{(n)}(x) &= (-1)^n \frac{1 \cdot 3 \cdots (2n-1)}{2^n}x^{-(2n+1)/2} \\
f^{(n)}(1) &= (-1)^n \frac{1 \cdot 3 \cdots (2n-1)}{2^n} \\
&= \frac{(-1)^n 1 \cdot 2 \cdots 2n}{2^n \cdot 2 \cdot 4 \cdots (2n)} \\
&= \frac{(-1)^n (2n)!}{2^n \cdot 2^n \cdot n!} \\
f(x) &= \sum_{n=0}^{\infty} \frac{f^{(n)}(a)}{n!}(x-a)^n \\
&= \sum_{n=0}^{\infty} \frac{f^{(n)}(1)}{n!}(x-1)^n \\
&= \sum_{n=0}^{\infty} \frac{(-1)^n (2n)!}{4^n (n!)^2}(x-1)^n
\end{aligned}
$$

455

- Use the Maclaurin series for $\cos(x)$ to find the Maclaurin series for

$$f(x) = x\cos(x^3).$$

$\cos(x) =$

$$1 - \frac{x^2}{2!} + \frac{x^4}{4!} - \cdots$$

$$= \sum_{n=0}^{\infty} (-1)^n \frac{x^{2n}}{(2n)!}.$$

$$f(x) = x\cos(x^3)$$

$$= x\left(\sum_{n=0}^{\infty}(-1)^n\frac{(x^3)^{2n}}{(2n)!}\right)$$

$$= x\left(\sum_{n=0}^{\infty}(-1)^n\frac{x^{6n}}{(2n)!}\right)$$

$$= \sum_{n=0}^{\infty}(-1)^n\frac{x^{6n+1}}{(2n)!}$$

$$= x - \frac{x^7}{2!} + \frac{x^{13}}{4!} - \frac{x^{19}}{6!} + \cdots$$

- Use Maclaurin series to compute

$$\int_0^1 x\cos(x^3)\,dx$$

to three decimal places.

The Maclaurin series for the integrand was computed in the preceding worked out example.

$$\int_0^1 x\cos(x^3)\,dx$$

$$= \int_0^1\left(\sum_{n=0}^{\infty}(-1)^n\frac{x^{6n+1}}{(2n)!}\right)dx$$

$$= \sum_{n=0}^{\infty}\frac{(-1)^n}{(2n)!}\int_0^1 x^{6n+1}\,dx$$

$$= \sum_{n=0}^{\infty}\frac{(-1)^n}{(2n)!}\frac{x^{6n+2}}{(6n+2)}\bigg]_0^1$$

$$= \sum_{n=0}^{\infty}\frac{(-1)^n}{(6n+2)(2n)!}$$

$$= \frac{1}{2} - \frac{1}{8\cdot 2} + \frac{1}{14\cdot 4!} - \frac{1}{20\cdot 6!} + \cdots$$

$$= 0.5 - 0.0625 + 0.0030 - 0.00007 + \cdots$$

The Alternating Series Estimation Theorem shows that the error involved in approximating the sum of this series with the sum of the first 3 terms is less than 0.00007.

$$\int_0^1 x\cos(x^3)\,dx \approx 0.5 - 0.0625 + 0.0030$$

$$\approx 0.4405$$

• Find the limit using Maclaurin series.

$$\lim_{x \to \infty} \frac{e^{-x^2} - 1 + x^2}{x^4}$$

$e^{-x^2} - 1 + x^2 =$

$(\sum_{n=0}^{\infty}(-x^2)^n/n!)$
$-1 + x^2 =$

$\sum_{n=2}^{\infty}(-1)^n x^{2n}/n!$

$$\lim_{x \to \infty} \frac{e^{-x^2} - 1 + x^2}{x^4}$$

$$= \lim_{x \to \infty} \frac{\sum_{n=2}^{\infty}(-1)^n x^{2n}/n!}{x^4}$$

$$= \lim_{x \to \infty} \frac{x^4 \sum_{n=2}^{\infty}(-1)^n x^{2n-4}/n!}{x^4}$$

$$= \lim_{x \to \infty} \sum_{n=2}^{\infty}(-1)^n x^{2n-4}/n!$$

$$= \lim_{x \to \infty} \frac{1}{2!} - \frac{x^2}{3!} + \frac{x^4}{4!} - \cdots$$

$$= \frac{1}{2!} - \frac{0^2}{3!} + \frac{0^4}{4!} - \cdots$$

$$= \frac{1}{2} = 0.5$$

457

Section 8.8 – The Binomial series

Key Concepts:

- The Binomial Theorem and the Binomial Series

Skills to Master:

- Use the Binomial Series to find power series representations for functions.

Discussion:

This section gives other examples of functions represented by a power series. The functions are of the form
$$(1+x)^k \text{ for } |x| < 1.$$

Key Concept: The Binomial Theorem and the Binomial Series

The Binomial Theorem states that
$$(a+b)^k = \sum_{n=0}^{k} \binom{k}{n} a^{k-n} b^n \text{ where } \binom{k}{n} = \frac{k(k-1)(k-1)\cdots(k-n+1)}{n!}.$$
(If $k=0$, $\binom{n}{k}$ is defined to be 1.) You should be familiar with this in the simple cases where $k=2$ or $k=3$. In the above formula, k is a fixed positive integer.

There is a generalization of this when k is any real number. The Binomial Series for $(1+x)^k$ is
$$\sum_{n=0}^{\infty} \binom{k}{n} x^n \text{ where } \binom{k}{n} = \frac{k(k-1)(k-1)\cdots(k-n+1)}{n!}.$$
If k is not an integer, there are an infinite number of terms in this series. This series is equal to $(1+x)^k$ if $|x| < 1$.

SkillMaster 8.17: Use the Binomial Series to find
power series representations for functions.

page 621.

Any function that is a product of a power of x and an expression of the form $(1 + x)^k$ can be written as a power series by using the Binomial Series. *Examples 1 and 2* in this section show how to do this.

Worked Examples

For each of the following examples, first try to find the solution without looking at the middle or right columns. Cover the middle and right columns with a piece of paper. If you need a hint, uncover the middle column. If you need to see the worked solution, uncover the right column.

Example	Tip	Solution
SkillMaster 8.17. • Use the binomial series to expand the following function as a power series. State the radius of convergence. $f(x) = \dfrac{x^3}{\sqrt{4+x}}$	Rewrite $\sqrt{4+x}$ as $\sqrt{4}\sqrt{1+\dfrac{x}{4}} =$ $2(1+x/4)^{1/2}$.	Using the tip, $\dfrac{x^3}{\sqrt{4+x}} = \dfrac{x^3}{2(1+x/4)^{1/2}}$ $= \dfrac{x^3}{2}(1+x/4)^{-1/2}$ The Binomial Series for $(1+x/4)^{-1/2}$ is $$\sum_{n=0}^{\infty} \binom{-\frac{1}{2}}{n}\left(\frac{x}{4}\right)^n$$ and converges when $\left|\dfrac{x}{4}\right| < 1$, or when $|x| < 4$. So a power series expansion for $f(x)$ is $$\frac{x^3}{2}\sum_{n=0}^{\infty}\binom{-\frac{1}{2}}{n}\left(\frac{x}{4}\right)^n =$$ $$\sum_{n=0}^{\infty}\binom{-\frac{1}{2}}{n}\frac{x^{n+3}}{2\cdot 4^n} =$$ $$\left[\frac{1}{2}x^3 - \frac{1}{2}\cdot\frac{1}{8}x^4 + \cdots\right]$$ The radius of convergence is 4.

SkillMasters for Chapter 8

SkillMaster 8.1: Find a defining equation for a sequence.

SkillMaster 8.2: Use the laws of limits together with known examples to determine if a sequence is divergent or convergent, and if convergent, to find the limit.

SkillMaster 8.3: Determine if a sequence is monotone and use the MonotonicSequence Theorem to show some sequences are convergent.

SkillMaster 8.4: Determine if a geometric series is convergent, and if convergent, find the sum of the series.

SkillMaster 8.5: Use the laws of series together with known examples to determine if a series is divergent or convergent, and if convergent, find the sum.

SkillMaster 8.6: Use the Integral Test to determine if a series with positive decreasing terms is convergent.

SkillMaster 8.7: Use the Comparison Test to determine whether a series is convergent or divergent.

SkillMaster 8.8: Estimate the error in approximating the sum of a series.

SkillMaster 8.9: Use the Alternating Series Test to determine if an alternating series is convergent.

SkillMaster 8.10: Use the Alternating Series Estimation Theorem to estimate the error in approximating the sum of an alternating series by a finite sum.

SkillMaster 8.11: Determine whether a series is absolutely convergent.

SkillMaster 8.12: Use the ratio test to determine whether a series is absolutely convergent.

SkillMaster 8.13: Find the radius of convergence and the interval of convergence of a power series.

SkillMaster 8.14: Represent a function as a power series using the known example $1/(1-x)$.

SkillMaster 8.15: Represent a function as a power series by differentiating and integrating known examples.

SkillMaster 8.16: Find the Taylor and Maclaurin series for a function using the definition.

SkillMaster 8.17: Use the Binomial Series to find power series representations for functions.

Chapter 9 - Vectors and the Geometry of Space

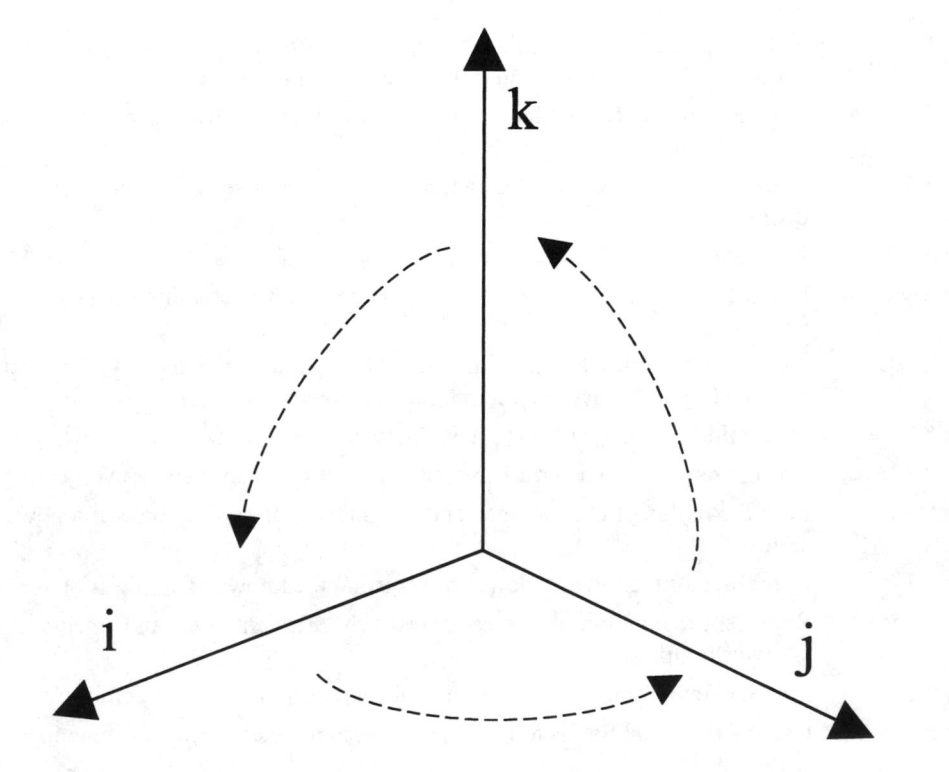

Section 9.1 – Three-Dimensional Coordinate Systems

Key Concepts:

- Distance Formula
- Graphs of Planes, Spheres, and other Sets in Three-Dimensions

Skills to Master:

- Describe and sketch points and planes in three dimensions.
- Use the distance formula to solve problems in three dimensions.
- Describe and sketch regions involving spheres in three dimensions.

Discussion:

This section introduces the standard three-dimensional coordinate system. Just as two coordinates are required to specify the location of a point in the plane, three coordinates are required to specify a point in three-dimensional space. If a point P in space has coordinates (x, y, z), the x and y represent the coordinates of the point in the xy plane that P lies above or below. The z coordinate represents how far above or below the xy plane P lies. Pay careful attention to *Figure 3* and make sure that you understand where the three coordinate planes are.

page 646.

Key Concept: The Distance Formula

The distance formula in three dimensions for the distance $|P_1P_2|$ between points P_1 with coordinates (x_1, y_1, z_1) and P_2 with coordinates (x_2, y_2, z_2) is

$$|P_1P_2| = \sqrt{(x_2 - x_1)^2 + (y_2 - y_1)^2 + (z_2 - z_1)^2}.$$

page 647.

Read through the *explanation* for this formula to see how the formula arises from two applications of the Pythagorean Theorem. Note the similarity to the distance formula in two dimensions. In fact, if the points P_1 and P_2 both lie in the xy plane (so that $z_1 = z_2 = 0$), the formula becomes

$$|P_1P_2| = \sqrt{(x_2 - x_1)^2 + (y_2 - y_1)^2},$$

the usual distance formula in the plane.

Key Concept: Graphs of Planes, Spheres and Other Sets in Three-Dimensions

The equation of a sphere with center $C(h, k, l)$ and radius r is

$$(x - h)^2 + (y - k)^2 + (z - l)^2 = r^2.$$

page 649.

Make sure you understand *Example 4* in the text which shows where this formula comes from. Pay careful attention to the other examples in this section explaining how certain equations correspond to graphs of planes in three-dimensional space.

SkillMaster 9.1: Sketch points and planes and in three-dimensions.

page 647.

page 648.

To sketch a point $P(x, y, z)$ in three-dimensions, first find the projection of P onto the xy plane as in *Figure 5* , then find the point by going above or below the xy plane. The distance to go above or below the plane is determined by the z coordinate. To sketch planes in three-dimensions, use geometric reasoning to determine where the plane is as was done in *Examples 1 and 2*. You will see later in Section 9.5 a more general method of sketching and determining planes in three-dimensions.

SkillMaster 9.2: Use the distance formula to solve problems in three dimensions.

If a subset of three-dimensional space is described in terms of the distance from certain points, you should use the distance formula to write down an equation describing the situation. Then you should solve this equation to solve the problem.

SkillMaster 9.3: Describe and sketch regions involving spheres in three dimensions.

page 650.

By rewriting equations so that the equations look like equations of spheres, the region described by the original equation can be determined. Pay careful attention to *Example 6* to see how this works.

Worked Examples

For each of the following examples, first try to find the solution without looking at the middle or right columns. Cover the middle and right columns with a piece of paper. If you need a hint, uncover the middle column. If you need to see the worked solution, uncover the right column.

Example	Tip	Solution
SkillMaster 9.1.		
• In a video game you have a spaceship at the location $(4, 4, 4)$. You navigate the ship 2 units toward the yz-plane and then you navigate upwards 2 units. What are the present coordinates of the ship?	Navigating the ship 2 units toward the yz-plane is the same as adjusting the x-coordinate 2 units closer to zero. Navigating the ship upward is the same as increasing the z-coordinate.	$(4 - 2, 4, 4 + 2) = (2, 4, 6)$
8• Describe in words and sketch the points in three dimensional space satisfying $y = 4$.	This is the set of all points with second coordinate 4.	This is the plane parallel to the xz-plane and 4 units to the right.

Example	**Tip**	**Solution**
• Describe in words and sketch the points in three dimensional space satisfying $x - z = 1$	This is a plane. First graph the line in which this plane intersects the xz plane. In three dimensions the plane will be the plane through this line in the xz plane and perpendicular to the xz-plane.	Solving for z gives $z = x - 1$. The sketch of this line in the xy plane is shown. 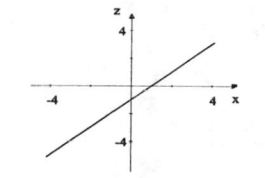 The sketch shows a plane perpendicular to the xz-plane, whose intersection with xz-plane has a slope of 1 and passes through the point $(1, 0, 0)$. 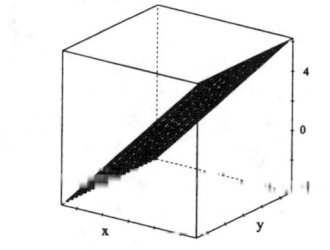

• Describe in words and sketch the points in three dimensional space satisfying
$$0 < x < z.$$

Describe the inequalities $0 < x$ and $x < z$ separately.

The inequality $0 < x$ is the set of points in front of the yz-plane. The inequality $x < z$ is the set of points above the plane through the origin making a $45°$ angle with the xy-plane (this is the plane $y = z$). Combining these inequalities together gives a wedge-shaped region whose edge is the y-axis.

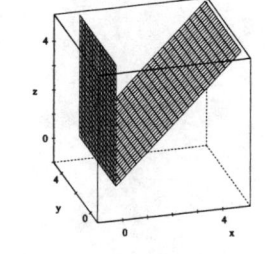

SkillMaster 9.2.

• Consider the triangle whose corners are the points $P(1, 1, -1)$, $Q(2, -1, 1)$, $R(4, 1, 2)$. Find the lengths of the sides. Is the triangle isosceles? Is it a right triangle?

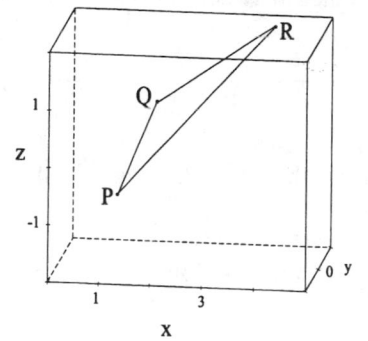

Use the distance formula to find the lengths. A triangle is isosceles if and only if it has two sides of equal length and it is a right triangle if and only if the Pythagorean Theorem holds.

The lengths are

$$|PQ| =$$
$$\sqrt{(2-1)^2 + (-1-1)^2 + (1-(-1))^2}$$
$$= \sqrt{1^2 + 2^2 + 2^2} = \sqrt{9} = 3,$$

$$|QR| =$$
$$\sqrt{(4-2)^2 + (1-(-1))^2 + (2-1)^2}$$
$$= \sqrt{2^2 + 2^2 + 1^2}$$
$$= \sqrt{9}$$
$$= 3,$$

$$|PR| =$$
$$\sqrt{(4-1)^2 + (1-1)^2 + (2-(-1))^2}$$
$$= \sqrt{3^2 + 0^2 + 3^2}$$
$$= \sqrt{18}$$
$$= 3\sqrt{2}.$$

The triangle is isosceles because it has two sides (PQ and QR) of equal length. PR is the longest side. The triangle is a right triangle if and only if the Pythagorean Theorem holds,

$$|PQ|^2 + |QR|^2 = |PR|^2.$$

This is true since

$$3^2 + 3^2 = (3\sqrt{2})^2 = 18.$$

• Find the point in R^3 that lies in the xz-plane, whose x-coordinate is 3, whose z-coordinate is positive, and is a distance of 5 from the origin.

The point has the form $(3, 0, z)$. Use the distance formula to solve for z.

Using the distance formula to solve for z we get $\sqrt{3^2 + 0^2 + z^2} = 5$ or $3^2 + 0^2 + z^2 = 25$, or $z^2 = 25 - 9 = 16$. So $z = \pm 4$. Since z is positive we must have $z = 4$ and the point is $(3, 0, 4)$.

SkillMaster 9.3.

• Find the equation of a sphere with radius 3 and center $(2, -1, 0)$. Is the origin inside this sphere? What is the intersection of this sphere with the yz-plane.

Recall that a sphere of radius r and center (h, k, l) is the set of points whose distance from (h, k, l) is r. the equation of this sphere is
$(x - h)^2 + (y - k)^2$
$+(z - l)^2 = r^2$.

The equation of the sphere is

$$(x - 2)^2 + (y - (-1))^2 + (z - 0)^2 = 3^2$$
$$(x - 2)^2 + (y + 1)^2 + z^2 = 9.$$

The distance between the center of the sphere and the origin is

$$\sqrt{2^2 + (-1)^2 + 0^2} = \sqrt{4 + 1 + 0}$$
$$= \sqrt{5} \approx 2.24.$$

This is less than 3 so the origin is inside the sphere.

The yz-plane is the set of (x, y, z) with $x = 0$. Setting $x = 0$ in the equation of the sphere gives

$$(0 - 2)^2 + (y + 1)^2 + z^2 = 9$$
$$4 + (y + 1)^2 + z^2 = 9$$
$$(y + 1)^2 + z^2 = 5.$$

This is a circle of radius $\sqrt{5}$ and center $(-1, 0)$ in the yz-plane. Below is a picture of the yz plane, the circle of intersection, and part of the sphere. The dark point in the center has coordinates $(-1, -1, 0)$.

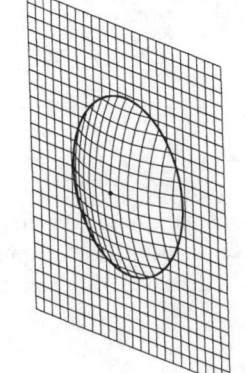

• Show that the equation below is an equation of a sphere and find its center and radius.
$x^2 - 2x + y^2 + z^2 + 4z = 11$

Complete the squares.

$$
\begin{aligned}
x^2 - 2x + y^2 + z^2 + 4z &= 11 \\
x^2 - 2x + 1 + y^2 + z^2 + 4z + 4 &= 11 + 1 + \\
(x - 1)^2 + y^2 + (z + 2)^2 &= 16 = (4)^2
\end{aligned}
$$

This is the equation of a sphere with radius 4 and center $(1, 0, -2)$.

• Show that the origin is on the sphere of radius 3 and center $(1, 2, 2)$. Find the other endpoint of the diameter of this sphere with the origin as one endpoint.

Show that the origin has distance 3 from $(1, 2, 2)$ using the distance formula.

The distance from $(1, 2, 2)$ to the origin is

$$
\begin{aligned}
&\sqrt{(1 - 0)^2 + (2 - 0)^2 + (2 - 0)^2} \\
&= \sqrt{1 + 4 + 4} \\
&= \sqrt{9} \\
&= 3.
\end{aligned}
$$

The sphere of radius 3 and center $(1, 2, 2)$ is the set of points whose distance from $(1, 2, 2)$ is 3. So the origin is on the sphere. The diameter with the origin as one endpoint passes through the origin $(0, 0, 0)$ and the center $(1, 2, 2)$. The other endpoint is on the line defined by these points and a distance 3 further from the origin. The endpoint is

$$(2, 4, 4).$$

• Show that the set of points that is twice as far from the origin as $(3,3,3)$ is a sphere and find its center and radius.

The distance between (x,y,z) and the origin is $\sqrt{x^2+y^2+z^2}$ and the distance between (x,y,z) and the point $(3,3,3)$ is $((x-3)^2+(y-3)^2+(z-3)^2)^{1/2}$.

A point (x,y,z) is twice as far from the origin as it is from the point $(3,3,3)$ if

(distance from origin) $= 2$(distance from $(3,3,3)$)

This is true when
$$\sqrt{x^2+y^2+z^2} = 2\sqrt{(x-3)^2+(y-3)^2+(z-3)^2},$$
or $x^2+y^2+z^2 = 4((x^2-6x+9)+(y^2-6y+9)+(z^2-6z+9)).$

This simplifies to
$x^2+y^2+z^2 = 4x^2-24x+4y^2-24y+4z^2-24z+108,$
or $0 = 3x^2-24x+3y^2-24y+3z^2-24z+108,$
or $0 = x^2-8x+y^2-8y+z^2-8z+36.$

Completing the square gives
$0 = x^2-8x+16+y^2-8y+16+z^2 -8z+16-12,$
or $0 = (x-4)^2+(y-4)^2+(z-4)^2-12.$
$(x-4)^2+(y-4)^2+(z-4)^2=12$
This is an equation of a sphere with radius $\sqrt{12}$ and center $(4,4,4)$.

Section 9.2 – Vectors

Key Concepts:

- Vectors and Representations of Vectors
- Properties of Vectors: Addition and Scalar Multiplication
- Horizontal and Vertical Components of a Vector

Skills to Master:

- Describe vectors and vector representations algebraically, graphically, and in words.
- Compute with vectors using the properties of vectors and vector operations.
- Solve problems involving force and velocity using the horizontal and vertical components of vectors.

Discussion:

A vector is a quantity that has both magnitude and direction. Vectors are extensively used in many areas of science and mathematics. It is important that you become familiar with the vector concepts presented in this section. Much of the material in the succeeding sections in the text depends on vector concepts. In this section, you will learn how to represent vectors, and how to work with them both algebraically and geometrically.

Key Concept: Vectors and Representations of Vectors

A vector is a quantity that has both magnitude and direction. A two dimensional vector can be thought of as an ordered pair $\mathbf{a} = \langle a_1, a_2 \rangle$ of real numbers. From any starting point $A(x, y)$ in the plane, if you proceed a_1 units in the x direction,, and a_2 units in the y direction, you end up at the point $B(x + a_1, y + a_2)$. The vector $\mathbf{a} = \langle a_1, a_2 \rangle$ has magnitude equal to the length of the line segment AB and direction equal to the direction from A to B. The directed line segment $\overrightarrow{AB}$ is called a representation of the vector $\mathbf{a}$. The magnitude or length of the vector $\mathbf{a}$, $|\mathbf{a}|$ is given by

$$|\mathbf{a}| = \sqrt{a_1^2 + a_2^2}.$$

The situation for three dimensional vectors is similar, except that three coordinates are used instead of two. Pay careful attention to the *explanation* of vectors and representations of vectors given in the text.

page 653.

Key Concept: Properties of Vectors: Addition and Scalar   Multiplication

To add two vectors together, you add their components. To multiply a vector by a scalar, you multiply each component of the vector by the scalar. The process of vector addition can be thought of geometrically as placing the initial point of one vector at the terminal point of the other. See *Figure 4* for an illustration of this. The process of multiplying a vector by a positive scalar can be thought of geometrically as keeping the direction of the vector the same, but multiplying the length by the scalar. Make sure that you understand these operations both algebraically and geometrically. Multiplying a vector by a negative scalar first reverses the direction of the vector, and then multiplies the length by the absolute value of the scalar.

page 654.

Key Concept: Components of a Vector

Every vector $\mathbf{a} = \langle a_1, a_2, a_3 \rangle$ can be represented as

$$a_1 \mathbf{i} + a_2 \mathbf{j} + a_3 \mathbf{k}$$

where

$$\mathbf{i} = \langle 1, 0, 0 \rangle \qquad \mathbf{j} = \langle 0, 1, 0 \rangle \qquad \text{and } \mathbf{k} = \langle 0, 0, 1 \rangle$$

page 656.

are the standard *basis vectors* in the directions of the x, y, and z axes. The vectors $a_1 \mathbf{i}$, $a_2 \mathbf{j}$, and $a_3 \mathbf{k}$ are said to be the components of the vector $\mathbf{a}$ since they add together to yield $\mathbf{a}$.

SkillMaster 9.4: Describe vectors and vector representations algebraically, graphically, and in words.

Make sure that you understand what a vector is, what a representation of a vector is, and how to find the vector joining two points. If you are given points $A(x_1, y_1, z_1)$ and $B(x_2, y_2, z_2)$, then the vector **a** with representation $\overrightarrow{AB}$ is the vector

$$\mathbf{a} = \langle x_2 - x_1, y_2 - y_1, z_2 - z_1 \rangle.$$

To graphically represent this vector, you can draw a directed line segment from the point A to the point B, or you can draw a directed line segment from the origin to the point $(x_2 - x_1, y_2 - y_1, z_2 - z_1)$. Both of these directed line segments have the same direction and the same magnitude.

SkillMaster 9.5: Compute with vectors using the properties of vectors and vector operations.

page 655.

The *properties of vectors* are listed in this section. Using these properties and the properties of vector addition and scalar multiplication, you can perform computations with vectors. The problems listed below under **SkillMaster 9.5** give examples of how to do this.

SkillMaster 9.6: Solve problems involving force and velocity using the horizontal and vertical components of vectors.

Force and velocity can be viewed as vectors since they have both direction and magnitude. By writing a force or velocity vector in terms of its components, you can solve problems involving force and velocity. For example, if an object is not moving, then the sum of the force vectors acting on the object must add up to the 0 vector. By writing the sum of the forces in terms of horizontal and vertical components, you will arrive at equations setting the sum of the horizontal components equal to 0 and setting the sum of the vertical components equal to 0. These equations can then be solved to give information about the original problem.

Worked Examples

For each of the following examples, first try to find the solution without looking at the middle or right columns. Cover the middle and right columns with a piece of paper. If you need a hint, uncover the middle column. If you need to see the worked solution, uncover the right column.

Example	Tip	Solution
SkillMaster 9.4.		
• Which of the following are vectors and which are scalars? Explain why. (a) The temperature at the Chicago Airport at 12:45 P.M. January 10, 2001. (b) The force pressing on a table exerted by a person leaning on it with her elbow. (c) The velocity of an asteroid heading toward earth. (d) The total mass of the observable universe.	A vector has direction and magnitude (unless it is the zero vector). A scalar is a single quantity that may be represented by a single number.	(a) The temperature is a scalar, it is a quantity. (b) The force is a vector, the elbow exerts force on the table which has a direction and a magnitude. (c) The velocity of the asteroid is a vector, it has a direction and a magnitude (the speed). (d) The total mass of the observable universe is a quantity and therefore a scalar.
• Find the vector $\overrightarrow{AB}$ where the point $A = A(1, -2, 1)$ and $B = B(0, 1, 2)$.	Recall that the vector from $A = A(x_1, y_1, z_1)$ to $B = B(x_2, y_2, z_2)$ is the vector $\langle x_2 - x_1, y_2 - y_1, z_2 - z_1 \rangle$. That is, we subtract the components of the initial point from the components of the terminal point.	$\begin{aligned} \overrightarrow{AB} &= \langle 0 - 1, 1 - (-2), 2 - 1 \rangle \\ &= \langle -1, 3, 1 \rangle \end{aligned}$

• Sketch the vector **a** whose initial point is $A = A(2, -1)$ and whose terminal point is $B = B(-3, 1)$. Express **a** in terms of the standard basis vectors **i**, **j**, **k**.

Plot the points A and B and draw an arrow between them. If you have computed the components of **a** then these components are the coefficients of **i**, **j**, **k**.

$$\vec{AB} = \langle -3 - 2, 1 - (-1) \rangle$$
$$= \langle -5, 2 \rangle$$
$$= -5\mathbf{i} + 2\mathbf{j}$$

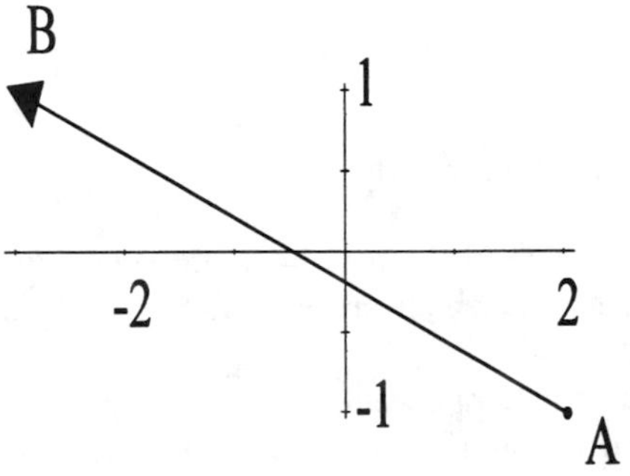

SkillMaster 9.5.

• Sketch **a** + **b** and **a**−2**b** where **a** and **b** are shown.

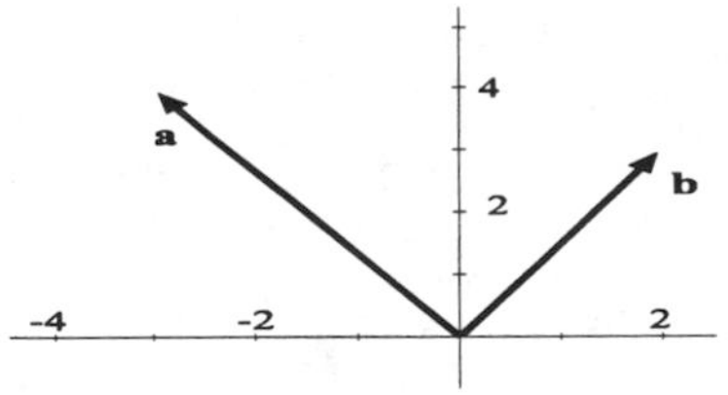

Find the coordinates of the endpoints of **a** and **b**.

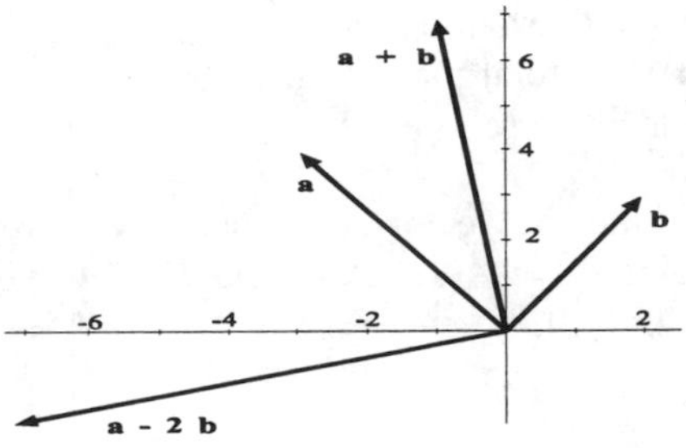

- Suppose that $\mathbf{a} = \overrightarrow{AB} = \overrightarrow{BC}$ and that $\mathbf{b} = \overrightarrow{AD}$. Express $\overrightarrow{CD}$ and $\overrightarrow{DB}$ in terms of $\mathbf{a}$ and $\mathbf{b}$. Explain why $\overrightarrow{CD} + \overrightarrow{DB} = -\mathbf{a}$.

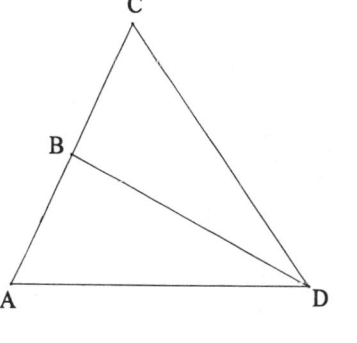

Use the Triangle Law with different triangles. The Triangle Law also implies that $\overrightarrow{CD} + \overrightarrow{DB} = \overrightarrow{CB}$.

Using the Triangle Law and following the diagram around the big triangle gives

$$\begin{aligned}
\overrightarrow{CD} &= \overrightarrow{CA} + \overrightarrow{AD} \\
&= \overrightarrow{CB} + \overrightarrow{BA} + \overrightarrow{AD} \\
&= -\overrightarrow{BC} - \overrightarrow{AB} + \overrightarrow{AD} \\
&= -\mathbf{a} - \mathbf{a} + \mathbf{b} \\
&= -2\mathbf{a} + \mathbf{b} \\
\overrightarrow{DB} &= \overrightarrow{DA} + \overrightarrow{AB} \\
&= -\overrightarrow{AD} + \overrightarrow{AB} \\
&= -\mathbf{b} + \mathbf{a} \\
&= \mathbf{a} - \mathbf{b}
\end{aligned}$$

We can compute $\overrightarrow{CD} + \overrightarrow{DB}$ in two ways. First we use the Triangle Law (as in the Tip) and second we use the expressions for $\overrightarrow{CD}$ and $\overrightarrow{DB}$ we have just computed in terms of $\mathbf{a}$ and $\mathbf{b}$.

$$\begin{aligned}
\overrightarrow{CD} + \overrightarrow{DB} &= \overrightarrow{CB} \\
&= -\overrightarrow{BC} \\
&= -\mathbf{a}
\end{aligned}$$

As a check, we can compute this in the second way.

$$\begin{aligned}
\overrightarrow{CD} + \overrightarrow{DB} &= (-2\mathbf{a} + \mathbf{b}) + (\mathbf{a} - \mathbf{b}) \\
&= (-2\mathbf{a} + \mathbf{a}) + (\mathbf{b} - \mathbf{b}) \\
&= -\mathbf{a}
\end{aligned}$$

- If $\mathbf{a} = \mathbf{i} + 2\mathbf{j} - \mathbf{k}$ and $\mathbf{b} = 3\mathbf{i} - 2\mathbf{j} + 2\mathbf{k}$ compute $3\mathbf{a} - 2\mathbf{b}$ and find its magnitude $|3\mathbf{a} - 2\mathbf{b}|$.

$$\begin{aligned}
3\mathbf{a} - 2\mathbf{b} &= 3(\mathbf{i} + 2\mathbf{j} - \mathbf{k}) - 2(3\mathbf{i} - 2\mathbf{j} + 2\mathbf{k}) \\
&= (3\mathbf{i} - 6\mathbf{i}) + (6\mathbf{j} + 4\mathbf{j}) + (-3\mathbf{k} - 4\mathbf{k} \\
&= -3\mathbf{i} + 10\mathbf{j} + -7\mathbf{k}
\end{aligned}$$

$$\begin{aligned}
|3\mathbf{a} - 2\mathbf{b}| &= |-3\mathbf{i} + 10\mathbf{j} + -7\mathbf{k}| \\
&= \sqrt{6^2 + 10^2 + 7^2} \\
&= \sqrt{36 + 100 + 49} \\
&= \sqrt{185}
\end{aligned}$$

• Find a vector of length 5 in the same directions as $\mathbf{a} = \langle 2, 1, -2 \rangle$

First compute the unit vector $\mathbf{u} = \mathbf{a}/|\mathbf{a}|$. Then $5\mathbf{u}$ is a vector of length 5 in the same direction.

First compute the magnitude of $\mathbf{a}$.
$$\begin{aligned} |a| &= \sqrt{2^2 + 1^2 + (-2)^2} \\ &= \sqrt{4 + 1 + 4} \\ &= \sqrt{9} \\ &= 3 \end{aligned}$$
Next compute the unit vector in the direction of $\mathbf{a}$.
$$\begin{aligned} \mathbf{u} &= \mathbf{a}/|\mathbf{a}| \\ &= \mathbf{a}/5 \\ &= \frac{1}{3}\langle 2, 1, -2 \rangle \\ &= \langle 2/3, 1/3, -2/3 \rangle \end{aligned}$$
Then the desired vector is
$$\begin{aligned} 5\mathbf{u} &= 5\langle 2/3, 1/3, -2/3 \rangle \\ &= \langle 10/3, 5/3, -10/3 \rangle. \end{aligned}$$

SkillMaster 9.6.

• Androcles is in an elevator descending a speed of 10 feet per second. Because of an asteroid that has hit the earth there is an earthquake which is making the building move due north at a speed of 2 feet per second. The asteroid has also caused a tidal wave that is pushing the building due east at a speed of 20 feet per second. How fast is Androcles moving?

Imagine that the standard basis vector $\mathbf{i}$ has the north direction, the basis vector $\mathbf{j}$ has the east direction, and the basis vector $\mathbf{k}$ points straight up. Then the velocity of Androcles may be written in terms of the standard basis vectors.

The velocity is
$$v = 2i + 20j - 10k.$$
The speed is the magnitude of the velocity vector.
$$\begin{aligned} |v| &= \sqrt{2^2 + 20^2 + 10^2} \\ &= \sqrt{4 + 400 + 100} \\ &= \sqrt{504} \\ &\approx 22.45 \text{ ft/s} \end{aligned}$$

• Suppose that 1 mule force is the force that a particular mule, named Henry, can exert on a chain when pulling forward. Suppose that Henry is pulling a chain attached to two chains which are themselves attached to fence posts which are near the corner of the south 40 of a farm. The chain attached to the mule is very strong but the chains attached to the fence posts will break if subjected to a force of magnitude 0.70 mule or greater. Will the chain break and the mule escape?

In the diagram the three force vectors must add to $\mathbf{0}$. Since the angles are 45° the horizontal and vertical components of $\mathbf{T}_1, \mathbf{T}_2$ have the same absolute value. This could be analyzed just using the diagram and what we know about isosceles right triangles. Instead use trigonometry and get $\mathbf{T}_1 = -\sin(45°)|\mathbf{T}_1|\,\mathbf{i} + \cos(45°)|\mathbf{T}_1|\,\mathbf{j} = -(\sqrt{2}/2)|\mathbf{T}_1|\,\mathbf{i} + (\sqrt{2}/2)|\mathbf{T}_1|\,\mathbf{j}$.

We have seen in the Tip that
$$\mathbf{T}_1 = -(\sqrt{2}/2)|\mathbf{T}_1|\,\mathbf{i} + (\sqrt{2}/2)|\mathbf{T}_1|\,\mathbf{j}.$$
Similarly
$$\begin{aligned}\mathbf{T}_2 &= \cos(45°)|\mathbf{T}_2|\,\mathbf{i} + \sin(45°)|\mathbf{T}_2|\,\mathbf{j}\\ &= (\sqrt{2}/2)|\mathbf{T}_2|\,\mathbf{i} + (\sqrt{2}/2)|\mathbf{T}_2|\,\mathbf{j}.\end{aligned}$$
For the forces to balance we must have
$$\begin{aligned}\mathbf{T}_1 + \mathbf{T}_2 + (-\mathbf{j}) &= \mathbf{0}\\ \mathbf{T}_1 + \mathbf{T}_2 &= \mathbf{j}\\ &= 0\mathbf{i} + \mathbf{j}.\end{aligned}$$
Equating the coefficients of $\mathbf{i}$ gives
$$\begin{aligned}-(\sqrt{2}/2)|\mathbf{T}_1| + (\sqrt{2}/2)|\mathbf{T}_2| &= 0\\ -|\mathbf{T}_1| + |\mathbf{T}_2| &= 0\\ |\mathbf{T}_1| &= |\mathbf{T}_2|.\end{aligned}$$
This means that the magnitude of the tensions on each of the chains attached to fence posts are equal. Now equate the coefficients of $\mathbf{j}$ and substitute $|\mathbf{T}_1|$ for $|\mathbf{T}_2|$ to get
$$\begin{aligned}(\sqrt{2}/2)|\mathbf{T}_1| + (\sqrt{2}/2)|\mathbf{T}_1| &= 1\\ \sqrt{2}|\mathbf{T}_1| &= 1\\ |\mathbf{T}_1| &= 1/\sqrt{2}\\ &\approx 0.7071.\end{aligned}$$
Since 0.7071 is greater than 0.7 we see that the force exerted by Henry puts too much tension on the chain. So the chain will break and Henry will escape.

Section 9.3 – The Dot Product

Key Concepts:

- The Dot Product of Two Vectors
- The Angle Between Two Vectors
- Scalar and Vector Projections of One Vector onto Another

Skills to Master:

- Compute dot products of vectors.
- Use dot products to find angles between vectors.
- Find scalar and vector projections.

Discussion:

The dot product of two vectors is introduced in this section. The dot product gives a way of multiplying two vectors together. However, it is important to remember that the dot product gives a scalar as a result, *NOT* a vector. The concept of dot product was first introduced to give a way of measuring the work done by a force that acts on an object in a direction different from the direction that the object moves. The dot product can also be used to measure the angle between two vectors and to find the scalar and vector projection of one vector onto another.

Key Concept: The Dot Product of Two Vectors

---

page 660.

page 662.

Pay careful attention to the explanation in the text of how the dot product can be used to calculate the *work* done by a force. Make sure that you are familiar with the definition of the dot product of two vectors and with the way to compute dot products in *component form*.

The dot product of two nonzero vectors **a** and **b** is the number
$$\mathbf{a} \cdot \mathbf{b} = |\mathbf{a}|\,|\mathbf{b}|\cos\theta$$

where θ is the angle between **a** and **b** with $0 \le \theta \le \pi$. If either vector is the zero vector, then the dot product is defined to be 0.

The dot product of $\mathbf{a} = \langle a_1, a_2, a_3 \rangle$ and $\mathbf{b} = \langle b_1, b_2, b_3 \rangle$ is
$$\mathbf{a} \cdot \mathbf{b} = a_1 b_1 + a_2 b_2 + a_3 b_3.$$

Key Concept: The Angle Between Two Vectors

By using the definition of dot product and solving for θ, you can find the angle between two nonzero vectors. First solve for $\cos\theta$.
$$\cos\theta = \frac{\mathbf{a} \cdot \mathbf{b}}{|\mathbf{a}|\,|\mathbf{b}|}$$
Then apply the inverse cosine function to both sides to get
$$\theta = \arccos\left(\frac{\mathbf{a} \cdot \mathbf{b}}{|\mathbf{a}|\,|\mathbf{b}|}\right).$$
Note that this gives you an angle between 0 and π. This formula shows that $\theta = \pi/2$ only when $\mathbf{a} \cdot \mathbf{b} = 0$, so this gives an easy way to check whether two vectors are orthogonal.

Key Concept: Scalar and Vector Projections of One Vector onto Another

page 664.

Study *Figure 4* and *Figure 5* in this section to gain a geometric understanding of the vector and scalar projection of one vector onto another. The scalar projection is the magnitude of the vector projection. The vector projection of **b** onto **a** is denoted $\text{proj}_\mathbf{a}\mathbf{b}$ and the scalar projection of **b** onto **a** is denoted $\text{comp}_\mathbf{a}\mathbf{b}$.
$$\text{proj}_\mathbf{a}\mathbf{b} = \frac{\mathbf{a} \cdot \mathbf{b}}{\mathbf{a} \cdot \mathbf{a}}\mathbf{a} \qquad \text{comp}_\mathbf{a}\mathbf{b} = \frac{\mathbf{a} \cdot \mathbf{b}}{|\mathbf{a}|}$$
You should understand how these formulas are derived. Do not just memorize them. Why is the formula for $\text{proj}_\mathbf{a}\mathbf{b}$ the same as that in the text?

SkillMaster 9.7: Compute dot products of vectors.

You should be able to compute dot products either by using the definition or by using the component form of the dot product. If you are given information about the lengths of vectors and the angle between them, use the definition to compute dot products. If the vectors are given in component form, use the component form of the dot product to compute dot products.

SkillMaster 9.8: Use dot products to find angles between vectors.

Given two vectors in component form, make sure that you know how to use the dot product to compute the angle between the vectors. The formula in the key concept above shows how to do this. This formula also gives you and easy way to check if two vectors are orthogonal.

SkillMaster 9.9: Find scalar and vector projections.

It is often necessary in applications to determine the projection of one vector onto another. Again, rather than just memorizing the formulas, make sure that you understand the geometric motivation for the formulas.

Worked Examples

For each of the following examples, first try to find the solution without looking at the middle or right columns. Cover the middle and right columns with a piece of paper. If you need a hint, uncover the middle column. If you need to see the worked solution, uncover the right column.

Example	Tip	Solution
SkillMaster 9.7. • Suppose the vector $\mathbf{a} = \langle 1, -1, 1 \rangle$ makes an angle of $30°$ with a vector $\mathbf{b}$, and that you know that $|\mathbf{b}| = 4$. Calculate the dot product, $\mathbf{a} \cdot \mathbf{b}$.	Recall the geometric definition of the dot product, $\mathbf{a} \cdot \mathbf{b} = |\mathbf{a}|\,|\mathbf{b}|\cos\theta$, where θ is the angle between $\mathbf{a}$ and $\mathbf{b}$.	$\begin{aligned} \mathbf{a} \cdot \mathbf{b} &= |\mathbf{a}|\,|\mathbf{b}|\cos\theta \\ &= \sqrt{1^2 + (-1)^2 + 1^2}(4)\cos 30° \\ &= 4\sqrt{3}(\sqrt{3}/2) \\ &= 6 \end{aligned}$
• Suppose that $\mathbf{a}$ has magnitude 2, $\mathbf{b}$ is a unit vector and that $\mathbf{a} \cdot \mathbf{b} = -2$. Compute the magnitude of $\mathbf{a} + 3\mathbf{b}$.	The magnitude $|\mathbf{a} + 3\mathbf{b}|$ is the square root of the dot product $(\mathbf{a} + 3\mathbf{b}) \cdot (\mathbf{a} + 3\mathbf{b})$. Use the laws of dot products to reduce this to a form so that the information from the problem is useful.	$\begin{aligned} &|\mathbf{a} + 3\mathbf{b}|^2 \\ &= (\mathbf{a} + 3\mathbf{b}) \cdot (\mathbf{a} + 3\mathbf{b}) \\ &= (\mathbf{a} \cdot (\mathbf{a} + 3\mathbf{b})) + (3\mathbf{b} \cdot (\mathbf{a} + 3\mathbf{b})) \\ &= (\mathbf{a} \cdot \mathbf{a} + \mathbf{a} \cdot 3\mathbf{b}) + (3\mathbf{b} \cdot (\mathbf{a} + 3\mathbf{b})) \\ &= (\mathbf{a} \cdot \mathbf{a} + \mathbf{a} \cdot 3\mathbf{b}) + (3\mathbf{b} \cdot \mathbf{a} + 3\mathbf{b} \cdot 3\mathbf{b}) \\ &= (\mathbf{a} \cdot \mathbf{a} + 3\mathbf{a} \cdot \mathbf{b}) + (3\mathbf{a} \cdot \mathbf{b} + 9\mathbf{b} \cdot \mathbf{b}) \\ &= \mathbf{a} \cdot \mathbf{a} + 6\mathbf{a} \cdot \mathbf{b} + 9\mathbf{b} \cdot \mathbf{b} \\ &= |\mathbf{a}|^2 + 6\mathbf{a} \cdot \mathbf{b} + 9\,|\mathbf{b}|^2 \\ &= 2^2 + 6(-2) + 9\left(1^2\right) \\ &= 4 - 12 + 9 \\ &= 1 \\[6pt] &|\mathbf{a} + 3\mathbf{b}| \\ &= \sqrt{(\mathbf{a} + 3\mathbf{b}) \cdot (\mathbf{a} + 3\mathbf{b})} \\ &= \sqrt{1} \\ &= 1 \end{aligned}$

- .A mass is dragged up an incline of 38° for 2 m by a force of 5.8 N that is directed at an angle of 54° to the horizontal as shown in the diagram. What is the work done?

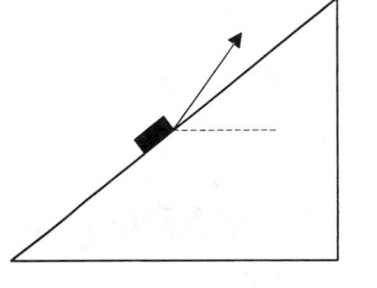

The angle between the force vector and the direction of the mass displacement is $54° - 38° = 16°$. Use the geometric definition of the dot product (which was inspired by the work problem).

Call the force vector $\mathbf{F}$ and the displacement vector $\mathbf{D}$. We know that the angle between these vectors is 16° and we know their magnitudes are $|\mathbf{F}| = 5.8$ and $|\mathbf{D}| = 2$. The work done is the dot product

$$
\begin{aligned}
\mathbf{F} \cdot \mathbf{D} &= |\mathbf{F}|\,|\mathbf{D}| \cos 16° \\
&= (5.8)(2)(0.9613) \\
&\approx 11.15 \text{ Nm} \\
&= 11.15 \text{ J.}
\end{aligned}
$$

SkillMaster 9.8.

- Find the angle between the two vectors $\mathbf{F}_1$ and $\mathbf{F}_2$ where $\mathbf{F}_1 = \mathbf{j} - \mathbf{k}$ and $\mathbf{F}_2 = 2\mathbf{i} - \mathbf{j} + 2\mathbf{k}$.

Equate the dot product formulas from the geometric and algebraic definitions. Solve the formula $\mathbf{F}_1 \cdot \mathbf{F}_2 = |\mathbf{F}_1|\,|\mathbf{F}_2| \cos \theta$ for $\cos \theta$ then solve for θ using the inverse cosine function.

$$
\begin{aligned}
|\mathbf{F}_1| &= \sqrt{\mathbf{F}_1 \cdot \mathbf{F}_1} \\
&= \sqrt{(\mathbf{j} - \mathbf{k}) \cdot (\mathbf{j} - \mathbf{k})} \\
&= \sqrt{0\,(0) + 1\,(1) + (-1)(-1)} \\
&= \sqrt{0 + 1 + 1} \\
&= \sqrt{2} \\
|\mathbf{F}_2| &= \sqrt{\mathbf{F}_2 \cdot \mathbf{F}_2} \\
&= \sqrt{(2\mathbf{i} - \mathbf{j} + 2\mathbf{k}) \cdot (2\mathbf{i} - \mathbf{j} + 2\mathbf{k})} \\
&= \sqrt{2(2) + (-1)(-1) + (2)(2)} \\
&= \sqrt{4 + 1 + 4} \\
&= \sqrt{9} \\
&= 3
\end{aligned}
$$

$$
\begin{aligned}
\mathbf{F}_1 \cdot \mathbf{F}_2 &= |\mathbf{F}_1|\,|\mathbf{F}_2| \cos \theta \\
(\mathbf{j} - \mathbf{k}) \cdot (2\mathbf{i} - \mathbf{j} + 2\mathbf{k}) &= \sqrt{2}(3) \cos \theta \\
(0)(1) + (1)(-1) + (-1)(2) &= 3\sqrt{2} \cos \theta \\
-3 &= 3\sqrt{2} \cos \theta \\
\cos \theta &= -1/\sqrt{2} \\
\theta &= 3\pi/4 = 135°
\end{aligned}
$$

• Find a nonzero vector that is orthogonal to the vector $\langle 2, 1 \rangle$. Then try to generalize your approach to find a formula for a non-zero vector orthogonal to $\langle a, b \rangle$ which is assumed to be a non-zero vector (that is, at least one of a and b are not 0).

First notice that there are many vectors that are orthogonal (perpendicular) to $< 2, 1 >$. Any scalar multiple of a vector that is orthogonal to $< 2, 1 >$ is also orthogonal to $< 2, 1 >$. Let $\langle x, y \rangle$ be a vector that is perpendicular (orthogonal) to $< 2, 1 >$. Simplify the equation $< 2, 1 > \cdot < x, y >= 0$ and choose a simple solution as possible. Then generalize the pattern that you see.

If $< x, y >$ is orthogonal to $< 2, 1 >$ then

$$0 = <2, 1> \cdot <x, y>$$
$$0 = 2x + y$$
$$y = -2x.$$

The simplest way to get a non-zero solution is to let $1 = x$ so that $y = -2$. A non-zero vector orthogonal to $< 2, 1 >$ is $< x, y >=< 1, -2 >$. Notice that this vector takes the original vector $< 2, 1 >$ and interchanges the components (to get $< 1, 2 >$) then puts a negative sign on the last component (to get $< 1, -2 >$). Copying this pattern to get a vector orthogonal to $< a, b >$ would give us $< b, -a >$. As a check that these vector are orthogonal compute $< a, b > \cdot < b, -a >= ab - ab = 0$. Notice that we could also have used $< -b, a >$ as the orthogonal vector which has the same length as $< b, -a >$ but points in the opposite direction.

SkillMaster 9.9.

• Given the vectors $\mathbf{a} = 2\mathbf{i} + \mathbf{j}$ and $\mathbf{b} = 3\mathbf{i} - 2\mathbf{j}$ find the scalar and the vector projections of $\mathbf{b}$ onto $\mathbf{a}$. Explain your reasoning (do NOT just look up the formula).

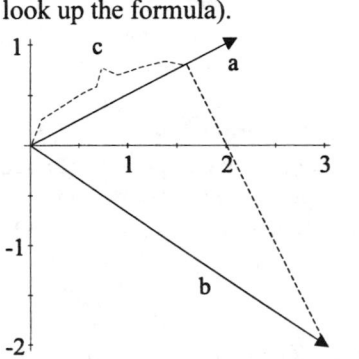

First remember that if $\mathbf{a}$ is a nonzero vector then the vector $\mathbf{u} = \mathbf{a}/|\mathbf{a}|$ is a unit vector (so $|\mathbf{u}| = 1$) with the same direction as $\mathbf{a}$. If $c > 0$ is a scalar then $c\mathbf{u}$ is a vector in the same direction as $\mathbf{a}$ with length c.

In the diagram the scalar projection of $\mathbf{b}$ onto $\mathbf{a}$ is the length c of the adjacent side in the right triangle. The definition of the cosine function tells us that the scalar projection of $\mathbf{b}$ onto $\mathbf{a}$ is $c = |\mathbf{b}| \cos \theta$. The vector projection of $\mathbf{b}$ onto $\mathbf{a}$ is the vector with magnitude c and direction $\mathbf{a}$ or $c\mathbf{u}$.

The scalar projection of $\mathbf{b}$ onto $\mathbf{a}$ is

$$
\begin{aligned}
c &= |\mathbf{b}| \cos \theta \\
&= |\mathbf{u}| \, |\mathbf{b}| \cos \theta \\
&= \mathbf{u} \cdot \mathbf{b} \\
&= (\mathbf{a}/|\mathbf{a}|) \cdot \mathbf{b} \\
&= \frac{\mathbf{a} \cdot \mathbf{b}}{|\mathbf{a}|} \\
&= \frac{(2\mathbf{i} + \mathbf{j}) \cdot (3\mathbf{i} - 2\mathbf{j})}{\sqrt{(2\mathbf{i} + \mathbf{j}) \cdot (2\mathbf{i} + \mathbf{j})}} \\
&= \frac{2(3) + 1(-2)}{\sqrt{2(2) + 1(1)}} \\
&= \frac{4}{\sqrt{5}}
\end{aligned}
$$

The vector projection of $\mathbf{b}$ onto $\mathbf{a}$ is the vector with magnitude c and direction $\mathbf{a}$ is

$$
\begin{aligned}
c\mathbf{u} &= c\frac{\mathbf{a}}{|\mathbf{a}|} \\
&= \frac{\mathbf{a} \cdot \mathbf{b}}{|\mathbf{a}|} \frac{\mathbf{a}}{|\mathbf{a}|} \\
&= \frac{\mathbf{a} \cdot \mathbf{b}}{|\mathbf{a}|^2}\mathbf{a} \\
&= \frac{\mathbf{a} \cdot \mathbf{b}}{\mathbf{a} \cdot \mathbf{a}}\mathbf{a} \\
&= \frac{(2\mathbf{i} + \mathbf{j}) \cdot (3\mathbf{i} - 2\mathbf{j})}{(2\mathbf{i} + \mathbf{j}) \cdot (2\mathbf{i} + \mathbf{j})}(2\mathbf{i} + \mathbf{j}) \\
&= \frac{4}{5}(2\mathbf{i} + \mathbf{j}) \\
&= \frac{8}{5}\mathbf{i} + \frac{4}{5}\mathbf{j}.
\end{aligned}
$$

• Given $\mathbf{a} = \langle 2, -6, 3 \rangle$ and $\mathbf{b} = \langle 1, -2, -2 \rangle$, find the vector projection of $\mathbf{b}$ onto $\mathbf{a}$.

See the hint for the previous problem.

The vector projection of $\mathbf{b}$ onto $\mathbf{a}$ is

$$\frac{\mathbf{a} \cdot \mathbf{b}}{\mathbf{a} \cdot \mathbf{a}} \mathbf{a}$$

$$= \frac{<2, -6, 3> \cdot <1, -2, -2>}{<2, -6, 3> \cdot <2, -6, 3>} <2, -6, 3>$$

$$= \frac{2(1) + (-6)(-2) + 3(-2)}{2(2) + (-6)(-6) + 3(3)} <2, -6, 3>$$

$$= \frac{2 + 12 - 6}{4 + 36 + 9} <2, -6, 3>$$

$$= \frac{8}{49} <2, -6, 3>$$

$$= <\frac{16}{49}, \frac{-48}{49}, \frac{24}{49}> .$$

• Suppose that $\mathbf{a} = \overrightarrow{AB}$ has length 3 and that $\mathbf{b} = \overrightarrow{AC}$, and that the angle ABC is a right angle. Compute the dot product $\mathbf{a} \cdot \mathbf{b}$.

Use the fact that the scalar projection of $\mathbf{b}$ onto $\mathbf{a}$ is the magnitude of $\mathbf{a}$.

$$|\mathbf{a}| = \text{comp}_\mathbf{a}\mathbf{b} = \frac{\mathbf{a} \cdot \mathbf{b}}{|\mathbf{a}|};$$
$$\text{So, } \mathbf{a} \cdot \mathbf{b} = |\mathbf{a}|^2$$
$$= 3^2$$
$$= 9$$

Note this only is true because we are dealing with a right triangle. We could also have used the trigonometric fact that $|\mathbf{b}| \cos \theta = |\mathbf{a}|$, we have
$$\mathbf{a} \cdot \mathbf{b} = |\mathbf{a}||\mathbf{b}| \cos \theta$$
$$= |\mathbf{a}||\mathbf{a}|$$
$$= |\mathbf{a}|^2$$
$$= 9.$$

Section 9.4 – The Cross Product

Key Concepts:

- Cross Product of Two Vectors
- Scalar and Vector Triple Products

Skills to Master:

- Compute and use cross products.
- Compute and use scalar and vector triple products.

Discussion:

page 666.

This section introduce the cross product of two vectors. In contrast to the dot product of two vectors (which yields a scalar), the cross product of two vectors yields a third vector that is perpendicular to the first two. One physical motivation for the cross product is *torque*. Read the *explanation* in the text to see how the cross product arises in computations of torque. You can only take the cross product of three-dimensional vectors. This section also introduces the scalar and vector triple products of three three-dimensional vectors.

Key Concept: Cross Product of Two Vectors

page 667.

The cross product of two three-dimensional vectors **a** and **b**, **a** × **b**, is a vector that is perpendicular to both **a** and **b** and that has magnitude
$$|\mathbf{a}||\mathbf{b}| \sin \theta$$
where θ is the angle between **a** and **b**. Since there are two vectors that satisfy this, we need to specify in addition that the cross product point in the direction determined by the *right hand rule*.

The cross product can be computed in component form also. If $\mathbf{a} = \langle a_1, a_2, a_3 \rangle$ and $\mathbf{b} = \langle b_1, b_2, b_3 \rangle$, then

$$\mathbf{a} \times \mathbf{b} = \langle a_2 b_3 - a_3 b_2, a_3 b_1 - a_1 b_3, a_1 b_2 - a_2 b_1 \rangle = \begin{vmatrix} \mathbf{i} & \mathbf{j} & \mathbf{k} \\ a_1 & a_2 & a_3 \\ b_1 & b_2 & b_3 \end{vmatrix}.$$

page 670.

Study the discussion of *determinants* in the text to see an easy way to remember this formula.

Key Concept: Scalar and Vector Triple Products

The scalar triple product of three three-dimensional vectors $\mathbf{a}$, $\mathbf{b}$, and $\mathbf{c}$ is the scalar quantity

$$\mathbf{a} \cdot (\mathbf{b} \times \mathbf{c}).$$

A physical motivation for the definition of the scalar triple product is that the magnitude is the volume of the parallelepiped determined by the vectors $\mathbf{a}$, $\mathbf{b}$, and $\mathbf{c}$. See *Figure 7* in the text.

page 671.

The vector triple product of three vectors $\mathbf{a}$, $\mathbf{b}$, and $\mathbf{c}$ is $\mathbf{a} \times (\mathbf{b} \times \mathbf{c})$.

SkillMaster 9.10: Compute and use cross products.

You should be able to use both the definition of cross product and the component form of cross product to compute the cross product of two three-dimensional vectors. Make sure that you understand how to apply the right hand rule. You should be able to use cross products in applications.

SkillMaster 9.11: Compute and use scalar and vector triple products.

In computing scalar triple products, you will need to use the definition of both dot product and cross product. It will sometimes be useful to use

$$\mathbf{a} \cdot (\mathbf{b} \times \mathbf{c}) = (\mathbf{a} \times \mathbf{b}) \cdot \mathbf{c}.$$

Note that on both sides of the above equation, the cross product needs to be computed first.

The following formula from the text shows how to compute vector triple products by using dot products and scalar multiplication.

$$\mathbf{a} \times (\mathbf{b} \times \mathbf{c}) = (\mathbf{a} \cdot \mathbf{c})\mathbf{b} - (\mathbf{a} \cdot \mathbf{b})\mathbf{c}$$

You should express both sides of this equation using the components of $\mathbf{a}$, $\mathbf{b}$, and $\mathbf{c}$ to convince yourself that it is true.

Worked Examples

For each of the following examples, first try to find the solution without looking at the middle or right columns. Cover the middle and right columns with a piece of paper. If you need a hint, uncover the middle column. If you need to see the worked solution, uncover the right column.

Example	Tip	Solution

SkillMaster 9.10.

- Suppose that $a = \langle -1, 3, 1 \rangle$ and $b = \langle 2, 5, 4 \rangle$. Find two unit vectors that are orthogonal to both a and b. Find the area of the parallelogram determined by a and b.

The cross product of a and b is a vector that is orthogonal to both a and b. The unit vector in this direction and the negative of this unit vector are both orthogonal to a and b. The magnitude of the cross product of a and b is the area of the parallelogram determined by a and b.

First compute the cross product

$$a \times b = \langle -1, 3, 1 \rangle \times \langle 2, 5, 4 \rangle$$

$$= \begin{vmatrix} i & j & k \\ -1 & 3 & 1 \\ 2 & 5 & 4 \end{vmatrix}$$

$$= \begin{vmatrix} 3 & 1 \\ 5 & 4 \end{vmatrix} i - \begin{vmatrix} -1 & 1 \\ 2 & 4 \end{vmatrix} j + \begin{vmatrix} -1 & 3 \\ 2 & 5 \end{vmatrix} k$$

$$= (12 - 5)i - (-4 - 2)j + (-5 - 6)k$$

$$= 7i + 6j + -11k.$$

The magnitude of the cross product is the area of the parallelogram determined by a and b which is

$$|a \times b| = |\langle 7, 6, -11 \rangle|$$

$$= \sqrt{7^2 + 6^2 + (-11)^2}$$

$$= \sqrt{49 + 36 + 121} = \sqrt{206} \approx 14.35$$

A unit vector that is orthogonal to both a and b is

$$n = \frac{a \times b}{|a \times b|} = \frac{1}{\sqrt{206}} \langle 7, 6, -11 \rangle$$

$$= \left\langle \frac{7}{\sqrt{206}}, \frac{6}{\sqrt{206}}, \frac{-11}{\sqrt{206}} \right\rangle.$$

Another unit vector orthogonal to both a and b is $-n$ which points in the opposite direction,

$$-n = \left\langle \frac{-7}{\sqrt{206}}, \frac{-6}{\sqrt{206}}, \frac{11}{\sqrt{206}} \right\rangle.$$

• Consider the triangle formed by the points $P = P(3, 0, 1)$, $Q = Q(7, 2, 4)$, and $R = R(4, 2, 5)$. Find a vector that is orthogonal to each of the edges of this triangle, then find the area.

Pick any of these three points as a "base point" to be the initial point of vectors coinciding with the two edges of the triangle touching the base point. The cross product of these two vectors is orthogonal to the plane containing the triangle. Also, the area of the triangle is one half the area or the parallelogram determined by these two edge vectors.

We pick the point $P = P(3, 0, 1)$ (but you could have chosen Q or R just as well). Let $\mathbf{a} = \vec{PQ} =< 7 - 3, 2 - 0, 4 - 1 >=< 4, 2, 3 >$ and $\mathbf{b} = \vec{PR} =< 4 - 3, 2 - 0, 5 - 1 >=< 1, 2, 4 >$. Then, a vector orthogonal to $\mathbf{a}$ and $\mathbf{b}$ (and thus to the plane containing P, Q, and R) is the cross product

$$\mathbf{a} \times \mathbf{b} = < 4, 2, 3 > \times < 1, 2, 4 >$$
$$= \begin{vmatrix} \mathbf{i} & \mathbf{j} & \mathbf{k} \\ 4 & 2 & 3 \\ 1 & 2 & 4 \end{vmatrix}$$
$$= \begin{vmatrix} 2 & 3 \\ 2 & 4 \end{vmatrix} \mathbf{i} - \begin{vmatrix} 4 & 3 \\ 1 & 4 \end{vmatrix} \mathbf{j} + \begin{vmatrix} 4 & 2 \\ 1 & 2 \end{vmatrix} \mathbf{k}$$
$$= (8 - 6)\mathbf{i} - (16 - 3)\mathbf{j} + (8 - 2)\mathbf{k}$$
$$= 2\mathbf{i} - 13\mathbf{j} + 6\mathbf{k}.$$

To find the area of the triangle formed by the three points, we need only find the area of the parallelogram formed by $\mathbf{a}$ and $\mathbf{b}$ and divide by two.

$$\begin{aligned} \text{Area} &= \frac{1}{2} |\mathbf{a} \times \mathbf{b}| \\ &= \frac{1}{2} \sqrt{2^2 + (-13)^2 + 6^2} \\ &= \frac{1}{2} \sqrt{4 + 169 + 36} \\ &= \frac{1}{2} \sqrt{209} \\ &\approx 7.23 \end{aligned}$$

SkillMaster 9.11.

• Suppose that $a = 2i + 3j - k$, $b = 5i$, $c = i + j$. Calculate the volume of the parallelepiped determined by these vectors. Also, using the cross product find the angle between b and c.

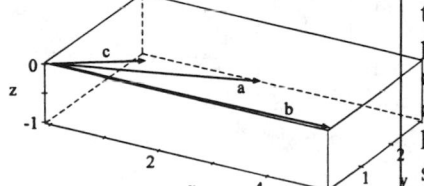

Recall the volume is given by the triple scalar product $V = |a \cdot (b \times c)|$. You could use the determinant formula to compute the cross product but in this case it is easier to do directly. The right hand rule may be simplified according to the diagram in the solution column below. The cross product of any two standard basis vectors is positive (and equal to the other standard basis vector) if they are ordered counter-clockwise in the diagram and the cross product is negative (and equal to the other standard basis vector) if they are ordered in the clockwise direction. For example $i \times j = k$ because j follows i counter-clockwise in the diagram.

$$
\begin{aligned}
b \times c &= (5i) \times (i + j) \\
&= 5i \times i + 5i \times j \\
&= 5(0) + 5k \\
&= 5k
\end{aligned}
$$

So
$$
\begin{aligned}
a \cdot (b \times c) &= (2i + 3j - k) \times (5k) \\
&= 2(0) + 3(0) + (-1)(5) \\
&= -5
\end{aligned}
$$
Therefore,
$$
V = |-5| = 5.
$$
To find the angle between b and c we use the geometric form of the magnitude of the cross product:
$$
\begin{aligned}
|b \times c| &= |b||c| \sin\theta \\
|5k| &= |5i| \, |i + j| \sin\theta \\
5 &= 5\sqrt{2} \sin\theta \\
\sin\theta &= 1/\sqrt{2} \\
\theta &= \pi/4 \text{ radians}
\end{aligned}
$$

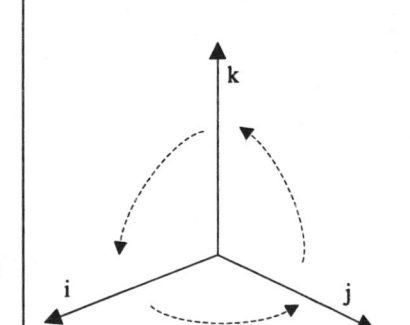

495

• Suppose that x is a real number that determines the vectors

$$\mathbf{a} = \mathbf{i} - x\mathbf{j} + \mathbf{k}$$
$$\mathbf{b} = x\mathbf{i} - \mathbf{j} + \mathbf{k}$$
$$\mathbf{c} = -\mathbf{i} + x\mathbf{j} + \mathbf{k}.$$

Determine the values of x that make these vectors coplanar. Find all values of x for which the volume of parallelepiped formed by $\mathbf{a}$, $\mathbf{b}$, and $\mathbf{c}$ equals 6.

Recall the area of the parallelepiped determined by $\mathbf{a}$, $\mathbf{b}$, and $\mathbf{c}$ is given by the scalar triple product. The volume is 0 if and only if the vectors all lie on a plane. Find all values of x for which the scalar triple product is 0.

$$0 = \pm V = \mathbf{a} \cdot (\mathbf{b} \times \mathbf{c})$$

$$= \begin{vmatrix} 1 & -x & 1 \\ x & -1 & 1 \\ -1 & x & 1 \end{vmatrix}$$

$$= (1)\begin{vmatrix} -1 & 1 \\ x & 1 \end{vmatrix} - (-x)\begin{vmatrix} x & 1 \\ -1 & 1 \end{vmatrix}$$
$$+ (1)\begin{vmatrix} x & -1 \\ -1 & x \end{vmatrix}$$

$$= (-1 - x) + x(x - (-1)) + (x^2 - (-1)(-1))$$
$$= -1 - x + x^2 + x + x^2 - 1$$
$$= 2x^2 - 2$$
$$= 2(x^2 - 1)$$
$$= 2(x - 1)(x + 1)$$

The vectors will be coplanar if and only if $x = 1$ or $x = -1$.

To find all x such that the volume equals 6 use the fact that $V = |2x^2 - 2|$. There are two possible cases: (a) $2x^2 - 2 = 6$ or (b) $2x^2 - 2 = -6$. In case (a)

$$\begin{aligned} 2x^2 - 2 &= 6 \\ x^2 - 1 &= 3 \\ x^2 &= 4 \\ x &= \pm 2. \end{aligned}$$

In case (b)

$$\begin{aligned} 2x^2 - 2 &= -6 \\ 2x^2 &= -4 \\ x^2 &= -2. \end{aligned}$$

There is no solution. Therefore all the values of x that give a volume of 6 are $x = 2$ and $x = -2$.

- If $\mathbf{a} = \langle a_1, a_2, a_3 \rangle$ express $\mathbf{i} \times (\mathbf{a} \times \mathbf{k})$ in component form.

Use the diagram for cross products above and the fact that $\mathbf{a} = a_1\mathbf{i} + a_2\mathbf{j} + a_3\mathbf{k}$ to first find $\mathbf{a} \times \mathbf{k}$. Then continue like this to find the vector triple product.

$$\mathbf{a} \times \mathbf{k} = (a_1\mathbf{i} + a_2\mathbf{j} + a_3\mathbf{k}) \times \mathbf{k}$$
$$= -a_1\mathbf{j} + a_2\mathbf{i} + 0$$
$$= a_2\mathbf{i} - a_1\mathbf{j}$$
$$\mathbf{i} \times (\mathbf{a} \times \mathbf{k}) = \mathbf{i} \times (a_2\mathbf{i} - a_1\mathbf{j})$$
$$= 0 - a_1\mathbf{k}$$
$$= -a_1\mathbf{k}$$
$$= \ <0, 0, -a_1>.$$

- Suppose that
$\mathbf{a} = < 1, -2, 3 >,$
$\mathbf{b} = < 4, 3, -2 >,$
$\mathbf{c} = < 2, 5, -1 >.$
Calculate the triple vector product.

You could use the determinant formula for the cross product twice, but simpler is to use the formula in the text,
$\mathbf{a} \times (\mathbf{b} \times \mathbf{c}) = (\mathbf{a} \cdot \mathbf{c})\mathbf{b} - (\mathbf{a} \cdot \mathbf{b})\mathbf{c}$.
This has an equivalent form that is easier to remember.
$\mathbf{a} \times (\mathbf{b} \times \mathbf{c}) = (\mathbf{a} \cdot \mathbf{c})\mathbf{b} - (\mathbf{a} \cdot \mathbf{b})\mathbf{c}$
$=$
$(\mathbf{c} \cdot \mathbf{a})\mathbf{b} - (\mathbf{b} \cdot \mathbf{a})\mathbf{c}$
This can be remembered by thinking "cab-back".

$$\mathbf{a} \times (\mathbf{b} \times \mathbf{c}) = (\mathbf{c} \cdot \mathbf{a})\mathbf{b} - (\mathbf{b} \cdot \mathbf{a})\mathbf{c}$$
Make some side calculations first.
$$(\mathbf{c} \cdot \mathbf{a}) = \ <2, 5, -1> \cdot <1, -2, 3>$$
$$= 2(1) + 5(-2) + (-1)(3)$$
$$= 2 - 10 - 3$$
$$= -11$$

$$(\mathbf{b} \cdot \mathbf{a}) = \ <4, 3, -2> \cdot <1, -2, 3>$$
$$= 4(1) + 3(-2) + (-2)(3)$$
$$= 4 - 6 - 6$$
$$= -8$$

$$\mathbf{a} \times (\mathbf{b} \times \mathbf{c})$$
$$= (\mathbf{c} \cdot \mathbf{a})\mathbf{b} - (\mathbf{b} \cdot \mathbf{a})\mathbf{c}$$
$$= -11\mathbf{b} - (-8)\mathbf{c}$$
$$= -11\mathbf{b} + 8\mathbf{c}$$
$$= (-11) < 4, 3, -2 > + (8) < 2, 5, -1 >$$
$$= \ < -44, -33, 22 > + < 16, 40, -8 >$$
$$= \ < -28, 7, 14 >$$

497

Section 9.5 – Equations of Planes and Lines

Key Concepts:

- Equations of Lines
- Equations of Planes

Skills to Master:

- Compute and use vector and parametric equations of lines.
- Compute and use equations of planes.

Discussion:

This section introduces equations of lines and planes in three-dimensional space. A line can be given by a vector equation, by parametric equations, or by symmetric equations. A plane can be given by a vector equation or by a scalar equation. The key to understanding the various forms of equations for lines is to realize that a line is determined by a point on the line and a direction vector. The key to understanding the various equations for planes is to realize that a plane is determined by a point on the plane and by a normal vector to the plane.

Key Concept: Equations of Lines

A line L can be determined by specifying a point $P_0(x_0, y_0, z_0)$ on the line and specifying a vector $\mathbf{v} = \langle a, b, c \rangle$ parallel to the line. The vector equation of the line is then

$$\mathbf{r} = \mathbf{r}_0 + t\mathbf{v}$$

 page 675.
where $\mathbf{r}_0 = \langle x_0, y_0, z_0 \rangle$ and where $\mathbf{r} = \langle x, y, z \rangle$ is the position vector to a point $P(x, y, z)$ on the line. See *Figure 1* for a geometric picture of the quantities involved. As t varies over all real numbers, the line is traced out.

498

Section 9.5 – Equations of Planes and Lines

If the above vector equation is rewritten in component form, the parametric equations of the line are obtained.

$$x = x_0 = at \qquad y = y_0 + bt \qquad z = z_0 + ct$$

Again, $P_0(x_0, y_0, z_0)$ is a point on the line and $\mathbf{v} = \langle a, b, c \rangle$ is a vector parallel to the line.

If the above parametric equations are solved for t and the resulting quantities are set equal, one obtains the symmetric equations of the line.

$$\frac{x - x_0}{a} = \frac{y - y_0}{b} = \frac{z - z_0}{c}$$

Key Concept: Equations of Planes

A plane can be determined by specifying a point $P_0(x_0, y_0, z_0)$ on the plane and by specifying a normal vector $\mathbf{n} = \langle a, b, c \rangle$ to the plane. The vector equation of the plane is then

$$\mathbf{n} \cdot (\mathbf{r} - \mathbf{r}_0) = 0 \ \text{ or } \ \mathbf{n} \cdot \mathbf{r} = \mathbf{n} \cdot \mathbf{r}_0$$

where $\mathbf{r}_0 = \langle x_0, y_0, z_0 \rangle$ and where $\mathbf{r} = \langle x, y, z \rangle$ is the position vector to a point $P(x, y, z)$ on the plane. See *Figure 6* for a geometric picture of these quantities.

page 678

If the above vector equation is written in component form, the scalar equation of the plane is obtained,

$$a(x - x_0) + b(y - y_0) + c(z - z_0) = 0 \ \text{ or } \ ax + by + cz = d$$

where $d = ax_0 + by_0 + cz_0$.

SkillMaster 9.12: Compute and use vector and parametric equations of lines.

page 677.

If you are given two points on a line, or if you are given a point on a line and a direction vector for the line, you should be able to determine an equation for the line. You should also be able to go from one form of the equation for a line to another. To determine whether two lines intersect, work with the symmetric equations of the lines as in *Example 3*. To determine where a line intersects a plane, work with the parametric equations of the line and the scalar equation of the plane.

SkillMaster 9.13: Compute and use equations of planes.

You should be able to determine an equation of a plane if you are given a point on the plane and a normal vector to the plane. If, instead, you are given three points on the plane, first determine a normal vector to the plane by taking the cross product of two vectors in the plane as in *Example 5*.

page 679.

To determine the angle between two planes, find the angle between normal vectors to the planes. To determine the line of intersection of two planes, follow the method described in *Example 7*.

page 680.

Worked Examples

For each of the following examples, first try to find the solution without looking at the middle or right columns. Cover the middle and right columns with a piece of paper. If you need a hint, uncover the middle column. If you need to see the worked solution, uncover the right column.

Example	Tip	Solution

SkillMaster 9.12.

- Find a vector equation, parametric equations, and the symmetric equations for the line that passes through the points $A(1, 2, 3)$ and $B(-2, 0, 5)$.

$\mathbf{v} = \overrightarrow{AB}$ is a vector parallel to the line. Use the vector representation $\mathbf{r} = \mathbf{r}_0 + t\mathbf{v}$ where the components of $\mathbf{r}_0$ are the components of $A(1, 2, 3)$. The parametric equations may be read off from the components of the vector equation. To find the symmetric equations solve each of the parametric equations for t and set the expressions in x, y, or z equal to each other.

$$\begin{aligned} \mathbf{v} &= \overrightarrow{AB} \\ &= <-2-1, 0-2, 5-3> \\ &= <-3, -2, 2>. \\ \mathbf{r}_0 &=< 1, 2, 3 > \end{aligned}$$

The vector equations are

$$\begin{aligned} \mathbf{r} &= \mathbf{r}_0 + t\mathbf{v} \\ \mathbf{r} &= <1, 2, 3> + t<-3, -2, 2> \\ \mathbf{r} &= <1-3t, 2-2t, 3+2t>. \end{aligned}$$

The parametric equations are the read off the components of the vector in the vector equation.

$$\begin{aligned} x &= 1 - 3t \\ y &= 2 - 2t \\ z &= 3 + 2t \end{aligned}$$

To obtain the symmetric equations solve the parametric equations for t.

$$\begin{aligned} t &= \frac{1-x}{3} \\ t &= \frac{2-y}{2} \\ t &= \frac{z-3}{2} \end{aligned}$$

Set all the expressions on the right hand side equal to each other to get the symmetric equations.

$$\frac{1-x}{2} = \frac{2-y}{2} = \frac{3-z}{3}$$

• Find a vector equation, parametric equations, and the symmetric equations for the line that is parallel to the vector $\langle 0, 1, -3 \rangle$ passes through the point $(3, -2, 0)$

Use the vector equation form $\mathbf{r} = \mathbf{r}_0 + t\mathbf{v}$ where $\mathbf{r}_0 = \langle 3, -2, 0 \rangle$ and $\mathbf{v} = \langle 0, 1, -3 \rangle$. Then proceed as in the previous worked out example.

The vector equations are
$$\begin{aligned}\mathbf{r} &= \mathbf{r}_0 + t\mathbf{v} \\ \mathbf{r} &= <3, -2, 0> + t < 0, 1, -3 > \\ \mathbf{r} &= <3, -2 + t, -3t > .\end{aligned}$$
The parametric equations are
$$\begin{aligned} x &= 3 \\ y &= -2 + t \\ z &= -3t.\end{aligned}$$
The equation   means that the line is contained in the plane parallel to the $y, z-$coordinate plane through $x = 3$. To get the symmetric equations solve for t whenever possible.
$$\begin{aligned} x &= 3 \\ t &= y + 2 \\ t &= z/(-3)\end{aligned}$$
The symmetric equations are
$$\begin{aligned} y + 2 &= z/(-3) \\ x &= 3.\end{aligned}$$

• Find parametric equations, and the symmetric equations for the line that passes through the point $(1, -2, 5)$ and is perpendicular to the plane $3x - 4y + 7z = 9$. Then find the point at which the line intersects the plane.

In the vector equation for the line $\mathbf{r} = \mathbf{r}_0 + t\mathbf{v}$, the vector $\mathbf{v}$ must be perpendicular to the plane. Since we are given the equation of the plane we can read off the components of such a vector to get $\mathbf{v} = <3, -4, 7>$.

The vector equation is the easiest way to get to the parametric equations.
$$\begin{aligned} \mathbf{r} &= \mathbf{r}_0 + t\mathbf{v} \\ \mathbf{r} &= <1, -2, 5> + t < 3, -4, 7 > \\ \mathbf{r} &= <1 + 3t, -2 - 4t, 5 + 7t > .\end{aligned}$$
The parametric equations are read off of the components of the vector equation.
$$\begin{aligned} x &= 1 + 3t \\ y &= -2 - 4t \\ z &= 5 + 7t\end{aligned}$$
(Solution continued on next page.)

Now, as always, solve for t to get the symmetric equations.

$$t = \frac{x-1}{3}$$

$$t = \frac{y+2}{-4}$$

$$t = \frac{z-5}{7}$$

$$\frac{x-1}{3} = \frac{y+2}{-4} = \frac{z-5}{7}$$

To find the point of intersection (x_0, y_0, z_0) notice that this point must satisfy the equation of the plane, $3x_0 - 4y_0 + 7z_0 = 9$ and be on the line. This means that there must be some t_0 value of the parametric equations so that:

$$x_0 = 1 + 3t_0$$
$$y_0 = -2 - 4t_0$$
$$z_0 = 5 + 7t_0.$$

Substitute these expressions in t_0 for x_0, y_0, z_0 in the equation of the plane.

$$3(1 + 3t_0) - 4(-2 - 4t_0) + 7(5 + 7t_0) = 9$$
$$3 + 9t_0 + 8 + 16t_0 + 35 + 49t_0 = 9$$
$$9t_0 + 16t_0 + 49t_0 + 46 = 9$$
$$9t_0 + 16t_0 + 49t_0 = -37$$
$$74t_0 = -37$$

So $t_0 = -1/2$.

Now that we know t_0 we can recover the point of intersection (x_0, y_0, z_0) by re-substituting.

$$x_0 = 1 + 3(-1/2)$$
$$y_0 = -2 - 4(-1/2)$$
$$z_0 = 5 + 7(-1/2)$$

$$x_0 = -1/2 \quad y_0 = 0 \quad z_0 = 3/2$$

The point of intersection is $(-1/2, 0, 3/2)$.

• Suppose L_1 is the line that passes through the points $(1, 5, -1)$ and $(5, 1, 4)$ and that L_2 is the line given by the symmetric equations $\dfrac{x - 2}{8} = \dfrac{y - 4}{-3} = \dfrac{z - 5}{1}$. Are these lines parallel?

Two lines are parallel if and only if their direction vectors are multiples of each other. It is easy to get the direction numbers without having to find the entire equations for the lines.

In the vector equation for line L_1,
$\mathbf{r} = \mathbf{r_0} + t\mathbf{v}$, $\mathbf{v}$
$=<5 - 1, 1 - 5, 4 - (-1) >$
$=< 4, -4, 5 >$
is the direction vector. The direction vector for L_2 is $< 8, -3, 1 >$. These vectors are not multiples of each other (for example any multiple of $< 4, -4, 5 >$ has the second component equal to the negative of the first component). Therefore the lines are not parallel.

SkillMaster 9.13.

• Find an equation of the plane that contains the points $A(0, 2, 1), B(1, -1, 2)$, and $C(-2, 1, 1)$.

First find a normal vector by taking the cross product of $\overrightarrow{BA}$ and $\overrightarrow{CA}$. Then use the scalar equation of the plane through $A(0, 2, 1)$.

$\overrightarrow{AB} = < 1 - 0, -1 - 2, 2 - 1 >$
$= < 1, -3, 1 >$ and
$\overrightarrow{AC} =$
$< -2 - 0, 1 - 2, 1 - 1 >$
$= < -2, -1, 0 >$. A normal vector is
$\mathbf{n} = \overrightarrow{AB} \times \overrightarrow{AC}$

$$= \begin{vmatrix} \mathbf{i} & \mathbf{j} & \mathbf{k} \\ 1 & -3 & 1 \\ -2 & -1 & 0 \end{vmatrix}$$

$$= \begin{vmatrix} -3 & 1 \\ -1 & 0 \end{vmatrix} \mathbf{i} - \begin{vmatrix} 1 & 1 \\ -2 & 0 \end{vmatrix} \mathbf{j} + \begin{vmatrix} 1 & -3 \\ -2 & -1 \end{vmatrix} \mathbf{k}$$

$$= (0 - (-1))\mathbf{i} - (0 + 2)\mathbf{j} + (-1 - 6)\mathbf{k}$$

$$= \mathbf{i} - 2\mathbf{j} - 7\mathbf{k}$$

$$= < 1, -2, -7 >$$

The equation of the plane is
$$1(x - 0) - 2(y - 2) - 7(z - 1) = 0$$
$$x - 2y - 7z = -11.$$

• Consider the two planes P_1 and P_2 where P_1 contains the point $(7, 4, -1)$ and is parallel to the plane $4x + 2y - 5z = 1$ and P_2 contains the points $A(1, 2, -1), B(2, 3, 1)$, and $C(3, -1, 2)$. Are P_1 and P_2 parallel? If not, find the angle of intersection between them.

First find the normal vector to each plane. If these vectors are multiples of each other then they are parallel. The angle between the planes is the equal to the angle between the normal vectors which may be found using the dot product.

A normal vector to plane P_1 can be read of the equation of the parallel plane's equation and is $\mathbf{n}_1 = <4, 2, -5>$. To find a normal vector to plane P_2 find the cross product to $\mathbf{a} = \overrightarrow{AB} = <1, 1, 2>$ and $\mathbf{b} = \overrightarrow{AC} = <2, 3, -3>$ that is set $\mathbf{n}_2 = \mathbf{a} \times \mathbf{b}$.

$$\mathbf{n}_2 = \mathbf{a} \times \mathbf{b} = \begin{vmatrix} \mathbf{i} & \mathbf{j} & \mathbf{k} \\ 1 & 1 & 2 \\ 2 & 3 & -3 \end{vmatrix}$$

$$= \begin{vmatrix} 1 & 2 \\ 3 & -3 \end{vmatrix} \mathbf{i} - \begin{vmatrix} 1 & 2 \\ 2 & -3 \end{vmatrix} \mathbf{j} + \begin{vmatrix} 1 & 1 \\ 2 & 3 \end{vmatrix} \mathbf{k}$$

$$= 9\mathbf{i} + \mathbf{j} - 5\mathbf{k}$$

The angle between $\mathbf{n}_2$ and $\mathbf{n}_1$ may be found from the dot product formula

$$\cos\theta$$
$$= \frac{\mathbf{n}_1 \cdot \mathbf{n}_2}{|\mathbf{n}_1||\mathbf{n}_2|}$$
$$= \frac{4(9) + 2(1) + (-5)(-5)}{\sqrt{16 + 4 + 25}\sqrt{81 + 1 + 25}}$$
$$= \frac{63}{\sqrt{45}\sqrt{107}}$$
$$\approx 0.9079$$
$$\theta \approx \cos^{-1}(0.9079)$$
$$\approx 0.4325 \text{ radian}$$

505

• Consider the intersection of the planes with equations $x+y+z = 1$ and $x+2y-z = 1$. Find the symmetric equations of their line of intersection.

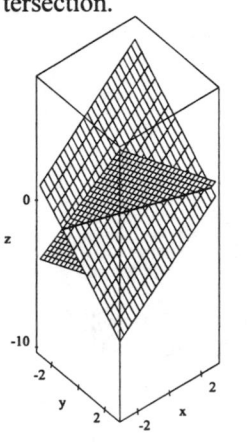

There are many ways to answer this. One method is to imitate the Example 7, on page 680 in the text. Another is to eliminate the x variable from by subtracting the first equation from the second. Then y may be expressed in terms of z and vice versa. Substitute the expression for y in the first equation. Solve the two final equations for z and set them equal to get a version of the symmetric equations.

$$x + y + z = 1$$
$$x + 2y - z = 1$$

Subtract the first equation from the second.

$$y - 2z = 0$$
$$y = 2z$$

Substitute for y in the first equation.

$$x + (2z) + z = 1$$
$$x + 3z = 1$$

Solve both two variable equations for z.

$$z = \frac{y}{2} \qquad z = \frac{-x+1}{3}$$

Set these equal to get the symmetric equations

$$\frac{-x+1}{3} = \frac{y}{2} = z$$

• Assume $A > 0, B > 0,$ and $C > 0$ are fixed positive numbers. Find the equation of the plane containing the points $(A,0,0), (0,B,0),$ and $(0,0,C)$. Sketch the tetrahedron enclosed by this plane and the coordinate planes.

There are two ways you might approach this. First you could use the method of finding two vectors in the plane by subtracting endpoints of the triangle formed by $(A,0,0), (0,B,0),$ and $(0,0,C)$. Then use the cross product to find a normal vector. Use the vector equation of a plane to get the scalar equation.

First method: Let $\mathbf{a} = < 0,B,0 > - < A,0,0 > = < -A,B,0 >$ and $\mathbf{b} = < 0,0,C > - < A,0,0 > = < -A,0,C >$. A normal vector to the plane is the cross product

$$\mathbf{n} = \mathbf{a} \times \mathbf{b} = \begin{vmatrix} \mathbf{i} & \mathbf{j} & \mathbf{k} \\ -A & B & 0 \\ -A & 0 & C \end{vmatrix}$$
$$= BC\mathbf{i} + AC\mathbf{j} + AB\mathbf{k} = < BC, AC, AB >.$$

Then

$$0 = \mathbf{n} \cdot (\mathbf{r} - < A,0,0 >)$$
$$= \mathbf{n} \cdot < x - A, y, z >$$
$$= < BC, AC, AB > \cdot < x - A, y, z >$$
$$= BC(x - A) + ACy + ABz$$
$$= BCx + ACy + ABz = ABC$$

The other method is to use the fact that the equation of the plane must look like $ax + by + cz = 1$ and that this plane contains the point $(A, 0, 0)$, $(0, B, 0)$, and $(0, 0, C)$ which must then satisfy the equation. Substituting them into $ax + by + cz = d$ gives expressions which may be solved for a, b, c in terms of A, B, C.

Second method: Substitute each of $< A, 0, 0 >, < 0, B, 0 >, < 0, 0, C >$ into the equation $ax + by + cz = 1$.

$$
\begin{aligned}
a(A) + b(0) + c(0) &= 1 \\
aA &= 1 \\
a &= 1/A
\end{aligned}
$$

$$
\begin{aligned}
a(0) + b(B) + c(0) &= 1 \\
bB &= 1 \\
b &= 1/B
\end{aligned}
$$

$$
\begin{aligned}
a(0) + b(0) + c(C) &= 1 \\
cC &= 1 \\
c &= 1/C
\end{aligned}
$$

Now we have expressions for the unknown parameters a, b, c in terms of A, B, C. Substitute these to get the equation of the plane in terms of the given A, B, C.

$$\frac{x}{A} + \frac{y}{B} + \frac{z}{C} = 1$$

Why are the two equations obtained by these methods the same? (Hint divide the first equation by ABC.)

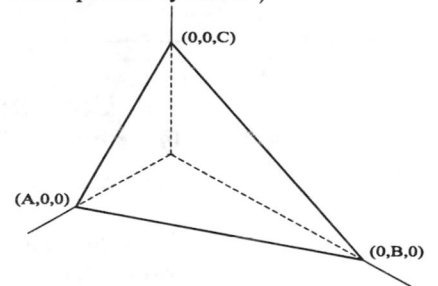

507

Section 9.6 – Functions and Surfaces

Key Concepts:

- Functions of Two Variables and their Graphs
- A Collection of Important Surfaces

Skills to Master:

- Sketch and describe properties of functions of two variables.
- Work with quadric surfaces.

Discussion:

You should be familiar with the concept of a function that assigns to each real number another real number. Polynomials, trigonometric functions and exponential functions are examples of these types of functions. In this section, functions of two variable are introduced and the graphs of such functions are described. The graph of such a function can be thought of as a surface in three-dimensional space.

Key Concept: Functions of Two Variables and their Graphs

page 686.

A function f of two variables can be thought of as a rule that assigns to each pair of real numbers (x, y) in a set D in the plane a unique real number denoted $f(x, y)$. The graph of the function f is the set of all points (x, y, z) in three-dimensional space such that $z = f(x, y)$. Study *Examples 4 and 5* to become more familiar with graphs of functions of two variables.

Key Concept: A Collection of Quadric and Other Important Surfaces

page 690.

The graph of a second degree equation in the variables x, y, and z is called a quadric surface. Examples of such surfaces include spheres, ellipsoids, paraboloids and hyperboloids. In general, such surfaces are not the graphs of single functions, but are the graphs of two functions as in *Figure 13* . Study *Figures 14, 15, and 16* to see the general form of ellipsoids, cones, and hyperboloids of one and two sheets.

SkillMaster 9.14: Sketch and describe properties of functions of two variables.

page 687.

In sketching surfaces, it is useful to first determine the cross sections or traces obtained by intersecting the surfaces with the vertical planes $x = k$ or $y = k$ or with the horizontal planes $z = k$. These traces can then be used to obtain information about the shape of the surface. Study *Example 7* to see how this works.

SkillMaster 9.15: Work with quadric surfaces.

In sketching quadric surfaces, you should again work with traces. Once you work with a few examples, you will begin to recognize the type of surface from the equation describing it. You should also determine where the surface intersects the coordinate axes before sketching it.

Worked Examples

For each of the following examples, first try to find the solution without looking at the middle or right columns. Cover the middle and right columns with a piece of paper. If you need a hint, uncover the middle column. If you need to see the worked solution, uncover the right column.

Example	Tip	Solution

SkillMaster 9.14.

• Find the domain of the function $z = f(x, y) = \dfrac{1}{\sqrt{\ln(10 - x^2 - y^2)}}$.

For a point (x, y) to be in the domain certain rules must not be violated. You are not allowed to divide by zero, you cannot take the square root of a negative number, and you cannot take the logarithm of a negative number. Gradually "unfold" this function. First let $u = \ln(10 - x^2 - y^2)$ and notice that we must have $u > 0$. Then continue.

Let $u = \ln(10 - x^2 - y^2)$. For $z = \frac{1}{\sqrt{u}}$ to be defined we need to have

$$u > 0$$
$$\ln(10 - x^2 - y^2) > 0.$$

Take the exponential of each side.

$$e^{\ln(10 - x^2 - y^2)} > e^0$$

(The inequality still holds because the exponential function is increasing.)

$$10 - x^2 - y^2 > 1$$
$$x^2 + y^2 < 9$$

Since $x^2 + y^2 = 9$ is a circle of radius 3 with center at the origin, the domain must be the disk inside this circle (and not including the boundary). The domain of f is

$$D = \{(x, y) | x^2 + y^2 < 9\}.$$

Below is a sketch of the graph of f lying over the domain.

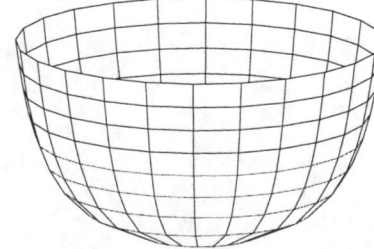

- Consider the function
$$z = f(x,y) = 2\sqrt{1-y^2}.$$
Find the domain, the range and give a rough sketch of the function.

To be in the domain we must have $1-y^2 \geq 0$. Notice that there is no restriction on x. Notice also that $z \geq 0$. Since x does not appear in the expression the surface is a kind of cylinder.

Now $z = 2\sqrt{1-y^2}$ makes sense for any x (since it does not appear it can cause no harm). On the other hand $z = 2\sqrt{1-y^2}$ is only defined for $1-y^2 \geq 0$ or $1 \geq y^2$ or $-1 \leq y \leq 1$. Thus the domain is
$$D + \{(x,y)| -\infty < x < \infty, -1 \leq y \leq 1\}$$
which is a strip as in the sketch.

The range of f is the set of possible output values $z = 2\sqrt{1-y^2}$ for $-1 \leq y \leq 1$. If $y = \pm 1$ then $z = 0$ and if $y = 0$ then $z = 2$. All intermediate values also can occur so the range is
$$R = \{z | 0 \leq z \leq 2\} = [0,2].$$
Since all cross-sections orthogonal to the $x - axis$ are the same we need to see what this cross-section looks like. The "profile" curve is
$$\begin{aligned} z &= 2\sqrt{1-y^2} \\ z^2 &= 4(1-y^2) \end{aligned}$$
$$\frac{y^2}{1} + \frac{z^2}{4} = 1 \text{ with } z \geq 0.$$
This is the upper half of the ellipse shown below.

This half ellipse is every cross-section so the surface looks like the cylindrical surface shown.

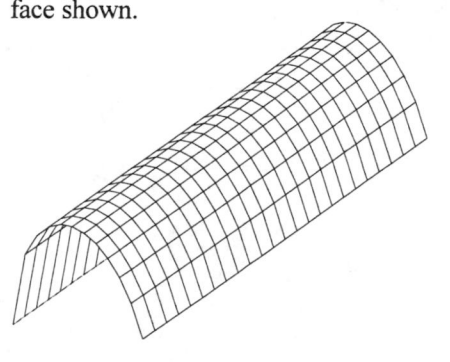

• Consider the function
$$z = f(x, y) = \frac{1}{x^2 + y^2}.$$
Find the domain and the range of this function. What are the traces of this function? Use the traces to sketch the graph of the function.

To find the domain observe that the only problem is the possibility of division by 0 or if $x^2 + y^2 = 0$.

Since $x^2 \geq 0$ and $y^2 \geq 0$ the only way this could happen is if $x = y = 0$. The domain is
$$D = \{(x, y) | (x, y) \neq (0, 0)\},$$
which is all points except the origin. To find the range notice that since $x^2 + y^2 > 0$, it is also the case that $z = \frac{1}{x^2+y^2} \geq 0$ and z achieves all positive number values. That is the range is
$$R = \{z | z > 0\} = (0, \infty).$$
To sketch the graph first find the trace on the $zy-$plane, by setting $x = 0$
$$x = 0 : z = \frac{1}{y^2}.$$
The trace in the $zx-$plane is similar
$$y = 0 : z = \frac{1}{x^2}.$$
Next sketch the traces $z = k$ for $k > 0$.
$$z = k : \qquad k = \frac{1}{x^2 + y^2}$$
$$k(x^2 + y^2) = 1 \qquad x^2 + y^2 = \frac{1}{k}$$
which is a circle of radius $1/\sqrt{k}$ centered at the origin. Thus the surface is an inverted funnel with tip becoming narrower and narrower at it stretches up toward infinity.

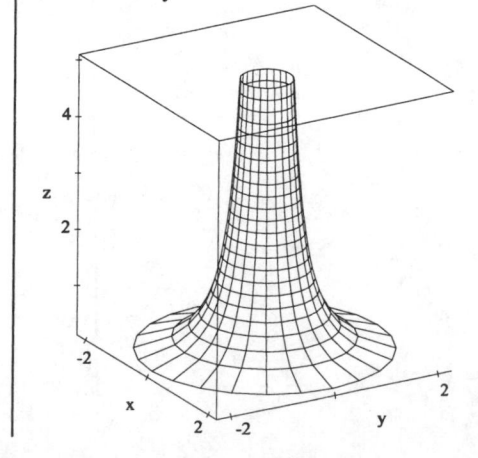

• Allegra is in a tree house with a water balloon. Allegra can drop the balloon somewhere between 12 and 17 feet from the ground. She can throw the balloon downward at a velocity between 10 and 0 feet per second. Suppose that $-v$ is the initial velocity and that p is the height from the ground. Let $t = f(p,v)$ be the time that it takes the balloon to hit the ground. What is the domain of this function? Give a formula that represents this function.

Find the equation for the time to splash on the ground as follows. Coordinatize the vertical axis by putting the origin at the ground level so that the acceleration due to gravity is negative, $a(t) = -32$ ft/s^2. Then integrate twice to get first the velocity, and then the position as a function of time. The length of time necessary to splash down may be found by setting the position equal to zero and solving for t.

$$a(t) = -32$$
$$v(t) = -32t - v$$
$$p(t) = -16t^2 - vt + p$$

To find the length of time it takes the balloon to get to the ground set $p(t) = 0$ and solve for t using the quadratic formula.

$$0 = -16t^2 - vt + p$$
$$t = \frac{v \pm \sqrt{v^2 - 4(-16)p}}{2(-16)}$$
$$= \frac{\sqrt{v^2 + 64p} - v}{32}$$

Simplify this by multiplying both the numerator and the denominator by $\sqrt{v^2 + 64} + v$.

$$t = \frac{\sqrt{v^2 + 64p} - v}{32}\left(\frac{\sqrt{v^2 + 64} + v}{\sqrt{v^2 + 64} + v}\right)$$
$$= \frac{(v^2 + 64p) - v^2}{32\sqrt{v^2 + 64} + v}$$
$$= \frac{64p}{32(\sqrt{v^2 + 64} + v)}$$
$$= \frac{2p}{\sqrt{v^2 + 64} + v}$$

$$f(p,v) = \frac{2p}{\sqrt{v^2 + 64} + v}$$

Notice that the larger v is the shorter the time until splash-down.
The domain is
$$D = \{(p,v) | 12 \le p \le 17, 0 \le v \le 10\}.$$

SkillMaster 9.15.

• Let $z = f(x, y) = \frac{2}{3}\sqrt{9 - x^2 - y^2}$. Find the domain of f, and sketch it. What type of quadric surface is this?

Put this into a more familiar form by squaring both sides the eliminate the square root sign. Don't forget that $z \geq 0$.

$$z = \frac{2}{3}\sqrt{9 - x^2 - y^2}$$

$$z^2 = \frac{4}{9}(9 - x^2 - y^2)$$

$$\frac{z^2}{4} = \frac{1}{9}(9 - x^2 - y^2) = 1 - \frac{x^2}{9} - \frac{y^2}{9}$$

$$1 = \frac{x^2}{3^2} + \frac{y^2}{3^2} + \frac{z^2}{4}$$

Since $z \geq 0$ this is the upper half of an ellipsoid with intercepts at $(\pm 3, 0, 0)$, $(0, \pm 3, 0)$, and $(0, 0, 2)$.

• Sketch the graphs of both $f(x, y) = x^2 - y^2 + 1$ and $g(x, y) = 5x^2 + 3y^2$. What types of quadric surfaces are these? Consider the points of intersection of these two graphs. Show that their projection onto the x, y-plane is a circle. (This intersection does not lie on a plane - it is sketched in the answer.)

Both surfaces have the form of paraboloids but of different types.

The xy-values of the intersection points may be found by eliminating z.

$$z = x^2 - y^2 + 1$$

$$z = 5x^2 + 3y^2$$

$$x^2 - y^2 + 1 = 5x^2 + 3y^2$$

$$1 = 4x^2 + 4y^2$$

$$x^2 + y^2 = (1/2)^2$$

The projection is a circle of radius $1/2$ and center at the origin.

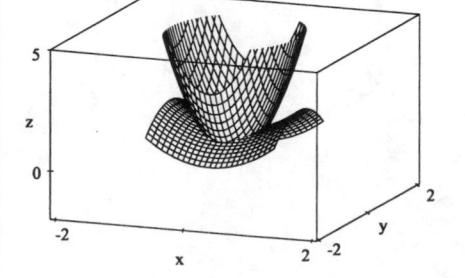

Section 9.7 – Cylindrical and Spherical Coordinates

Key Concepts:

- Cylindrical Coordinates
- Spherical Coordinates

Skills to Master:

- Convert from one coordinate system to another.
- Graph functions given in cylindrical and spherical coordinate systems.

Discussion:

page 695.

You should already be familiar with rectangular coordinates in two and three dimensions and with polar coordinates in two dimensions. Cylindrical coordinates are essentially polar coordinates with a third coordinate z added to represent the height above or below the $xy-$ plane. Spherical coordinates use the polar angle θ, and two additional coordinates. You will need to become familiar with these two new coordinate systems. Certain surfaces are best represented using these new coordinates. The graph below and in *Figure 9* shows the relation among the various coordinates.

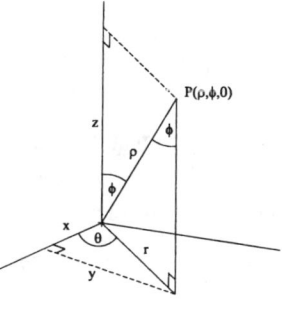

515

Key Concept: Cylindrical Coordinates

The cylindrical coordinates of a point P in three-dimensional space are represented by

$$(r, \theta, z)$$

page 692.

where r and θ are the polar coordinates of the projection of P onto the $xy-$plane and where z is the usual z coordinate of the point P. Study *Figure 1* to see a geometric representation of these coordinates.

Key Concept: Spherical Coordinates

The spherical coordinates of a point P in three dimensional space are represented by

$$(\rho, \theta, \phi)$$

where ρ is the distance from the origin O to P, θ is the same as in cylindrical coordinates, and where ϕ is the angle from the positive $z-$axis to OP. Note that ϕ takes on values between 0 and π.

SkillMaster 9.16: Convert one coordinate system to another.

To convert from cylindrical to rectangular coordinates, use the equations
$$x = r\cos\theta \qquad y = r\sin\theta \qquad z = z.$$
To convert from rectangular to cylindrical coordinates, use the equations
$$r^2 = x^2 + y^2 \qquad \tan\theta = \frac{y}{x} \qquad z = z.$$
To convert from spherical to rectangular coordinates, use the equations
$$x = \rho\sin\phi\cos\theta \qquad y = \rho\sin\phi\sin\theta \qquad z = \rho\cos\phi.$$
To convert from rectangular to spherical coordinates, use the equations
$$\rho^2 = x^2 + y^2 + z^2 \qquad \tan\theta = \frac{y}{x} \qquad \cos\phi = \frac{z}{\rho}$$

and the above equations.

SkillMaster 9.17: Graph functions given in cylindrical and spherical coordinate systems.

In graphing functions using cylindrical or spherical coordinate systems, you need to be familiar with what the coordinates represent. You will often also need to use spherical or cylindrical symmetry of certain objects. Sometimes, you will need to convert equations to rectangular form before graphing.

Worked Examples

For each of the following examples, first try to find the solution without looking at the middle or right columns. Cover the middle and right columns with a piece of paper. If you need a hint, uncover the middle column. If you need to see the worked solution, uncover the right column.

Example	Tip	Solution

SkillMaster 9.16.

● Find the rectangular and cylindrical coordinates of the point with spherical coordinates $(2, \pi/3, \pi/6)$. Reason from first principles - do not look up the formulas for cylindrical and spherical coordinates. Use the diagram below if needed.

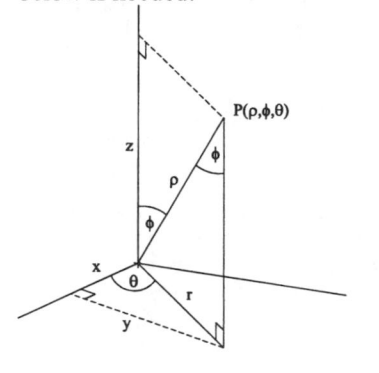

Identify $\rho = 2, \theta = \pi/3, \varphi = \pi/6$. Use the diagram that connects $x, y, z, r, \theta, \rho, \varphi$. There are two basic right triangles that connect these variables. The first right triangle has an edge on the z−axis, the second right triangle lies on the xy−plane and has an edge on the x−axis.

From the trigonometry of the first triangle we get
$$r = \rho \sin \varphi$$
$$z = \rho \cos \varphi.$$
From the second right triangle we get
$$x = r \cos \theta$$
$$y = r \sin \theta.$$
Plug in the values $\rho = 2, \theta = \pi/3, \varphi = \pi/6$ into the first set of equations
$$r = 2 \sin \pi/6 = 2(1/2) = 1$$
$$z = 2 \cos \pi/6 = 2(\sqrt{3}/2) = \sqrt{3}.$$
Plug $r = 1$ and the other values into the second set of equations to find x and y.
$$x = 1 \cos \pi/3 = 1/2$$
$$y = 1 \sin \pi/3 = \sqrt{3}/2.$$
The cylindrical coordinates are
$$(r, \theta, z) = (1, \pi/3, \sqrt{3})$$
and rectangular coordinates are
$$(x, y, z) = (1/2, \sqrt{3}/2, \sqrt{3}).$$

• Find the cylindrical coordinates of the point $(4, 3, -5)$.

Recall the formulas for cylindrical coordinates.

$$x = r\cos\theta = 4$$
$$y = r\sin\theta = 3$$
$$z = z = -5$$

and

$$r = \sqrt{x^2 + y^2}.$$

Solve for r first, then substitute this value for r in the first two equations and then use inverse trig functions.

$$\begin{aligned} r &= \sqrt{x^2 + y^2} \\ &= \sqrt{4^2 + 3^2} \\ &= 5 \end{aligned}$$

$$\begin{aligned} x &= 4 = r\cos\theta \\ &= 5\cos\theta \end{aligned}$$

$$\cos\theta = 4/5$$
$$\theta = \cos^{-1}(4/5) \approx 0.6435 \text{ radians}$$
$$(r, \theta, z) = \left(5, \cos^{-1}(4/5), -5\right)$$

• Find the spherical coordinates of the point with rectangular coordinates $\left(\frac{3}{2}, \frac{3}{2}\sqrt{3}, -3\right)$.

Recall the formulas for spherical coordinates.

$$\rho = \sqrt{x^2 + y^2 + z^2}$$
$$x =$$
$$\rho\sin\varphi\cos\theta = \frac{3}{2}$$
$$y =$$
$$\rho\sin\varphi\cos\theta = \frac{3}{2}\sqrt{3}$$

$$z = \rho\cos\varphi = -3$$

First find ρ then solve for φ in the last equation.

$$\begin{aligned} \rho &= \sqrt{x^2 + y^2 + z^2} \\ &= \sqrt{\left(\frac{3}{2}\right)^2 + \left(\frac{3\sqrt{3}}{2}\right)^2 + (-3)^2} \\ &= \sqrt{\frac{9}{4} + \frac{27}{4} + 9} \\ &= 3\sqrt{2} \end{aligned}$$

$$-3 = 3\sqrt{2}\cos\varphi$$
$$\cos\varphi = -1/\sqrt{2}$$
$$\varphi = \cos^{-1}\left(1/\sqrt{2}\right) - 3\pi/4$$
$$\sin\varphi = 1/\sqrt{2}$$

$$\frac{3}{2} = \rho\sin\varphi\cos\theta$$
$$\frac{3}{2} = 3\sqrt{2}\left(1/\sqrt{2}\right)\cos\theta$$
$$\cos\theta = \frac{1}{2}$$
$$\theta = \pi/3$$
$$(\rho, \theta, \varphi) = \left(3\sqrt{2}, \pi/3, 3\pi/4\right)$$

SkillMaster 9.17

• Describe the graph of each of the following in cylindrical coordinates.

(a) $r = \sqrt{3}$

Look again at the diagram for SkillMaster 9.16. The variables can be described as follows: r is the distance from the z-axis, θ is the angle counter-clockwise from the x-axis, φ is the angle down from the z-axis, and ρ is the distance from the origin.

(a) $r = \sqrt{3}$ means the set of point that are $\sqrt{3}$ from the $z-$axis. This is the circular cylinder of radius $\sqrt{3}$ Another way to see this is to square both sides of the equation

$$r = \sqrt{3}$$
$$r^2 = 3$$
$$x^2 + y^2 = 3$$

This is a circle of radius 1 in each cross-section parallel to the $xy-$plane.

• (b) $\theta = \pi/4$

Look again at the diagram for SkillMaster 9.16. Remember what θ represents.

(b) $\theta = \pi/4$ is the set of points that make an angle of $\pi/4$ measured counter-clockwise from the $x-$axis. This is the plane through the $z-$axis with this angle.

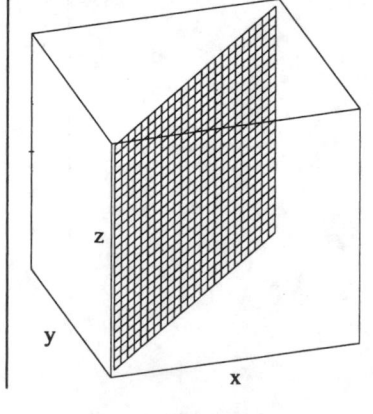

- (c) $z = 1 - r^2$

Look again at the diagram for SkillMaster 9.16. What is the relation between r and x and y?

(c)
$$z = 1 - r^2$$
$$z = 1 - x^2 - y^2$$
This is the inverted paraboloid with vertex at $(0, 0, 1)$.

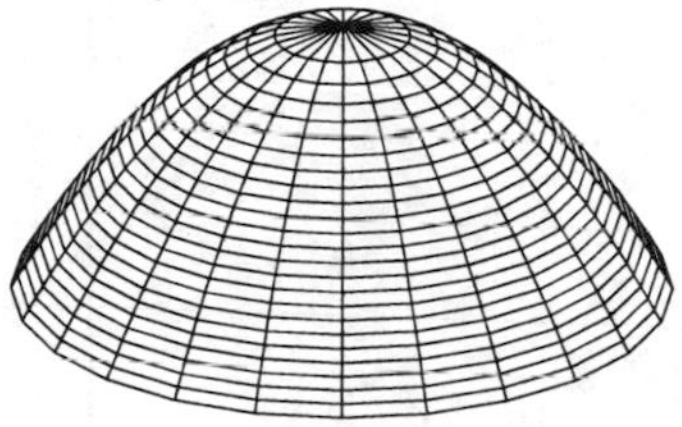

- (d) $r = 4 \sin \theta$

Look again at the diagram for SkillMaster 9.16. Multiply both sides of the equation by r.

(d)
$$r = 4 \sin \theta$$
$$r^2 = 4r \sin \theta$$
$$x^2 + y^2 = 4y$$
$$x^2 + y^2 - 4y = 0$$
$$x^2 + y^2 - 4y + 4 = 4$$
$$x^2 + (y - 2)^2 = 2^2$$
Since z is unrestricted this is a circular cylinder with central axis at $x = 0$ and radius 2.

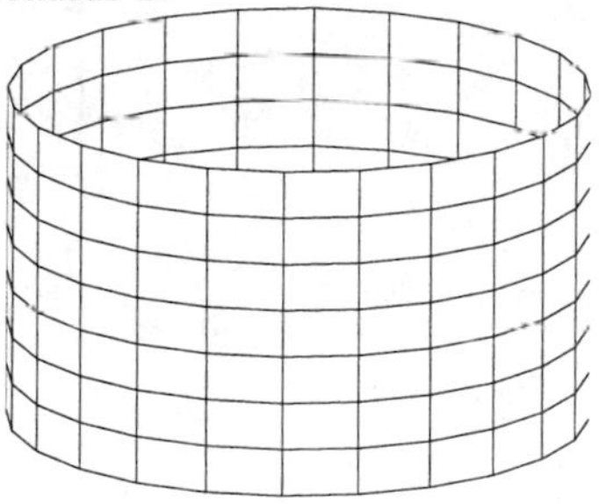

• (e) $z = ar,\ a > 0$

Look again at the diagram for SkillMaster 9.16. What happens for a fixed value of z?

(e)
$$z = ar$$
for $a > 0$ is the surface whose traces at $z = k$ are circles of radius k/a. This is an inverted cone with vertex at the origin.

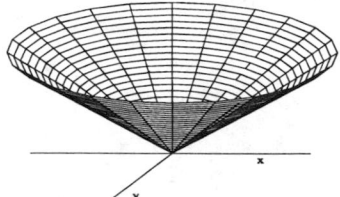

• Describe the graph of each of the following in spherical coordinates.
(a) $\rho = 2$

Look again at the diagram from SkillMaster 9.16. The variables can be described as follows: r is the distance from the z-axis, θ is the angle counter-clockwise from the x-axis, φ is the angle down from the z-axis, and ρ is the distance from the origin.

(a)
$$\rho = 2$$
is the set of points of that are a distance of 2 from the origin. This is the definition of a sphere of radius 2 with center the origin.

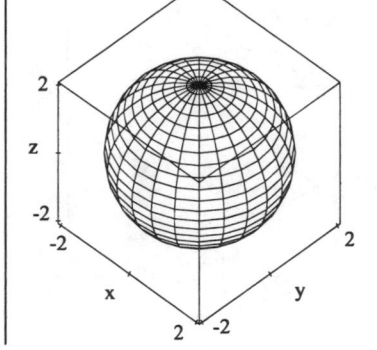

• (b) $\varphi = \pi/3$

Look again at the diagram for SkillMaster 9.16. Remember what φ represents.

(b) This is the set of points that make a angle of $\pi/3$ measured downward from the $z-$axis. This is an inverted cone with vertex at the origin. To see how this connects with part (e) of the previous worked out example notice from the basic diagram that

$$\frac{z}{r} = \cot \varphi = \sqrt{3}$$
$$z = \sqrt{3}r$$

which is the equation of (e) in the previous worked out example with $a = \sqrt{3}$.

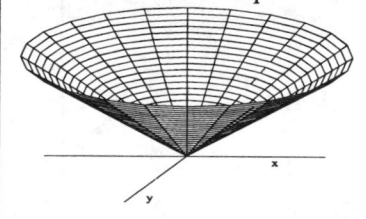

• (c) $\rho^2 = z^2 - 1$

Look again at the diagram for SkillMaster 9.16. Convert ρ^2 to rectangular form.

(c)

$$\rho^2 - z^2 = 1$$
$$(x^2 + y^2 + z^2) - z^2 = 1$$
$$x^2 + y^2 = 1$$

is a circular cylinder of radius 1 with the $x-$axis as central axis and unit circles as cross sections.

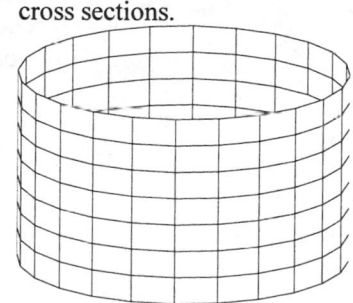

● (d) $\rho = 4\cos\varphi$

Look again at the diagram for SkillMaster 9.16. Multiply the equation by ρ and reduce to rectangular coordinates.

(d)
$$
\begin{aligned}
\rho &= 4\cos\varphi \\
\rho^2 &= 4\rho\cos\varphi \\
x^2 + y^2 + z^2 &= 4z \\
x^2 + y^2 + z^2 - 4z + 4 &= 4 \\
x^2 + y^2 + (z-2)^2 &= 2^2
\end{aligned}
$$
This is sphere of radius 2 and center at $(0,0,2)$.

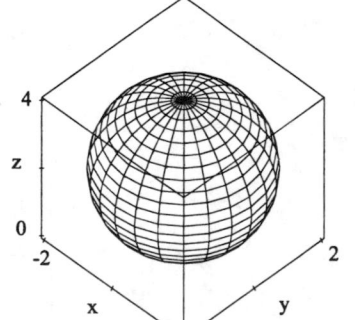

● (e) $\rho\sin^2\varphi = \cos\varphi$

Look again at the diagram for SkillMaster 9.16. Multiply the equation by ρ and reduce to rectangular coordinates.

(e)
$$
\begin{aligned}
\rho\sin^2\varphi &= \cos\varphi \\
\rho^2\sin^2\varphi &= \rho\cos\varphi \\
(\rho\sin\varphi)^2 &= \rho\cos\varphi \\
r^2 &= z \\
x^2 + y^2 &= z
\end{aligned}
$$
This is a circular paraboloid.

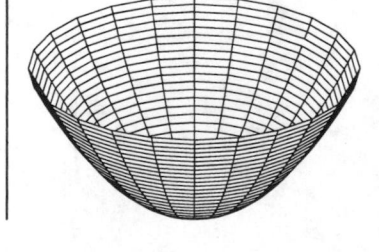

• Express the paraboloid $z = x^2 + y^2$ in both cylindrical and spherical coordinates. Which seems more natural?

Cylindrical:
$$z = x^2 + y^2$$
$$z = r^2$$
Spherical:
$$r^2 = z$$
$$(\rho \sin \varphi)^2 = \rho \cos \varphi$$
$$\rho^2 \sin^2 \varphi = \rho \cos \varphi$$
$$\rho \sin^2 \varphi = \cos \varphi$$
Cylindrical coordinates seem more natural and easier to use.

• Express the sphere $1 = x^2 + y^2 + z^2 + 2z$ in both cylindrical and spherical coordinates. Which seems more natural?

Cylindrical:
$$1 = x^2 + y^2 + z^2 + 2z$$
$$1 = r^2 + z^2 + 2z$$
Spherical:
$$1 = x^2 + y^2 + z^2 + 2z$$
$$1 = \rho^2 + 2\rho \cos \varphi$$
Both seem reasonably natural.

SkillMasters for Chapter 9

SkillMaster 9.1:   Use the distance formula to solve problems in three dimensions.

SkillMaster 9.2:   Use the distance formula to solve problems in three dimensions.

SkillMaster 9.3:   Describe and sketch regions involving spheres in three dimensions.

SkillMaster 9.4:   Describe vectors and vector representations algebraically, graphically, and in words.

SkillMaster 9.5:   Compute with vectors using the properties of vectors and vector operations.

SkillMaster 9.6:   Solve problems involving force and velocity using the horizontal and vertical components of vectors.

SkillMaster 9.7:   Compute dot products of vectors.

SkillMaster 9.8:   Use dot products to find angles between vectors.

SkillMaster 9.9:   Find scalar and vector projections.

SkillMaster 9.10:   Compute and use cross products.

SkillMaster 9.11:   Compute and use scalar and vector triple products.

SkillMaster 9.12:   Compute and use vector and parametric equations of lines.

SkillMaster 9.13:   Compute and use equations of planes.

SkillMaster 9.14:   Sketch and describe properties of functions of two variables.

SkillMaster 9.15:   Work with quadric surfaces.

SkillMaster 9.16:   Convert one coordinate system to another.

SkillMaster 9.17:   Graph functions given in cylindrical and spherical coordinate systems.

Chapter 10 - Vector Functions

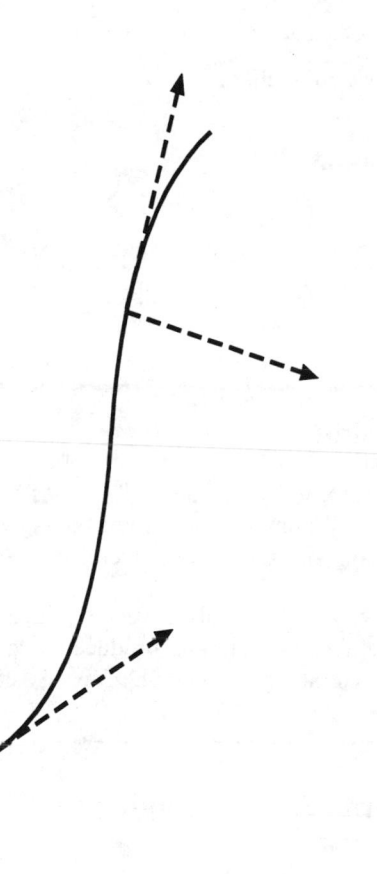

Section 10.1 – Vector Functions and Space Curves

Key Concepts:

- Vector functions and Representations of Space Curves
- Limits of Vector Functions

Skills to Master:

- Sketch and recognize graphs of space curves given parametrically.
- Find the domain, range, and limits of vector functions.

Discussion:

In the last chapter, you learned about functions of two variables. Those functions took as input 2 real numbers and produced as output a single real number. The graphs of functions of two variables were surfaces in three-dimensional space.

In this section, you will learn about vector valued functions. These functions take a single real number as input and produce a vector as output. You can think of these functions as tracing out a curve in space as the input varies.

Key Concept: Vector functions and Representations of Space Curves

---

A vector valued function is a function whose domain is a set of real numbers and whose range is a set of vectors. You can take limits of vector functions by taking limits of the components of the function. We represent vector functions as

$$\mathbf{r}(t) = \langle f(t), g(t), h(t) \rangle = f(t)\mathbf{i} + g(t)\mathbf{j} + h(t)\mathbf{k}.$$

The set C of points (x, y, z) in three-dimensional space with
$$x = f(t) \qquad y = g(t) \qquad z = h(t)$$

where t varies over the domain of the function, is called a space curve. Study *Figures 1 and 2* to see a geometric representation of certain space curves.

page 705.

Key Concept: Limits of Vector Functions

You can take limits of vector functions by taking limits of the components of the function. That is, if $\mathbf{r}(t) = \langle f(t), g(t), h(t) \rangle$, then
$$\lim_{t \to a} \mathbf{r}(t) = \left\langle \lim_{t \to a} f(t), \lim_{t \to a} g(t), \lim_{t \to a} h(t) \right\rangle.$$
Just as with real valued functions, a vector valued function is continuous at $\mathbf{a}$ if $\lim_{t \to a} \mathbf{r}(t) = \mathbf{r}(a)$.

SkillMaster 10.1: Sketch and recognize graphs of space curves given parametrically.

To sketch a graph of a space curve, you should first determine what each component of the vector function that gives the space curve is doing. If you have access to a computer that can draw representations of space curves, try viewing the space curve from different perspectives to get a better geometric understanding of the curve. Study *Example 5* to see how you can represent certain curves in space by vector functions.

page 706.

SkillMaster 10.2: Find the domain, range, and limits of vector functions.

Finding the domain, range and limits of vector functions is very similar to finding the domain, range and limit of ordinary real valued functions. To work with vector functions, work with each component separately and then combine the information that you get.

Worked Examples

For each of the following examples, first try to find the solution without looking at the middle or right columns. Cover the middle and right columns with a piece of paper. If you need a hint, uncover the middle column. If you need to see the worked solution, uncover the right column.

Example	Tip	Solution

SkillMaster 10.1.

- Consider the vector function
$\mathbf{r}(t) = <t^2 - 1, 2t^2, t^2 + 1>$.
Describe the space curve defined by this function.

Tip: Eliminate the variable t in the parametric equations for the vector function to get expressions for (x, y, z) .

Solution: The parametric equations are
$$\begin{aligned} x &= t^2 - 1 \\ y &= 2t^2 \\ z &= t^2 + 1 \end{aligned}$$

$$\begin{aligned} x + 1 &= t^2 \\ y/2 &= t^2 \\ z - 1 &= t^2 \end{aligned}$$

Since these expressions are all equal to t^2 they must be equal to each other, i.e.
$$x + 1 = \frac{y}{2} = z - 1.$$

We recognize these as the symmetric equations for a straight line with direction vector $(1, 2, 1)$ and passing through the point $(-1, 0, 1)$.

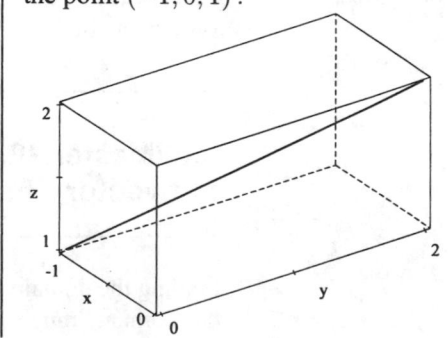

530

- Sketch the curve whose vector equation is

$$\mathbf{r}(t) = <t \cos t, t \sin t, t>.$$

First eliminate t in the parametric equation by using the relation $\cos^2 t + \sin^2 t = 1$. Then show that the curve lies on the inverted right circular cone with vertex at the origin.

$$x = t \cos t$$
$$y = t \sin t$$
$$z = t$$

$$x^2 + y^2 = t^2 \cos^2 t + t^2 \sin^2 t = t^2 = z^2$$
$$x^2 + y^2 = z^2$$

This is the right circular cone. $(\cos t, \sin t)$ describes a point rotating counter-clockwise around the unit circle. The x, y-coordinates of vector function also rotate counter-clockwise but spiral outward and upward, staying on the surface of the cone.

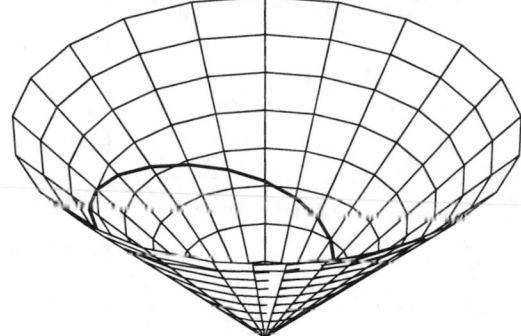

SkillMaster 10.2.

• Consider the vector function

$$r(t) = \sqrt{\frac{2}{3}\ln(t)}\,i + \sqrt{\frac{1}{3}\ln(t)}\,j + \frac{1}{t}k.$$

(a) Find the domain of $r(t)$.

(b) Does $\displaystyle\lim_{t\to 0^+} |tr(t)|^2$ exist? If it does then find the limit.

(c) Show that $r(t)$ moves on the "Gaussian bump" surface

$$z = e^{-(x^2+y^2)}$$

The domain of $z(t)$ is $t \neq 0$ since we may not divide by zero. Now the domain of the square root function is the set of non-negative numbers.

(a) $\ln(t)$ is nonnegative if and only if $t \geq 1$. The domain of the vector function is

$$D = [1, \infty) = \{t | t \geq 1\}.$$

(b) First try to simplify the expression.

$$|tr(t)|^2 = \left| t\sqrt{\frac{2}{3}\ln(t)}\,i + t\sqrt{\frac{1}{3}\ln(t)}\,j + t\frac{1}{t}k \right|$$

$$= \left(t\sqrt{\frac{2}{3}\ln(t)} \right)^2 + \left(t\sqrt{\frac{1}{3}\ln(t)} \right)^2 + \left(t\frac{1}{t} \right)^2$$

$$= t^2\frac{2}{3}\ln(t) + t^2\frac{1}{3}\ln(t) + 1 = t^2\ln(t) + 1.$$

Try to evaluate the limit, using L'Hospital's Rule,

$$\lim_{t\to 0^+} |tr(t)|^2 = \lim_{t\to 0^+} (t^2\ln(t) + 1)$$

$$= \lim_{t\to 0^+} (\frac{\ln(t)}{t^{-2}} + 1) = \lim_{t\to 0^+} (\frac{\frac{d}{dt}(\ln(t))}{\frac{d}{dt}(t^{-2})} + 1)$$

$$= \lim_{t\to 0^+} (\frac{1/t}{-2t^{-3}} + 1) = \lim_{t\to 0^+} (\frac{-1}{2}t^2 + 1)$$

$$= 0 + 1 = 1.$$

(c)

$$x^2 + y^2 = \left(\sqrt{\frac{2}{3}\ln(t)} \right)^2 + \left(\sqrt{\frac{1}{3}\ln(t)} \right)^2$$

$$= \frac{2}{3}\ln(t) + \frac{1}{3}\ln(t) = \ln(t)$$

$$e^{x^2+y^2} = e^{\ln(t)} = t$$

$$z = \frac{1}{t} = 1/e^{x^2+y^2} = e^{-(x^2+y^2)}$$

The surface below is this surface.

- Find a vector function that parametrizes the intersection of the elliptic cylinder

$$\frac{y^2}{9} + \frac{z^2}{4} = 1$$

and the plane

$$x + 2z = 0.$$

First parametrize the ellipse in the usual way. Then solve for z in terms of x to get an expression for z in terms of t.

The standard parametrization of the ellipse is

$$x = 2\cos t$$
$$y = 3\sin t.$$

$$
\begin{aligned}
x + 2z &= 0 \\
z &= -(1/2)x \\
&= -(1/2)2\cos t \\
& -\cos t
\end{aligned}
$$

The vector function is

$$\mathbf{r}(t) = <2\cos t, 3\sin t, -\cos t>.$$

Section 10.2 – Derivatives and Integrals of Vector Functions

Key Concepts:

- Derivatives of Vector Functions
- Unit Tangent Vector to a Space Curve
- Differentiation Rules
- Integral of a Vector Function

Skills to Master:

- Compute derivatives of vector functions.
- Compute tangent vectors and tangent lines for space curves.
- Compute integrals of vector functions.

Discussion:

In this section, derivatives and integrals of vector functions are introduced and the basic properties of these derivatives and integrals are discussed. Differentiating or integrating a vector function is similar to differentiating or integrating a real valued function. The differentiation or integration is done component by component, so there are no new differentiation or integration formulas to learn. Both differentiation and integration of vector valued functions yield another vector valued function. There are important physical interpretations of the derivative of a vector function.

Key Concept: Derivatives of Vector Functions and Tangents to Space Curves

The derivative of a vector function is defined in the same way that the derivative of a real valued function. If $\mathbf{r}(t)$ is a vector function,

$$\mathbf{r}'(t) = \lim_{h \to 0} \frac{\mathbf{r}(t+h) - \mathbf{r}(t)}{h}.$$

page 711.

If $\mathbf{r}(t) = \langle f(t), g(t), h(t) \rangle$, then $\mathbf{r}'(t) = \langle f'(t), g'(t), h'(t) \rangle$. See *Examples 1 and 2* for computations of the derivative of vector functions.

If C is the space curve traced out by the vector function $\mathbf{r}(t)$, the tangent line to C at a point $P = \mathbf{r}(t)$ is the line through P that is parallel to the vector $\mathbf{r}'(t)$. The vector $\mathbf{r}'(t)$ is called a tangent vector, and

$$\mathbf{T}(t) = \frac{\mathbf{r}'(t)}{|\mathbf{r}'(t)|}$$

page 710.

is called the unit tangent vector to C at P. See *Figure 1* for a geometric representation of this.

Key Concept: Differentiation Rules

page 712.

The differentiation rules are listed in *Theorem 3*. Notice the similarity to the differentiation rules that you are already familiar with. The product rule for differentiating products of real valued functions carries over to dot products or cross products of vector functions. It is important that you keep track of when you are dealing with vector functions and when you are dealing with scalar functions in these rules.

One rule that you should spend extra time understanding is the chain rule:

$$\frac{d}{dt}\mathbf{u}(f(t)) = f'(t)\mathbf{u}'(f(t)).$$

Here, $\mathbf{u}$ is a vector valued function and f is a real valued function.

Key Concept: Integrals of a Vector Function

The definite integral of a continuous vector function is defined in a similar way to the definite integral of a continuous real valued function. It can then be shown that if $\mathbf{r}(t) = \langle f(t), g(t), h(t) \rangle$, then

$$\int_a^b \mathbf{r}(t)\, dt = \left(\int_a^b f(t)\, dt \right) \mathbf{i} + \left(\int_a^b g(t)\, dt \right) \mathbf{j} + \left(\int_a^b h(t)\, dt \right) \mathbf{k}.$$

The Fundamental Theorem of calculus also holds in this setting. If $\mathbf{R}'(t) = \mathbf{r}(t)$, then

$$\int_a^b \mathbf{r}(t)\, dt = \mathbf{R}(b) - \mathbf{R}(a).$$

SkillMaster 10.3: Compute derivatives of vector functions and tangent vectors of space curves.

To compute derivatives of vector functions, differentiate component by component. Remember that the derivative is another vector function. To find a tangent vector at a point P on the space curve defined by $\mathbf{r}(t)$, take the derivative. This is a tangent vector to the curve at that point! This vector can then be used to find tangent lines and unit tangent vectors.

SkillMaster 10.4: Compute integrals of vector functions.

To compute integrals of vector functions, integrate component by component. Review the integration techniques that you learned in previous courses. Study *Example 6* to see a worked out example of this.

page 714.

Worked Examples

For each of the following examples, first try to find the solution without looking at the middle or right columns. Cover the middle and right columns with a piece of paper. If you need a hint, uncover the middle column. If you need to see the worked solution, uncover the right column.

Example	Tip	Solution
SkillMaster 10.3.		
• Sketch the curve $$\mathbf{r}(t) = \sqrt{t}\,\mathbf{i} + (t+1)\mathbf{j}.$$ Find the derivative $\mathbf{r}(t)$ and sketch $\mathbf{r}'(1)$.	To sketch the curve first eliminate the variable t.	$$\begin{aligned}\mathbf{r}(t) &= \sqrt{t}\,\mathbf{i} + (t+1)\mathbf{j} \\ \mathbf{r}(t) &= t^{1/2}\mathbf{i} + (t+1)\mathbf{j}\end{aligned}$$ $$\begin{aligned}\mathbf{r}'(t) &= (1/2)t^{-1/2}\mathbf{i} + \mathbf{j} \\ &= \frac{1}{2\sqrt{t}}\mathbf{i} + \mathbf{j}\end{aligned}$$ $$\mathbf{r}'(1) = (1/2)\mathbf{i} + \mathbf{j}$$

SkillMaster 10.4.

• Consider the vector function
$$\mathbf{r}(t) =< e^t, e^{-t}, t > .$$
Find $\mathbf{r}\prime(t)$, the unit tangent vector $\mathbf{T}(t)$, and the symmetric equations for the tangent line to $\mathbf{r}(0) =< 1, 1, 0 >$.

$$\mathbf{r}'(t) =< e^t, -e^{-t}, 1 >$$

$$\mathbf{T}(t) = \frac{\mathbf{r}'(t)}{|\mathbf{r}'(t)|}$$

$$= \frac{< e^t, -e^{-t}, 0 >}{\sqrt{(e^t)^2 + (e^{-t})^2 + 0^2}}$$

$$= \frac{< e^t, -e^{-t}, 0 >}{\sqrt{e^{2t} + e^{-2t}}}$$

$$\mathbf{r}'(0) =< 1, -1, 1 >$$
gives the direction numbers.
$$\mathbf{r}(0) =< 1, 1, 0 >$$
gives a point the tangent line passes thro.ugh

$$\frac{x-1}{1} = \frac{y-1}{-1} = \frac{z-0}{1}$$
$$x - 1 = 1 - y = z$$

give the symmetric equations of the tangent line.

• If $\mathbf{r}(t)$ is a smooth nonzero vector function such that $\mathbf{r}(t) \times \mathbf{r}'(t) = c$, a constant, show that $\mathbf{r}(t)$ and $\mathbf{r}''(t)$ are parallel.

Differentiate $\mathbf{r}(t) \times \mathbf{r}'(t) = c$ and use the rules of differentiation to show that the cross product of $\mathbf{r}(t)$ and $\mathbf{r}'(t)$ is zero.

$$\mathbf{r}(t) \times \mathbf{r}'(t) = c$$
$$0 = \frac{d}{dt}\mathbf{r}(t) \times \mathbf{r}'(t))$$
$$= \mathbf{r}'(t) \times \mathbf{r}'(t) + \mathbf{r}(t) \times \mathbf{r}''(t)$$
$$= \mathbf{0} + \mathbf{r}(t) \times \mathbf{r}''(t)$$
$$= \mathbf{r}(t) \times \mathbf{r}''(t)$$

Now $\mathbf{r}(t)$ is not zero by assumption and $\mathbf{r}'(t)$ is not zero since the vector function is smooth so $\mathbf{r}(t)$ and $\mathbf{r}"(t)$ are parallel since their cross product is zero.

| • If $\mathbf{r}(0) = 0$ and $\mathbf{r}'(t) =<\ 2t, \cos t, e^t >$ find $\mathbf{r}(t)$ for $t > 0$. | Use the Fundamental Theorem of Calculus to obtain $\mathbf{r}(t)$ by antidifferentiation. | $\begin{aligned} \mathbf{r}(t) &= \mathbf{r}(0) + \int_0^t \mathbf{r}'(u)du \\ &= 0+ < \int_0^t 2udu, \int_0^t \cos udu \int_0^t e^u du > \\ &= < \frac{2u^2}{2}|_0^t, \sin u|_0^t, e^u|_0^t > \\ &= < t^2, \sin t, e^t - 1 > . \end{aligned}$ |

Section 10.3 – Arc Length and Curvature

Key Concepts:

- Arc length and parametrization with respect to arc length
- Curvature
- Tangent, Normal and Binormal Vectors
- Normal and Osculating Planes

Skills to Master:

- Compute arc length and parametrize curves with respect to arc length.
- Compute the curvature of a curve.
- Compute the normal and binormal vectors and the associated planes of a curve.

Discussion:

In this section, you will investigate two properties associated with space curves: arc length and curvature. Arc length will be defined using a limiting process, and will turn out to be the definite integral of the magnitude of the tangent vector to the curve. There will be a useful parametrization of the curve in terms of arc length. Curvature is a measure of how fast the curve is turning. This will be defined in terms of a certain derivative.

Each space curve will have three mutually orthogonal vectors associated with it at each point, the tangent, normal and binormal vectors. There will also be two planes associated with the space curve at each point, the normal and osculating planes. A lot of new concepts are introduced in this section. Make sure that you set aside enough time to master these concepts.

Key Concept: Arc length and parametrization with respect to arc length.

page 716.

Study *Figure 1* to see the geometric motivation for the definition of arc length. Recall that arc length was defined for parametric curves in the plane in *Section 6.3*. For space curves, arc length is defined similarly. If $\mathbf{r}(t) = \langle f(t), g(t), h(t) \rangle$ is the vector equation of a space curve, then the arc length of the curve from the point corresponding to $t = a$ to the point corresponding to $t = b$ is defined to be

$$L = \int_a^b \sqrt{[f'(t)]^2 + [g'(t)]^2 + [h'(t)]^2}$$

$$= \int_a^b |\mathbf{r}'(t)|\, dt = \int_a^b \sqrt{\left(\frac{dx}{dt}\right)^2 + \left(\frac{dy}{dt}\right)^2 + \left(\frac{dz}{dt}\right)^2}.$$

Make sure that you understand why all three integrals above represent the same thing.

If $s(t)$ represents the arc length of a curve from $u = a$ to $u = t$,

$$s(t) = \int_a^t |\mathbf{r}'(u)|\, du,$$

it is sometimes possible to solve for t as a function of arc length s and reparametrize the curve in terms of arc length.

Key Concept: Curvature

Curvature measures how fast the unit tangent vector changes as you move along the curve. Specifically, the curvature is

$$\kappa = \left| \frac{d\mathbf{T}}{ds} \right|$$

where $\mathbf{T}$ is the unit tangent vector and s is arc length. Two additional useful formulations of curvature are

$$\kappa = \frac{|\mathbf{T}'(t)|}{|\mathbf{r}'(t)|} = \frac{|\mathbf{r}'(t) \times \mathbf{r}''(t)|}{|\mathbf{r}'(t)|^3}.$$

page 719.

Make sure that you understand why these are the same. Study *Theorem 10* to see how the cross product formula arises. For a plane curve given by $y = f(x)$, curvature is given by

$$\kappa = \frac{|f''(x)|}{\left|1 + (f'(x))^2\right|^{3/2}}.$$

Key Concept: Tangent, Normal and Binormal Vectors

The unit tangent vector $\mathbf{T}(t)$ for a space curve given by $\mathbf{r}(t)$ is
$$\mathbf{T}(t) = \frac{\mathbf{r}'(t)}{|\mathbf{r}'(t)|}.$$
The principal unit normal $\mathbf{N}(t)$, which is orthogonal to $\mathbf{T}(t)$, is given by
$$\mathbf{N}(t) = \frac{\mathbf{T}'(t)}{|\mathbf{T}'(t)|}.$$
The binormal vector $\mathbf{B}(t)$ is given by
$$\mathbf{B(t)} = \mathbf{T}(t) \times \mathbf{N}(t)$$

page 721.

and is orthogonal to both $\mathbf{T}$ and $\mathbf{N}$. Study *Figure 7* to see a geometric picture of these three mutually orthogonal vectors associated with the space curve.

Key Concept: Normal and Osculating Planes

The plane determined by $\mathbf{T}$ and $\mathbf{N}$ at a point P on the curve is called the osculating plane and is the plane closest to containing the part of the curve near P. The plane determined by $\mathbf{N}$ and $\mathbf{B}$ at a point P on the curve is called the normal plane to the curve at P. It contains all lines orthogonal to the tangent vector.

page 721.

Study *Figure 8* to get a geometric picture of the osculating plane.

SkillMaster 10.5: Compute arc length and parametrize curves with respect to arc length.

To compute arc length, use the definition. Take the derivative of the vector function giving the space curve, take the magnitude of this derivative, and integrate if possible. You will often need to numerically approximate the integral in applications to get approximations to the exact arc length.

To parametrize the curve in terms of arc length, start with the vector function $\mathbf{r}(t)$ giving the curve. Find a formula for $s(t)$, the arc length in terms of t, by the process described above. Then solve for t in terms of s if possible, and substitute this expression into $\mathbf{r}(t)$. Study *Example 2* to see how this is done.

page 716.

SkillMaster 10.6: Compute the curvature of a curve.

In computing the curvature of a curve, you have a number of choices. You can work directly from the definition, you can use the alternate form that uses the cross product of the first and second derivatives, or you can use the formula for curvature of a plane curve if you are in that situation. Before beginning, decide which approach will work best and keep careful track of the quantities that you are computing. Study *Examples 3 and 4* to see computations of curvature

page 718-719.

SkillMaster 10.7: Compute the normal and bi-normal vectors and the associated planes of a curve.

Finding the normal and binormal vectors, and the osculating and normal planes requires careful attention to detail. Starting with $\mathbf{r}(t)$, first compute $\mathbf{T}(t)$. Next, compute $\mathbf{N}(t)$ by differentiating $\mathbf{T}(t)$ and dividing by the magnitude of this derivative. Once you have $\mathbf{T}(t)$ and $\mathbf{N}(t)$, take the cross product to get $\mathbf{B}(t)$.

page 675.

To find the osculating plane, use $\mathbf{B}(t)$ as a normal vector to this plane and follow the method from *Section 9.5*. To find the normal plane, use $\mathbf{T}(t)$ as a normal vector to the plane.

Worked Examples

For each of the following examples, first try to find the solution without looking at the middle or right columns. Cover the middle and right columns with a piece of paper. If you need a hint, uncover the middle column. If you need to see the worked solution, uncover the right column.

Example	Tip	Solution

SkillMaster 10.5.

• Set up the integral to compute the arc length of $\mathbf{r}(t) = te^t\mathbf{i} + t^2\mathbf{j} + \tan t\mathbf{k}$ from $t = 1$ to $t = 3$. Do NOT attempt to compute this integral.

$$L = \int_1^3 \sqrt{\left(\frac{dx}{dt}\right)^2 + \left(\frac{dy}{dt}\right)^2 + \left(\frac{dz}{dt}\right)^2}$$

$$= \int_1^3 \sqrt{(e^t + te^t)^2 + (2t)^2 + (\sec^2 t)^2}$$

$$= \int_1^3 \sqrt{e^{2t} + 2te^{2t} + t^2 e^{2t} + 4t^2 + \sec^4 t}.$$

• Let $\mathbf{r}(t) = (\sin t - t\cos t)\mathbf{i} + (\cos t + t\sin t)\mathbf{j} - t^2\mathbf{k}$ for $t \geq 0$. Reparametrize with respect to arc length beginning at $t = 0$.

Tip: First find the arc length function $s = s(t) = \int_0^t |\mathbf{r}'(u)|\, du$. Then solve for t to get $t = t(s)$. The reparametrization is $\mathbf{r}(t(s))$.

Solution: First calculate
$$\mathbf{r}'(t)$$
$$= (\cos t - (\cos t - t\sin t)\mathbf{i}$$
$$+(-\sin t + (\sin t + t\cos t)\mathbf{j} - 2t\mathbf{k}$$
$$= t\sin t\mathbf{i} + t\cos t\mathbf{j} - 2t\mathbf{k}$$
$$= t(\sin t\mathbf{i} + \cos t\mathbf{j} - 2\mathbf{k})$$
$$|\mathbf{r}'(t)| = t\,|\sin t\mathbf{i} + \cos t\mathbf{j} - 2\mathbf{k}|$$
$$= t\sqrt{\sin^2 t + \cos^2 t + (-2)^2}$$
$$= t\sqrt{1 + 4}$$
$$= t\sqrt{5}$$

The arc length function is
$$s(t) = \int_0^t |\mathbf{r}'(u)|\, du$$
$$= \int_0^t u\sqrt{5}\,du$$
$$= \frac{t^2}{2}\sqrt{5}.$$

Now solve for t.

$$s = \frac{t^2}{2}\sqrt{5}$$

$$\frac{2s}{\sqrt{5}} = t^2$$

$$t = \left(\frac{2s}{\sqrt{5}}\right)^{1/2}$$

The reparametrization is

$$\mathbf{r}(t(s))$$

$$= (\sin t(s) - t(s)\cos t(s))\mathbf{i}$$
$$+ (\cos t(s) + t(s)\sin t(s))\mathbf{j} - t(s)^2\mathbf{k}$$

$$= \left(\sin\left(\frac{2s}{\sqrt{5}}\right)^{1/2}\right.$$

$$\left. - \left(\frac{2s}{\sqrt{5}}\right)^{1/2}\cos\left(\frac{2s}{\sqrt{5}}\right)^{1/2}\right)\mathbf{i}$$

$$+ \left(\cos\left(\frac{2s}{\sqrt{5}}\right)^{1/2}\right.$$

$$\left. + \left(\frac{2s}{\sqrt{5}}\right)^{1/2}\sin\left(\frac{2s}{\sqrt{5}}\right)^{1/2}\right)\mathbf{j}$$

$$- \left(\frac{2s}{\sqrt{5}}\right)\mathbf{k}$$

SkillMaster 10.6.

• Compute the curvature of $\mathbf{r}(t) =< 3t^2, 2t^3 >$ for $t \geq 0$. Describe in words what the curvature at $t = 1$ represents.

First compute the unit tangent vector $\mathbf{T}(t) = \mathbf{r}'(t)/|\mathbf{r}'(t)|$. The curvature is $\kappa(t) = \frac{|T'(t)|}{|\mathbf{r}'(t)|}$.

$$\mathbf{r}(t) = < 3t^2, 2t^3 >$$

$$\mathbf{r}'(t) = < 6t, 6t^2 >= 6t < 1, t >$$

$$|\mathbf{r}'(t)| = 6t| < 1, t > | = 6t\sqrt{1+t^2}$$

This uses the given fact that $t \geq 0$. Thus

$$\mathbf{T}(t) = \frac{\mathbf{r}'(t)}{|\mathbf{r}'(t)|} = \frac{6t < 1, t >}{6t\sqrt{1+t^2}}$$

$$= \left\langle \frac{1}{\sqrt{1+t^2}}, \frac{t}{\sqrt{1+t^2}} \right\rangle$$

$$= \left\langle (1+t^2)^{-1/2}, t(1+t^2)^{-1/2} \right\rangle$$

$$\mathbf{T}'(t)$$

$$= \frac{d}{dt}((1+t^2)^{-1/2})\mathbf{i} + \frac{d}{dt}(t(1+t^2)^{-1/2})\mathbf{j}$$

$$= (-1/2)(1+t^2)^{-3/2}(2t)\mathbf{i}$$
$$+((1+t^2)^{-1/2} + t(-1/2)(1+t^2)^{-3/2}(2t))\mathbf{j}$$

$$= -t(1+t^2)^{-3/2}\mathbf{i}$$
$$+((1+t^2)^{-1/2} - t^2(1+t^2)^{-3/2})\mathbf{j}$$

$$= (1+t^2)^{-3/2}[-t\mathbf{i} + ((1+t^2) - t^2)\mathbf{j}]$$

$$= (1+t^2)^{-3/2}[-t\mathbf{i} + \mathbf{j}]$$

So the curvature is

$$\kappa(t) = \frac{|\mathbf{T}'(t)|}{|\mathbf{r}'(t)|}$$

$$= \frac{(1+t^2)^{-3/2}\sqrt{(-t)^2 + 1^2}}{6t\sqrt{1+t^2}}$$

$$= \frac{(1+t^2)^{-3/2}\sqrt{1+t^2}}{6t\sqrt{1+t^2}}$$

$$= \frac{(1+t^2)^{-3/2}}{6t} = \frac{1}{6t(1+t^2)^{3/2}}.$$

At $t = 1$ we have

$$\kappa(1) = \frac{1}{6(1)(1+1^2)^{3/2}}$$

$$= \frac{1}{6(2)\sqrt{2}} = \frac{1}{12\sqrt{2}}.$$

This means that the circle that best approximates the shape of the curve very close to the point $t = 1$ has radius $\frac{1}{\kappa(t)} = 12\sqrt{2}$

• Find the curvature $\kappa(t)$ for the space curve
$$\mathbf{r}(t) = 4t\mathbf{i} + 5\ln(\cos t)\mathbf{j} - 3t\mathbf{k}$$
for $0 \le t \le \pi/4$.

Use $\kappa(t) = \dfrac{|\mathbf{T}'(t)|}{|\mathbf{r}'(t)|}$

$$
\begin{aligned}
\mathbf{r}'(t) &= 4\mathbf{i} - 5\frac{\sin t}{\cos t}\mathbf{j} - 3\mathbf{k} \\
&= 4\mathbf{i} - 5\tan t\mathbf{j} - 3\mathbf{k} \\
|\mathbf{r}'(t)| &= \sqrt{4^2 + 5^2 \tan^2 t + (-3)^2} \\
&= \sqrt{16 + 25\tan^2 t + 9} \\
&= \sqrt{25 + 25\tan^2 t} \\
&= 5\sqrt{1 + \tan^2 t} \\
&= 5\sqrt{\sec^2(t)} \\
&= 5\sec t
\end{aligned}
$$

since $\sec t$ is positive for $0 \le t \le \pi/4$.

$$
\begin{aligned}
\mathbf{T}'(t) &= \frac{\mathbf{r}'(t)}{|\mathbf{r}'(t)|} \\
&= \frac{4\mathbf{i} - 5\tan t\mathbf{j} - 3\mathbf{k}}{5\sec t} \\
&= \frac{4}{5}\cos t\mathbf{i} - \sin t\mathbf{j} - \frac{3}{5}\cos t\mathbf{k}
\end{aligned}
$$

$$
\begin{aligned}
|\mathbf{T}'(t)| \\
&= \left(\left(\frac{4}{5}\right)^2 \cos^2 t + \sin^2 t \right. \\
&\qquad \left. + \left(-\frac{3}{5}\right)^2 \cos^2 t\right)^{1/2} \\
&= \left(\frac{16}{25}\cos^2 t + \sin^2 t + \frac{9}{25}\cos^2 t\right)^{1/} \\
&= \sqrt{\cos^2 t + \sin^2 t} \\
&= 1
\end{aligned}
$$

So the curvature is
$$
\begin{aligned}
\kappa(t) &= \frac{|\mathbf{T}'(t)|}{|\mathbf{r}'(t)|} \\
&= \frac{1}{5\sec t} \\
&= \frac{1}{5}\cos t.
\end{aligned}
$$

SkillMaster 10.7.

• Consider the vector curve
$$\mathbf{r}(t) = <t, t^2, t^2 - 1>.$$
Find the unit tangent vector $\mathbf{T}(t)$, the unit normal vector $\mathbf{N}(t)$, and the binormal vector $\mathbf{B}(t)$.

Refer back to the definitions of these quantities if you need to.

$$\mathbf{r}'(t) = <1, 2t, 2t>$$

$$
\begin{aligned}
|\mathbf{r}'(t)| &= \sqrt{1^2 + (2t)^2 + (2t)^2} \\
&= \sqrt{1 + 4t^2 + 4t^2} \\
&= \sqrt{1 + 8t^2} = (1 + 8t^2)^{1/2}
\end{aligned}
$$

$$
\begin{aligned}
\mathbf{T}(t) &= \frac{\mathbf{r}'(t)}{|\mathbf{r}'(t)|} \\
&= \frac{1}{(1 + 8t^2)^{1/2}}\mathbf{i} \\
&\quad + \frac{2t}{(1 + 8t^2)^{1/2}}\mathbf{j} + \frac{2t}{(1 + 8t^2)^{1/2}}\mathbf{k} \\
&= (1 + 8t^2)^{-1/2}\mathbf{i} + 2t(1 + 8t^2)^{-1/2}\mathbf{j} \\
&\quad + 2t(1 + 8t^2)^{-1/2}\mathbf{k} \\
&= (1 + 8t^2)^{-1/2} <1, 2t, 2t>
\end{aligned}
$$

$$
\begin{aligned}
\mathbf{T}'(t) &= \frac{d}{dt}(1 + 8t^2)^{-1/2}\mathbf{i} \\
&\quad + \frac{d}{dt}(2t(1 + 8t^2)^{-1/2})\mathbf{j} \\
&\quad + \frac{d}{dt}(2t(1 + 8t^2)^{-1/2})\mathbf{k} \\
&= (-1/2)(1 + 8t^2)^{-3/2}(16t)\mathbf{i} \\
&\quad + (2(1 + 8t^2)^{-1/2} \\
&\quad + 2t(-1/2)(1 + 8t^2)^{-3/2})(16t)\mathbf{j} \\
&\quad + (2(1 + 8t^2)^{-1/2} \\
&\quad + 2t(-1/2)(1 + 8t^2)^{-3/2})(16t)\mathbf{k} \\[6pt]
&= -8t(1 + 8t^2)^{-3/2}\mathbf{i} \\
&\quad + 2(1 + 8t^2)^{-1/2} - 16t^2(1 + 8t^2)^{-3/2}\mathbf{j} \\
&\quad + 2(1 + 8t^2)^{-1/2} - 16t^2(1 + 8t^2)^{-3/2}\mathbf{k} \\
&= 2(1 + 8t^2)^{-3/2}[-4t\mathbf{i} \\
&\quad + ((1 + 8t^2) - 8t^2)\mathbf{j} \\
&\quad + ((1 + 8t^2) - 8t^2)\mathbf{k}] \\
&= 2(1 + 8t^2)^{-3/2}[-4t\mathbf{i} + \mathbf{j} + \mathbf{k}]
\end{aligned}
$$

$$\mathbf{N}(t) = \frac{\mathbf{T}'(t)}{|\mathbf{T}'(t)|}$$

$$= \frac{<-4t, 1, 1>}{|<-4t, 1, 1>|}$$

$$= \frac{<-4t, 1, 1>}{\sqrt{16t^2 + 2}}$$

$$= \frac{1}{\sqrt{2}} \frac{1}{\sqrt{1 + 8t^2}} <-4t, 1, 1>$$

$$\mathbf{B}(t)$$

$$= \mathbf{T}(t) \times \mathbf{N}(t)$$

$$= \left(\frac{(1 + 8t^2)^{-1/2}}{\sqrt{2}\sqrt{1 + 8t^2}} \right)$$

$$\cdot (< 1, 2t, 2t > \times < -4t, 1, 1 >)$$

$$= \left(\frac{1}{\sqrt{2}} \frac{1}{1 + 8t^2} \right)$$

$$\cdot (< 1, 2t, 2t > \times < -4t, 1, 1 >)$$

$$= \left(\frac{1}{\sqrt{2}} \frac{1}{1 + 8t^2} \right) \begin{vmatrix} \mathbf{i} & \mathbf{j} & \mathbf{k} \\ 1 & 2t & 2t \\ 4t & 1 & 1 \end{vmatrix}$$

$$= \left(\frac{1}{\sqrt{2}} \frac{1}{1 + 8t^2} \right)$$

$$\cdot \left((2t - 2t)\mathbf{i} - (1 + 8t^2)\mathbf{j} + (1 + 8t^2)\mathbf{k} \right)$$

$$= \left(\frac{1}{\sqrt{2}} \frac{1}{1 + 8t^2} \right)$$

$$\cdot \left(0\mathbf{i} - (1 + 8t^2)\mathbf{j} + (1 + 8t^2)\mathbf{k} \right)$$

$$= \left(\frac{1}{\sqrt{2}} \frac{1 + 8t^2}{1 + 8t^2} \right) < 0, -1, 1 >$$

$$= \frac{1}{\sqrt{2}} < 0, -1, 1 >$$

549

• In the worked out example above, why is $\mathbf{B}(t)$ constant?

Show that the curve lies in a plane.

From the vector equation
$$\mathbf{r}(t) = <t, t^2, t^2 - 1>$$
we have
$$
\begin{aligned}
x &= t \\
y &= t^2 \\
z &= t^2 - 1 = y - 1
\end{aligned}
$$
so the curve must lie on the plane
$$(0)x + (-1)y + (1)z = -1.$$
The direction numbers of this equation are given in the vector
$$< 0, -1, 1 >.$$
Thus the tangent vector and the normal vector must lie in this plane and the binormal vector must have the direction of $\langle 0, -1, 1 \rangle$.

• Find the equation of the osculating and normal plane at the point $(1, 1, 0)$ of the vector valued function of the previous two worked out examples.

Note that the point $(1, 1, 0) = (t, t^2, t^2 - 1)$ corresponds to $t = 1$.

The osculating plane was already found in the previous worked out example, since the curve lies within a fixed plane
$$-y + z = -1.$$
The normal plane has direction numbers given by the direction of
$$
\begin{aligned}
\mathbf{T}(1) &= \frac{1}{\sqrt{1 + 8(1)^2}} < 1, 2(1), 2(1) > \\
&= \frac{1}{3} < 1, 2, 2 >.
\end{aligned}
$$
For simplicity we work with the direction numbers $< 1, 2, 2 >$ (since any nonzero multiple of $\mathbf{T}(1)$ will do). The normal plane contains the point $(1, 1, 0)$ so its equation is
$$x - 1 = \frac{y - 1}{2} = \frac{z}{2}.$$

Section 10.4 –Motion in Space

Key Concepts:

- Space Motion: Position, Velocity, and Acceleration
- Tangential and Normal Components of Acceleration

Skills to Master:

- Compute the velocity and acceleration of a moving particle.
- Solve problems involving motion, acceleration, or force.

Discussion:

In this section, you apply much of what you have learned about space curves to the situation where points on the space curve represents the position of an object travelling through three-dimensional space. The first and second derivatives of the vector function giving the space curve then represent the velocity and acceleration of the object. It is then possible to break up the acceleration vector into the sum of two component vectors, one of which is in the direction of the tangent vector, and the other of which is in the direction of the normal vector to the curve.

Key Concept: Space Motion: Position, Velocity, and Acceleration

If a particle moves through space so that its position at time t is given by $\mathbf{r}(t)$, then the velocity of the particle, $\mathbf{v}(t)$ is given by

$$\mathbf{v}(t) = \mathbf{r}'(t)$$

and the acceleration of the particle, $\mathbf{a}(t)$ is given by

$$\mathbf{a}(t) = \mathbf{v}'(t) = \mathbf{r}''(t).$$

551

page 725.

Given any one of $\mathbf{r}(t)$, $\mathbf{v}(t)$, and $\mathbf{a}(t)$, and given some information about the initial conditions of the particle, you should be able to find the other two vectors by differentiation or integration. Study *Figures 2 and 3* to see geometric representations of velocity and acceleration vectors.

Key Concept: Tangential and Normal Components of Acceleration

If you represent the magnitude of the velocity vector $\mathbf{v}(t)$ by $v(t)$, the acceleration vector $\mathbf{a}(t)$ can be written as

$$\mathbf{a} = v'\mathbf{T} + \kappa v^2 \mathbf{N}$$

page 728.

where $\mathbf{T}$ is the unit tangent vector and $\mathbf{N}$ is the unit normal vector. Study *Figure 7* to see an example of how the acceleration vector can be broken down into these two components. The coefficient of $\mathbf{T}$, v', is called the tangential component of acceleration and is denoted a_T. The coefficient of $\mathbf{N}$, κv^2, is called the normal component of acceleration and is denoted a_N. The tangential component of acceleration speeds up or slows down the particle, and the normal component of acceleration turns the particle. Make sure that you understand the explanation in the text that shows how the formulas for the tangential and normal components acceleration are obtained.

SkillMaster 10.8: Compute the velocity and acceleration of a moving particle.

page 725.

Study *Examples 1, 2, and 3* to see worked out examples of how to find the position, velocity and acceleration of a moving particle. If you are given the position vector $\mathbf{r}(t)$, differentiating once gives velocity and differentiating again gives acceleration. If you are given the acceleration vector $\mathbf{a}(t)$, and some information about the velocity and position at some time, integrating once and applying the given information gives velocity, and integrating again and applying the information gives position.

SkillMaster 10.9: Solve problems involving motion, acceleration, or force.

To solve problems involving motion, acceleration or force, remember the relation between acceleration and force

$$\mathbf{F}(t) = m\mathbf{a}(t)$$

page 729.

where m is the mass of the moving object. This relation allows you to find $\mathbf{F}$ if given $\mathbf{a}$ and vice versa. You can then break the acceleration down into its tangential and normal components to get more information about how the object is moving. Study the discussion of *Kepler's Laws* to see how this is put into practice to derive important physical information.

Worked Examples

For each of the following examples, first try to find the solution without looking at the middle or right columns. Cover the middle and right columns with a piece of paper. If you need a hint, uncover the middle column. If you need to see the worked solution, uncover the right column.

Example	Tip	Solution

SkillMaster 10.8.

- Find the velocity, speed, and acceleration of the particle whose position is given by
$$\mathbf{r}(t) = 2t\mathbf{i} + (\tfrac{t^2}{2} - \ln(t))\mathbf{j} + (\tfrac{t^2}{2} + \ln(t))\mathbf{k} \text{ for } t \geq 1.$$

The velocity is the derivative of position. The speed is the magnitude of acceleration. The acceleration is the derivative of velocity.

$$
\begin{aligned}
\mathbf{v}(t) \\
&= \mathbf{r}'(t) \\
&= 2\mathbf{i} + (t - \frac{1}{t})\mathbf{j} + (t + \frac{1}{t})\mathbf{k} \\
|\mathbf{v}(t)| \\
&= \left| 2\mathbf{i} + (t - \frac{1}{t})\mathbf{j} + (t + \frac{1}{t})\mathbf{k} \right| \\
&= \sqrt{2^2 + (t - \frac{1}{t})^2 + (t + \frac{1}{t})^2} \\
&= \left(4 + (t^2 - 2t(1/t) + 1/t^2) \right. \\
&\quad \left. + (t^2 + 2t(1/t) + 1/t^2) \right)^{1/2} \\
&= \left(4 + (t^2 - 2 + 1/t^2) \right. \\
&\quad \left. + (t^2 + 2 + 1/t^2) \right)^{1/2} \\
&= \sqrt{2(t^2 + 2 + 1/t^2)} \\
&= \sqrt{2}\sqrt{(t + 1/t)^2} \\
&= \sqrt{2}(t + 1/t) \\
\mathbf{a}(t) \\
&= \mathbf{v}'(t) \\
&= (0)\mathbf{i} + (1 + \frac{1}{t^2})\mathbf{j} + (1 - \frac{1}{t^2})\mathbf{k}
\end{aligned}
$$

• Find the velocity, acceleration, and speed of a particle with the position function $\quad \mathbf{r}(t) =< e^t \cos t, e^t \sin t >$. Sketch the function from $t = -2\pi$ to $t = 2\pi$. Show that the acceleration is orthogonal to the position vector. Show that the acceleration always makes an angle of $\pi/2$ with the velocity. Thus the particle is always accelerated at a constant angle to the velocity with a magnitude proportional to the position. Illustrate these vectors for $t = 0$.	The velocity is $\mathbf{r}'(t)$.	$$\begin{aligned} \mathbf{v}(t) &= \mathbf{r}'(t) \\ &= \; < e^t \cos t - e^t \sin t, e^t \sin t + e^t \cos t > \\ &= \; < e^t(\cos t - \sin t), e^t(\sin t + \cos t) > . \end{aligned}$$
	The speed is the magnitude of $\mathbf{v}(t)$.	$$\begin{aligned} &|\mathbf{v}(t)| \\ &= \quad |< e^t(\cos t - \sin t), e^t(\sin t + \cos t) >| \\ &= \quad e^t \, |< (\cos t - \sin t), (\sin t + \cos t) >| \\ &= \quad e^t \sqrt{(\cos t - \sin t)^2 + (\sin t + \cos t)^2} \\ &= \quad e^t((\cos^2 t - 2\cos t \sin t + \sin^2 t) \\ &\qquad + (\sin^2 t + 2\sin t \cos t + \cos^2 t))^{1/2} \\ &= \quad e^t \sqrt{2\cos^2 + 2\sin^2 t} = e^t \sqrt{2} \\[4pt] &\frac{\mathbf{v}(t)}{|\mathbf{v}(t)|} = (1/\sqrt{2})(\cos t - \sin t)\mathbf{i} \\ &\qquad\qquad + (1/\sqrt{2})(\sin t + \cos t)\mathbf{j} \end{aligned}$$
	The acceleration is the derivative of $\mathbf{v}(t)$.	$$\begin{aligned} \mathbf{a}(t) &= \mathbf{v}'(t) \\ &= \; < e^t(\cos t - \sin t) + e^t(-\sin t - \cos t), \\ &\qquad e^t(\sin t + \cos t) + e^t(\cos t - \sin t) > \\ &= \; < e^t(-2\sin t), e^t(2\cos t) > \\ &= \; 2e^t < -\sin t, \cos t > \\ &\frac{\mathbf{a}(t)}{|\mathbf{a}(t)|} = -\sin t\mathbf{i} + \cos t\mathbf{j} \end{aligned}$$

To show that the position and the acceleration are orthogonal, compute the dot product and show that it is zero.

To find the angle between the acceleration and the velocity use the geometric definition of the dot product.

$$\cos \theta = \frac{\mathbf{v}(t) \cdot \mathbf{a}(t)}{|\mathbf{v}(t)| \, |\mathbf{a}(t)|}$$
$$= (1/\sqrt{2}) < \cos t - \sin t, \cos t + \sin t >$$
$$\cdot (< - \sin t, \cos t >)$$
$$= (1/\sqrt{2})(\cos t(-\sin t) - \sin t(-\sin t)$$
$$+ \cos t(\cos t) + \sin t(\cos y))$$
$$= \frac{(\sin^2 t + \cos^2 t)}{\sqrt{2}} = \frac{1}{\sqrt{2}}$$

Thus $\theta = \pi/4$.

SkillMaster 10.9.

• Suppose that the acceleration of a particle is given by
$\mathbf{a}(t) = 4\cos(2t)\mathbf{i} + 12\sin(2t)\mathbf{j} + 2\mathbf{k}$.
If the initial velocity and initial position vectors are

$$\mathbf{v}(0) = 6\mathbf{j} + \mathbf{k}$$
$$\mathbf{r}(0) = -\mathbf{k}$$

find the position vector of the particle for $t \geq 0$.

Use $\mathbf{a}(t) = \mathbf{v}'(t)$.

$$\mathbf{a}(t) = \mathbf{v}'(t)$$
$$= 4\cos(2t)\mathbf{i} + 12\sin(2t)\mathbf{j} + 2\mathbf{k}$$

$$\mathbf{v}(t)$$
$$= 4\int \cos(2t)dt\mathbf{i} + 12\int \sin(2t)dt\mathbf{j}$$
$$+ \int 2dt\mathbf{k}$$
$$= (4(1/2)(\sin 2t) + C_1)\mathbf{i}$$
$$+ (12(1/2)(-\cos(2t)) + C_2)\mathbf{j}$$
$$+ (2t + C_3)\mathbf{k}$$
$$= (2\sin(2t) + C_1)\mathbf{i} +$$
$$(-6\cos(2t) + C_2)\mathbf{j} + (2t + C_3)\mathbf{k}$$

To evaluate the constants use the initial condition.

$$\begin{aligned}
\mathbf{v}(0) \\
&= 6\mathbf{j} + \mathbf{k} \\
&= 0\mathbf{i} + 6\mathbf{j} + \mathbf{k} \\
&= (2\sin(2(0)) + C_1)\mathbf{i} + \\
&\quad (-6\cos(2(0)) + C_2)\mathbf{j} + (2(0) + C_3)\mathbf{k} \\
&= C_1\mathbf{i} + (-6 + C_2)\mathbf{j} + C_3\mathbf{k}
\end{aligned}$$

Equate these coefficients

$$\begin{aligned}
0 &= C_1 \\
6 &= -6 + C_2 \\
1 &= 2 + C_3
\end{aligned}$$

$$\begin{aligned}
C_1 &= 0 \\
C_2 &= 12 \\
C_3 &= 1
\end{aligned}$$

$$\begin{aligned}
\mathbf{v}(t) \\
&= 2\sin(2t)\mathbf{i} + (-6\cos(2t) + 12)\mathbf{j} \\
&\quad + (2t + 1)\mathbf{k}
\end{aligned}$$

557

Repeat this process of integration to find the position vector.

$$\mathbf{r}(t)$$
$$= \quad 2 \int \sin(2t)dt\mathbf{i}$$
$$+ \int 12 - 6\cos(2t)dt\mathbf{j} + \int 2t + 1dt\mathbf{k}$$
$$= \quad (-\cos(2t) + D_1)\mathbf{i}$$
$$+(12t - 3\sin(2t) + D_2)\mathbf{j}$$
$$+(t^2 + t + D_3)\mathbf{k}$$

$$\mathbf{r}(0)$$
$$= \quad -\mathbf{k}$$
$$= \quad (-1 + D_1)\mathbf{i} + D_2\mathbf{j} + D_3\mathbf{k}$$
$$-1 + D_1 \quad = \quad 0$$
$$D_2 \quad = \quad 0$$
$$D_3 \quad = \quad -1$$

$$\mathbf{r}(t)$$
$$= \quad (1 - \cos(2t))\mathbf{i}$$
$$+(12t - 3\sin(2t))\mathbf{j} + (t^2 + t - 1)\mathbf{k}$$

- Suppose the speed of a particle is constant and not zero. Show that the acceleration is orthogonal to the velocity. Thus the acceleration only changing the direction of the particle.

You need to show $\mathbf{v}(t) \cdot \mathbf{a}(t) = 0$. You are given that the speed $|\mathbf{v}(t)| = k$. This means that $\mathbf{v}(t) \cdot \mathbf{v}(t) = k^2$. Differentiate this equation, using the rules of differentiation.

$$k^2 = \mathbf{v}(t) \cdot \mathbf{v}(t)$$

Differentiate both sides.

$$0 = \mathbf{v}'(t) \cdot \mathbf{v}(t) + \mathbf{v}(t) \cdot \mathbf{v}'(t)$$
$$= \mathbf{v}(t) \cdot \mathbf{v}'(t) + \mathbf{v}(t) \cdot \mathbf{v}'(t)$$
$$= 2\mathbf{v}(t) \cdot \mathbf{v}'(t)$$
$$= 2\mathbf{v}(t) \cdot \mathbf{a}(t)$$

$$0 = \mathbf{v}(t) \cdot \mathbf{a}(t)$$

• Consider the path of a particle given by
$$r(t) = \ <2t^2, t^3, -1>.$$
for $t \geq 0$. Find the tangential and normal components of acceleration at the point $t = 1$.

The acceleration vector $\mathbf{a} = \mathbf{a}(t)$ lies in the osculating plane, which contains the tangent unit vector $\mathbf{T} = \mathbf{T}(t)$ and the normal vector $\mathbf{N} = \mathbf{N}(t)$. This means that
$$\mathbf{a} = a_\mathbf{T}\mathbf{T} + a_\mathbf{N}\mathbf{N}$$
where
$$a_\mathbf{T} = \frac{dv}{dt}$$
$$a_\mathbf{N} = \kappa v^2$$
where $v = |\mathbf{v}(t)| = |\mathbf{r}'(t)|$ is the speed and κ is the curvature.
Notice that since $\mathbf{T}$ and $\mathbf{N}$ are unit vectors which are orthogonal to each other
$$|\mathbf{a}|^2 = a_\mathbf{T}^2 + a_\mathbf{N}^2$$
so that $a_\mathbf{N}$ may be found if $|\mathbf{a}|^2$ and $a_\mathbf{T}$ have already been computed.

$$\begin{aligned} v &= |\mathbf{v}(t)| = |\mathbf{r}'(t)| \\ &= |<4t, 3t^2, 0>| \\ &= \sqrt{(4t)^2 + (3t^2)^2 + 0^2} \\ &= \sqrt{16t^2 + 9t^4} \end{aligned}$$
since $t \geq 0$. Therefore
$$\begin{aligned} a_\mathbf{T} &= \frac{dv}{dt} \\ &= \frac{(16t + 18t^3)}{\sqrt{16t^2 + 9t^4}} \\ &= \frac{2t(8 + 9t^2)}{t\sqrt{16 + 9t^2}} \\ &= \frac{2(8 + 9t^2)}{\sqrt{16 + 9t^2}} \\ &= \frac{2(8 + 9(1)^2)}{\sqrt{16 + 9(1)^2}} \\ &= \frac{2(17)}{5} = \frac{34}{5}. \end{aligned}$$
To use the Tip to compute
$$\begin{aligned} \mathbf{a} &= \ <4, 6t, 0> \\ |\mathbf{a}|^2 &= \mathbf{a} \cdot \mathbf{a} \\ &= 4^2 + (6t)^2 \\ &= 16 + 36t^2 \\ &= 16 + 36(1)^2 = 52 \end{aligned}$$
According to the Tip
$$\begin{aligned} a_\mathbf{N} &= \sqrt{|\mathbf{a}|^2 - (a_\mathbf{T})^2} \\ &= \sqrt{52 - (\frac{34}{5})^2} \\ &= \frac{1}{5}\sqrt{52(25) - (34)^2} \\ &= \frac{1}{5}\sqrt{144} = \frac{12}{5}. \end{aligned}$$
Thus
$$\mathbf{a} = \frac{34}{5}\mathbf{T} + \frac{12}{5}\mathbf{N}.$$

559

Section 10.5 – Parametric Surfaces

Key Concepts:

- Parametric Surfaces

Skills to Master:

- Graph and recognize parametric surfaces.
- Find parametric representations of surfaces.

Discussion:

Section 9.6.

In the *previous chapter*, you were introduced to surfaces that were graphs of functions of two variables, $f(x, y)$. These types of surfaces are examples of parametric surfaces: the third coordinate of a point on the surface, z, depends on the parameters x and y via the relation $z = f(x, y)$.

This section introduces a more general type of parametric surface in which each of the coordinates x, y, and z of a point on the surface can be viewed as a function of two parameters. Most of the surfaces that you have encountered can be viewed in this new way as parametric surfaces.

Key Concept: Parametric Surfaces

page 734.

A *parametric surface* is a vector function of two parameters that assigns to each pair of parameters (u, v) in some region in the $uv-$plane a vector $\mathbf{r}(u, v)$ in three-dimensional space.

$$\mathbf{r}(u, v) = x(u, v)\mathbf{i} + y(u, v)\mathbf{j} + z(u, v)\mathbf{k}$$

Here, $x(u, v)$, $y(u, v)$, and $z(u, v)$ are real valued functions of two variables. The

parametric surface is the set of points D in three-dimensional space satisfying
$$x = x(u,v), \ y = y(u,v), \text{ and } z = z(u,v).$$

The above three equations are called parametric equations to the surface. Notice the similarity to parametric curves.

SkillMaster 10.10: Graph and recognize parametric surfaces.

pages 735.

Graphing a parametric surface is best done using a computer algebra system or graphing utility. To gain insight into what the surface looks like, consider the various cross sections perpendicular to the x, y, and z axes. See *Example 1* for a worked out example on how to do this.

SkillMaster 10.11: Find parametric representations of surfaces.

pages 736-737.

Pay careful attention to *Examples 3, 4, and 5* to see how to find parametric representations of surfaces.

As an additional example, consider the ellipsoid $2x^2 + 3y^2 + z^2 = 4$. The cross sections in the planes $z = k$ are ellipses of the form
$$2x^2 + 3y^2 = 4 - k^2.$$
Such an ellipse can be parametrized by
$$x = \sqrt{\frac{4 - k^2}{2}} \cos u \qquad y = \sqrt{\frac{4 - k^2}{3}} \sin u$$
where $0 \leq u \leq 2\pi$. If the parameter v is used to represent z, $-2 \leq v \leq 2$, we obtain the parametrization
$$x = \sqrt{\frac{4 - v^2}{2}} \cos u \qquad y = \sqrt{\frac{4 - v^2}{3}} \sin u \qquad z = v$$
of the surface.

Worked Examples

For each of the following examples, first try to find the solution without looking at the middle or right columns. Cover the middle and right columns with a piece of paper. If you need a hint, uncover the middle column. If you need to see the worked solution, uncover the right column.

Example	Tip	Solution

SkillMaster 10.10.

• Identify and graph the surface
$\mathbf{r}(u, v) = 2\cos u\,\mathbf{i} + v\,\mathbf{j} + 3\sin u\,\mathbf{k}$.

Here
$$x = 2\cos u$$
$$y = v$$
$$z = 3\sin u.$$
Note that $\cos u = \frac{x}{2}$, $\sin u = \frac{z}{3}$, and $y = v$ (which has no connection to the variable u). Eliminate u and get an equation relating x and y.

Thus
$$\frac{x^2}{4} + \frac{z^2}{9} = 1$$
and y is any real number. This gives an elliptic cylinder along the y-axis.

$$x = 2\cos u$$
$$y = v$$
$$z = 3\sin u.$$

• Identify and graph the surface $\mathbf{r}(u, v)$
$= u\cos v\mathbf{i} + u\sin v j + (1 - u^2)\mathbf{k}.$

Use the identity $\cos^2 v + \sin^2 v = 1$ to get eliminate v.

$x^2 + y^2 = u^2\cos^2 v + u^2\sin^2 v = u^2$. So $z = 1 - u^2 = 1 - (x^2 + y^2)$.

The equation is $z = 1 - (x^2 + y^2)$.
Now $z = x^2 + y^2$ is the standard circular paraboloid, so $z = 1 - (x^2 + y^2)$ is the same paraboloid reflected across the xy-plane and then moved up one unit along the z-axis. The vertex is at $(0, 0, 1)$.

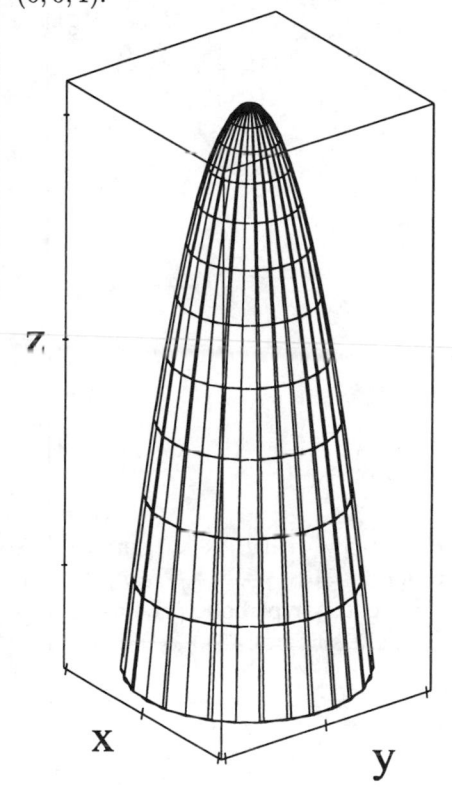

• Identify and graph the surface $\mathbf{r}(u,v)$ $=< u, -\sqrt{9-u^2-v^2}, v >$.

Notice that $y \leq 0$. First express y in terms of x and z.

We have $x = u$ and $z = v$ so
$$
\begin{aligned}
y &= -\sqrt{9-x^2-z^2} \\
y^2 &= 9-x^2-z^2 \\
x^2 + y^2 + z^2 &= 3^2
\end{aligned}
$$
This is a sphere of radius 3 and center at the origin. Since $y \leq 0$ the surface is the "left" hemisphere.

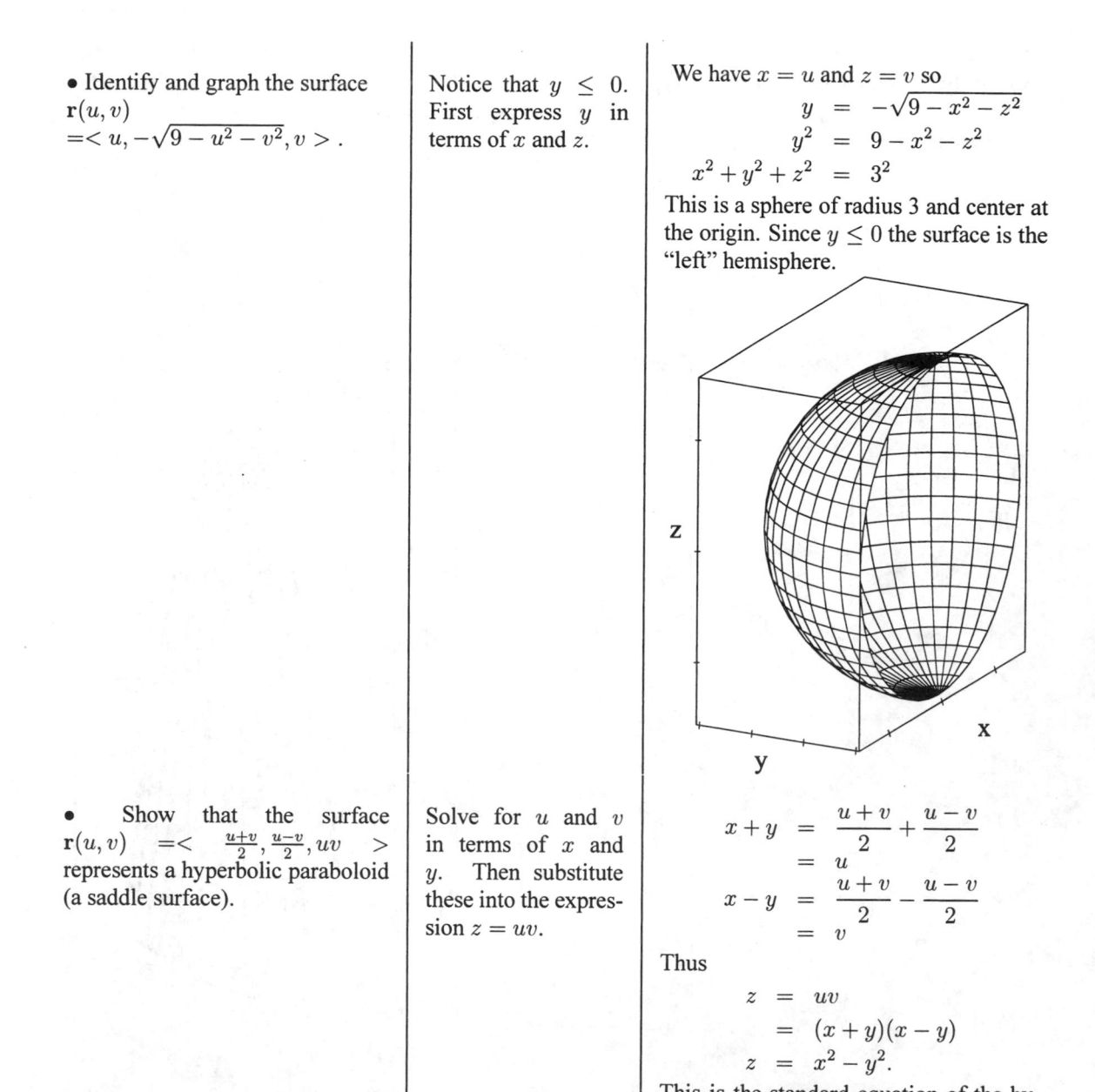

• Show that the surface $\mathbf{r}(u,v)$ $=< \frac{u+v}{2}, \frac{u-v}{2}, uv >$ represents a hyperbolic paraboloid (a saddle surface).

Solve for u and v in terms of x and y. Then substitute these into the expression $z = uv$.

$$
\begin{aligned}
x + y &= \frac{u+v}{2} + \frac{u-v}{2} \\
&= u \\
x - y &= \frac{u+v}{2} - \frac{u-v}{2} \\
&= v
\end{aligned}
$$

Thus
$$
\begin{aligned}
z &= uv \\
&= (x+y)(x-y) \\
z &= x^2 - y^2.
\end{aligned}
$$
This is the standard equation of the hyperbolic paraboloid.

SkillMaster 10.11.

- Find a parametrization for the "Gaussian-bump" surface $z = e^{-x^2-y^2}$.

There are many approaches. One is to use x and y as the parameters. Another is to use polar coordinates $x = r \cos \theta$ and $y = r \sin \theta$.

If x and y are used as parameters then the parametrization is $\mathbf{r}(x,y) = <x, y, e^{-x^2-y^2}>$.

If polar coordinates are used then we have instead $\mathbf{r}(r, \theta) = r \cos \theta \mathbf{i} + r \sin \theta \mathbf{j} + e^{-r^2} \mathbf{k}$.

SkillMasters for Chapter 10

SkillMaster 10.1:   Sketch and recognize graphs of space curves given parametrically.

SkillMaster 10.2:   Find the domain, range, and limits of vector functions.

SkillMaster 10.3:   Compute derivatives of vector functions and tangent vectors of space curves.

SkillMaster 10.4:   Compute integrals of vector functions.

SkillMaster 10.5:   Compute arc length and parameterize curves with respect to arc length.

SkillMaster 10.6:   Compute the curvature of a curve.

SkillMaster 10.7:   Compute the normal and binormal vectors and the associated planes of a curve.

SkillMaster 10.8:   Compute the velocity and acceleration of a moving particle.

SkillMaster 10.9:   Solve problems involving motion, acceleration, or force.

SkillMaster 10.10:   Graph and recognize parametric surfaces.

SkillMaster 10.11:   Find parametric representations of surfaces.

Chapter 11 - Partial Derivatives

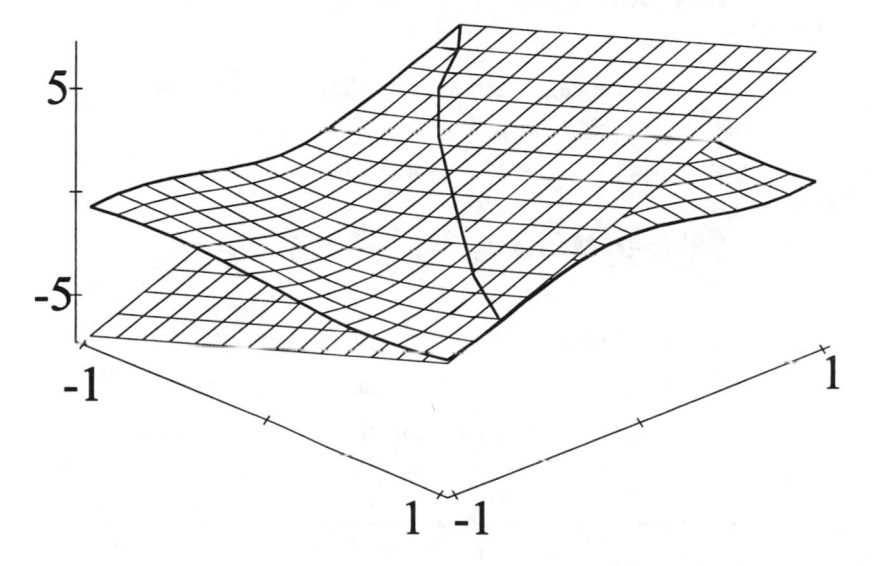

Section 11.1 – Functions of Several Variables

Key Concepts:

- Verbal, Numerical, Algebraic, and Visual Interpretations of Functions of Two Variables
- Functions of Three or More Variables

Skills to Master:

- Find the domain, and range of functions of two variables.
- Use different descriptions of functions.
- Describe the level surfaces of functions of three variables.

Discussion:

page 684.

In *Chapter 9*, you studied functions of two variables and their graphs. In this section, functions of two or more variables are studied in more detail from several points of view. You should think of a function of several variables in the following way.

$$\text{Input} \left\{ \begin{array}{c} (x,y) \\ \text{or} \\ (x,y,z) \end{array} \right\} \Longrightarrow \text{Output } \{\text{a real number}\}$$

The input to the function is a pair (x, y) or a triple (x, y, z) of real numbers. The output or result of applying the function is a unique real number. Whenever you have a situation where a quantity depends on two, three or more variables, you are dealing with a function of several variables. Study carefully the examples presented in this section.

Key Concept: Verbal, Numerical, Algebraic, and Visual Interpretations of Functions of Two Variables

The examples presented in this section illustrate the various ways of viewing a function of several variables. A function can be given by describing it in words. A function can also be described numerically by giving a table of values. *Examples 1 and 2* in this section are of functions given by tables of values. From previous mathematics courses, you are most familiar with functions given algebraically by an explicit formula. *Example 3* in the text is of a function given by a formula.

The graph of a function of two variables is a surface in three dimensional space. Visualizing the graph often leads to a better understanding of the function. Study *Examples 4 through 8* in this section to gain a better understanding of visual representations of functions.

page 748-749.

page 750.

page 750-753.

Key Concept: Functions of Three or More Variables

As indicated in the discussion above, a function of three variables is a rule that assigns a unique real number to each (x, y, z) in the domain of the function. You can't easily visualize the graph of such a function since the graph lies in four dimensional space. However you can get a better understanding of the function by considering its level surfaces, that is, surfaces with the equations

$$f(x, y, z) = k \text{ where } k \text{ is a constant.}$$

Study *Examples 11 and 12* to gain a better understanding of functions of three variables.

page 755.

SkillMaster 11.1: Find the domain and range of functions of two variables.

To find the domain of a function of two variables, find all points (x, y) in the plane at which the function is defined. If the function is given by a formula that includes a fraction, points where the denominator is zero will not be included in the domain. If the formula involves functions such as the logarithm or tangent, the domain of the function will not include points that would lead to values where the logarithm or tangent would not be defined. Study the examples in this section to gain a better understanding of how to find the domain of functions.

page 749.

The range of a function of two variables is the set of all values taken on by the function. If the function is given by a formula, use your understanding of the functions in the formula to determine the range. *Example 3* in the text gives an illustration of finding the domain and range of a function.

SkillMaster 11.2: Use different descriptions of functions.

page 750.

Using different descriptions of a function can give you a much better understanding of the function and its properties. Working with a formula for the function allows you to find the domain and range of the function. Using a visual representation of the function gives you a picture of what the graph looks like. Together, these representations allow you to find information about the function that might be of interest. *Example 4*, combined with Example 3, gives an illustration of this process.

SkillMaster 11.3: Describe the level surfaces of functions of three variables.

Since the graph of a function of three variables requires four dimensional space, information about functions of three variables is often obtained by using level surfaces which only require three dimensional space. Finding level surfaces involves solving equations of the form

$$f(x, y, z) = k.$$

You have already worked with equations of this form in previous chapters when you worked with equations of planes, spheres and quadric surfaces. Apply what you learned in Chapter 9 to this new situation.

Worked Examples

For each of the following examples, first try to find the solution without looking at the middle or right columns. Cover the middle and right columns with a piece of paper. If you need a hint, uncover the middle column. If you need to see the worked solution, uncover the right column.

Example	Tip	Solution
SkillMaster 11.1.		
• Let $f(x, y) = \ln(xy)$. (a) Find the domain and range of f.	Remember that $\ln u$ is defined if and only if $u > 0$.	(a) $f(x, y)$ is defined if and only if $\ln(xy)$ is defined if and only if $xy > 0$. The domain of f is $D = \{(x, y) \mid xy > 0\}$. If $xy > 0$ then either (i) $x > 0, y > 0$ or (ii) $x < 0, y < 0$. The domain of f consists of the first and third quadrants (without the axes). The range is the same as the range of $\ln u$ which is the set of all real numbers. 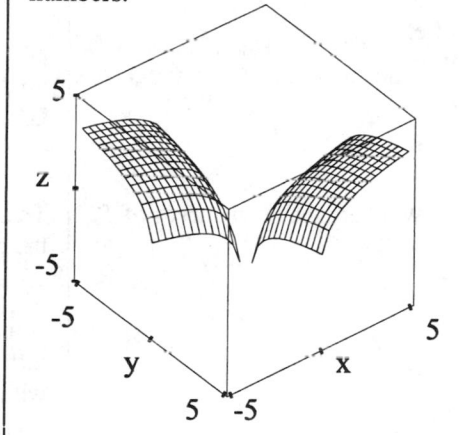
(b) Is it true that $f(x, y) = \ln x + \ln y$ for all (x, y) in the domain of f?	Is it possible that (x, y) is in the domain of f but that $\ln x$ is not defined?	(b) It is not true that $f(x, y) = \ln x + \ln y$ for all (x, y) in the domain of f. If both $x > 0$ and $y > 0$ then we have $f(x, y) = \ln(xy) = \ln x + \ln y$; however of (x, y) is in the third quadrant then both $x < 0$ and $y < 0$ then $\ln x$ and $\ln y$ are not even defined.

(c) Describe and sketch the points (x, y) such that $f(x, y) = 0$.

Recall that $\ln u = 0$ if and only if $u = 1$.

(c) $f(x, y) = 0$ if and only if $\ln(xy) = 0$ if and only if $xy = 1$. This is the equation of a hyperbola.

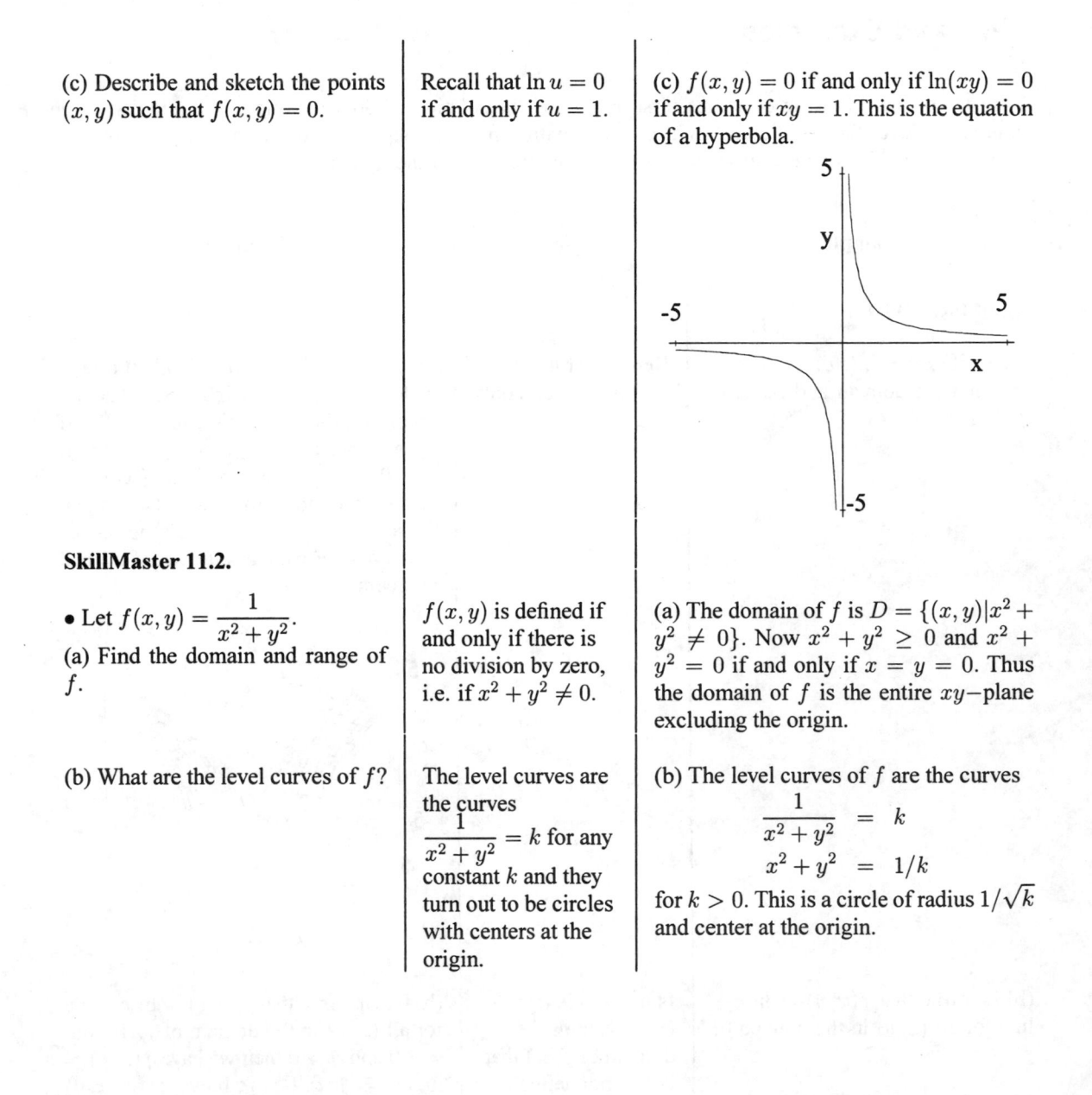

SkillMaster 11.2.

• Let $f(x, y) = \dfrac{1}{x^2 + y^2}$.

(a) Find the domain and range of f.

$f(x, y)$ is defined if and only if there is no division by zero, i.e. if $x^2 + y^2 \neq 0$.

(a) The domain of f is $D = \{(x, y)|x^2 + y^2 \neq 0\}$. Now $x^2 + y^2 \geq 0$ and $x^2 + y^2 = 0$ if and only if $x = y = 0$. Thus the domain of f is the entire xy-plane excluding the origin.

(b) What are the level curves of f?

The level curves are the curves
$$\frac{1}{x^2 + y^2} = k$$
for any constant k and they turn out to be circles with centers at the origin.

(b) The level curves of f are the curves
$$\frac{1}{x^2 + y^2} = k$$
$$x^2 + y^2 = 1/k$$
for $k > 0$. This is a circle of radius $1/\sqrt{k}$ and center at the origin.

(c) Sketch a contour graph of f and sketch the surface of f.

Since the level curves are circles with centers at the origin and the values of the function are positive the shape could be approximated by clay shaped at a potter's wheel.

(c)

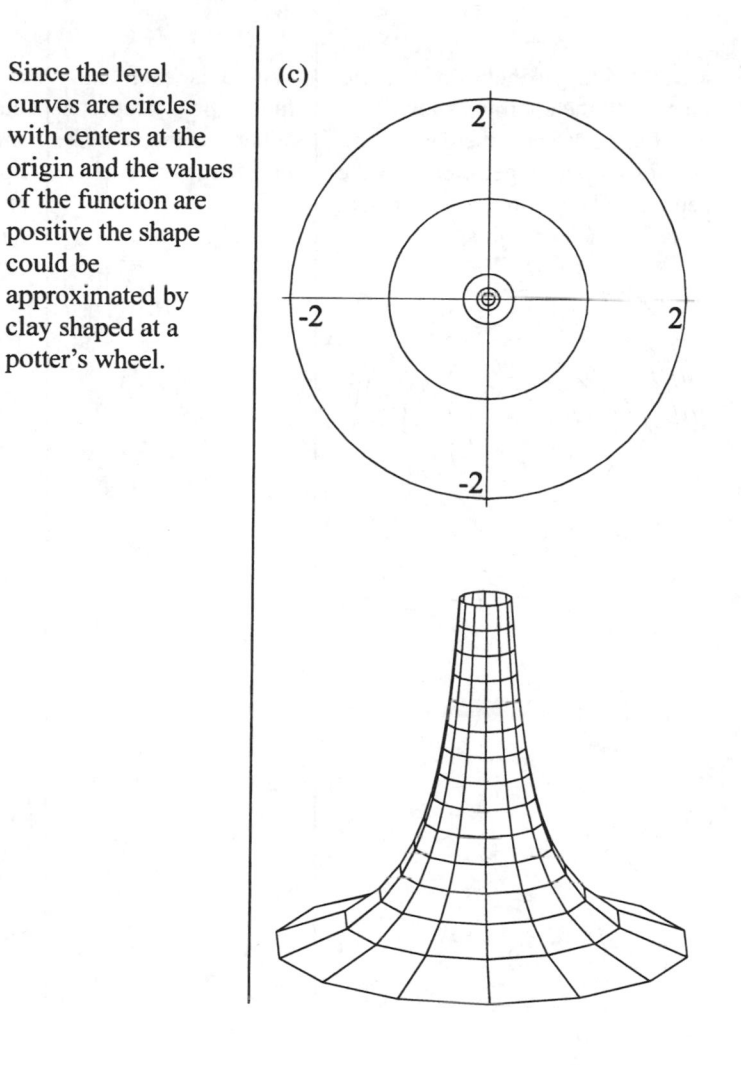

● Make rough sketches of the surfaces with the contours shown.
For both surface, the height increases as points get closer to the center. The contours represent equally spaced heights.

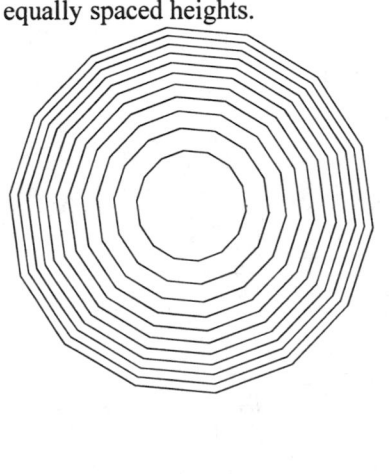

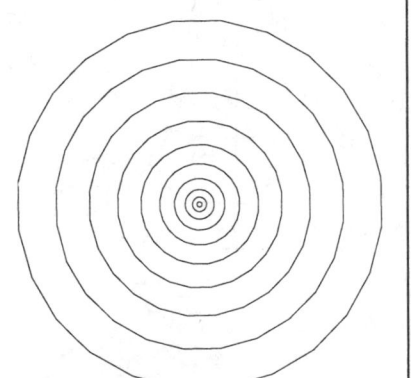

Where is each surface the steepest? Where is each surface the most level?

A surface is steepest when the level curves are closest together and is the most level when the level curves are far apart. The first surface is steepest away from the center and the second surface is steepest close to the center. The surfaces are pictured below.

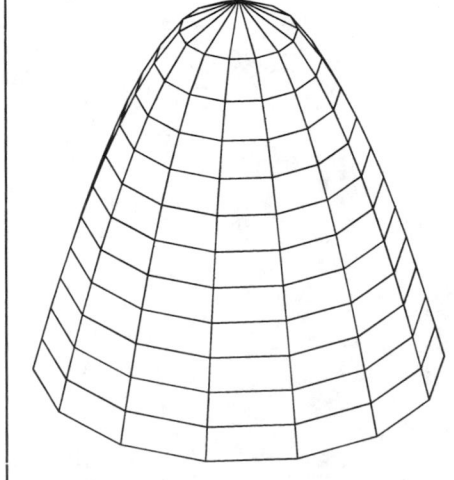

SkillMaster 11.3.

• Consider the function
$$f(x, y, z) = \frac{1}{\sqrt{1 - \ln(x^2 + y^2 + z^2)}}.$$
(a) Describe the domain of this function.

Notice that for (x, y, z) to be in the domain all the functions that make up $f(x, y, z)$ must be defined. In particular, one cannot divide by 0, one cannot take the square root of a negative number, and one cannot take the logarithm of a number that is not positive.

(a) First the inside of the square root must be positive:
$$1 - \ln(x^2 + y^2 + z^2) > 0$$
$$\ln(x^2 + y^2 + z^2) < 1$$
$$x^2 + y^2 + z^2 < e^1 = e.$$
Second, the inside of the logarithm must be positive:
$$x^2 + y^2 + z^2 > 0.$$
Putting this together and taking square roots we have
$$0 < \sqrt{x^2 + y^2 + z^2} < e^{1/2}.$$
The domain of f is the ball strictly inside the sphere of radius $e^{1/2}$ and with the origin excluded.

(b) Describe the level surfaces of this function.

Since $x^2 + y^2 + z^2 = \rho^2$ (in spherical coordinate) is the square of the distance of (x, y, z) from the origin, the level curves must be spheres centered at the origin.

(b) To describe the level surfaces fix $k > 0$ and set
$$f(x, y, z) = k$$
$$\frac{1}{\sqrt{1 - \ln(x^2 + y^2 + z^2)}} = k$$
$$\sqrt{1 - \ln(x^2 + y^2 + z^2)} = \frac{1}{k}$$
$$1 - \ln(x^2 + y^2 + z^2) = \frac{1}{k^2}$$
$$\ln(x^2 + y^2 + z^2) = 1 - \frac{1}{k^2}$$
$$x^2 + y^2 + z^2 = e^{1 - \frac{1}{k^2}}$$
$$\sqrt{x^2 + y^2 + z^2} = \sqrt{e^{1 - \frac{1}{k^2}}}.$$
The level surfaces are spheres centered at the origin with radius $\sqrt{e^{1 - \frac{1}{k^2}}}$.

Section 11.2 – Limits and Continuity

Key Concepts:

- The limit of a function $f(x, y)$ as (x, y) approached a fixed value
- Continuity of a function $f(x, y)$ at a point (a, b)

Skills to Master:

- Determine if a function $f(x, y)$ has a limit at (a, b).
- Determine the points of continuity of a function $f(x, y)$.

Discussion:

The definitions of limit and continuity for functions of several variables are identical to the definitions of these terms for functions of a single variable. However, deciding whether a limit exists is more complicated for functions of several variables because a point in the plane or in three dimensional space can be approached from many directions and along many paths. Contrast this with the situation for one variable where a point on the line can be only approached from two directions. Much of what you already know about limits and continuity can be used in this section. Make sure that you study the examples carefully to gain an understanding of the additional complexity for functions of several variables.

Key Concept: The limit of a function $f(x, y)$ as (x, y) approached a fixed value

The definition of the limit of $f(x, y)$ as (x, y) approaches a fixed point (a, b) is identical to the definition of the limit of a function of a single variable. That is, the limit is L if you can make the values of $f(x, y)$ as close to L as you like by taking the point (x, y) sufficiently close to the point (a, b), but not equal to

page 760.
the point (a, b). Study *Examples 1, 2, and 3* carefully to see how a function of several variables can fail to have a limit at certain points. The easiest way to show that the limit of $f(x, y)$ as (x, y) approaches (a, b) doesn't exist is to find two paths approaching the point (a, b) so that the function has different limits along the paths.

page 762.
To show that a limit does exist, you may use results such as the Squeeze Theorem and Limit Laws that you already know. Study *Example 4* to see how this works.

Key Concept: Continuity of a function $f(x, y)$ at a point (a, b)

A function f of two variables is continuous at (a, b) if
$$\lim_{(x,y)\to(a,b)} f(x, y) = f(a, b).$$
Note the similarity of this definition to the definition for continuity of functions of a single variable. The function f is continuous on a set D if it is continuous at every point in D. A function can fail to be continuous at a point for three reasons:

- $\lim_{(x,y)\to(a,b)} f(x, y)$ might not exist.

- f might not be defined at (a, b).

- Even if $\lim_{(x,y)\to(a,b)} f(x, y)$ exists and $f(a, b)$ is defined, these values might not be equal.

By using the Limit Laws that you already know, you can conclude that all polynomial and rational functions of two variables are continuous at each point where they are defined.

SkillMaster 11.4: Determine if a function $f(x, y)$ has a limit at (a, b).

If a function $f(x, y)$ is not a rational function or a polynomial and you want to know whether the function has a limit at a specific point (a, b), try taking different

page 760-761.

page 761.

paths to the point and evaluating limits as you approach the point along those paths. *Examples 1 and 2* show you how to do this. Refer back to the earlier examples in this section to see that the paths that you choose should not always be straight line paths. Reread *Example 3* to see how this might occur.

SkillMaster 11.5: Determine the points of continuity of a function $f(x, y)$.

Polynomial and rational functions of two variables are continuous at each point where they are defined. This follows by applying the Limit Laws to such functions. For example, the limit of a product is the product of the limits and the limit of a sum or difference is the sum or difference of the limits. All of the usual combinations of standard special functions like $\ln(xy^3)$ or $\sqrt{\tan^{-1}(xy)}$ are continuous at all points in their domains. *Examples 5 and 6 and 8* show how to apply this. Make sure that you understand these examples.

page 763-764.

Worked Examples

For each of the following examples, first try to find the solution without looking at the middle or right columns. Cover the middle and right columns with a piece of paper. If you need a hint, uncover the middle column. If you need to see the worked solution, uncover the right column.

Example	Tip	Solution

SkillMaster 11.4.

- Determine if the limit exists. If the limit does exist, find its value.

(a)

$$\lim_{(x,y)\to(1,-1)} \frac{e^{-xy}\cos(x+y)}{(x+1)^2+(y+1)^2-1}$$

Tip: First find the domain. If the limit point in question is not in the domain then use l'Hospital's rule to show the limit exists or else show the limit does not exist by finding different paths with different limits. Sometimes polar coordinates are useful.

Solution: (a) In this case $(1,-1)$ is in the domain and the limit may be evaluated by plugging in the values.

$$\lim_{(x,y)\to(1,-1)} \frac{e^{-xy}\cos(x+y)}{(x+1)^2+(y+1)^2-1}$$

$$= \frac{e^{-1(-1)}\cos(1+(-1))}{(1+1)^2+(-1+1)^2-1}$$

$$= \frac{e\cos 0}{4-1}$$

$$= e/3$$

(b)
$$\lim_{(x,y)\to(0,0)} \frac{\sqrt{x^2+y^2}}{\ln(1+\sqrt{x^2+y^2})}$$

Tip: Sometimes polar coordinates are useful.

Solution: (b) Here the point $(0,0)$ is not in the domain. Use polar coordinates, $r = \sqrt{x^2+y^2}$ and l'Hospital's Rule.

$$\lim_{(x,y)\to(0,0)} \frac{\sqrt{x^2+y^2}}{\ln(1+\sqrt{x^2+y^2})}$$

$$= \lim_{r\to 0} \frac{r}{\ln(1+r)}$$

$$= \lim_{r\to 0} \frac{d(r)/dr}{d(\ln(1+r))/dr}$$

$$= \lim_{r\to 0} \frac{1}{1/(1+r)}$$

$$= \lim_{r\to 0} (1+r)$$

$$= 1+0 = 1$$

(c) $\displaystyle\lim_{(x,y)\to(0,0)} \frac{\sqrt{xy}}{x-y}$

Consider different paths to the origin.

(c) The limit does not appear to exist. Try limits along different paths to the origin. One path might be along the axis $(x,0) \to (0,0)$, another along the line with slope 1/4, $(4y,y) \to (0,0)$.

$$\lim_{(x,0)\to(0,0)} \frac{\sqrt{x(0)}}{x-(0)}$$

$$= \lim_{(x,0)\to(0,0)} \frac{0}{x}$$

$$= \quad 0$$

$$\lim_{(4y,y)\to(0,0)^+} \frac{\sqrt{(4y)y}}{4y-y}$$

$$= \lim_{(4y,y)\to(0,0)^+} \frac{2y}{3y}$$

$$= \quad 2/3$$

Since these values are different the limit does not exist.

• Either evaluate the limit, or show that it does not exist.

Try spherical coordinates and l'Hospital's Rule.

(a) In spherical coordinates $\rho^2 = x^2 + y^2 + z^2$.

(a) $\displaystyle\lim_{(x,y,z)\to(0,0,0)} \frac{e^{(x^2+y^2+z^2)} - 1}{x^2 + y^2 + z^2}$

$$\lim_{(x,y,z)\to(0,0,0)} \frac{e^{(x^2+y^2+z^2)} - 1}{x^2 + y^2 + z^2}$$

$$= \lim_{\rho\to 0} \frac{e^{\rho^2} - 1}{\rho^2}$$

$$= \lim_{\rho\to 0} \frac{d(e^{\rho^2} - 1)/d\rho}{d(\rho^2)/d\rho}$$

$$= \lim_{\rho\to 0} \frac{2\rho e^{\rho^2}}{2\rho}$$

$$= \lim_{\rho\to 0} e^{\rho^2}$$

$$= e^0 = 1$$

(b) $\displaystyle\lim_{(x,y,z)\to(0,0,0)} \frac{(x+2y+3z)^2}{x^2+2y^2+3z^2}$

Try seeing if you get different values for limits along different coordinate axes. Try the paths $(x,0,0)\to(0,0,0)$ and $(0,y,0)\to(0,0,0)$.

(b)

$$\lim_{(x,0,0)\to(0,0,0)} \frac{(x+2y+3z)^2}{x^2+2y^2+3z^2}$$

$$= \lim_{(x,0,0)\to(0,0,0)} \frac{(x+2(0)+3(0))^2}{x^2+2(0)^2+3(0)^2}$$

$$= \lim_{(x,0,0)\to(0,0,0)} \frac{x^2}{x^2}$$

$$= 1$$

$$\lim_{(0,y,0)\to(0,0,0)} \frac{(x+2y+3z)^2}{x^2+2y^2+3z^2}$$

$$= \lim_{(0,y,0)\to(0,0,0)} \frac{((0)+2y+3(0))^2}{(0)^2+2y^2+3(0)^2}$$

$$= \lim_{(0,y,0)\to(0,0,0)} \frac{(2y)^2}{2y^2}$$

$$= \lim_{(0,y,0)\to(0,0,0)} \frac{4y^2}{2y^2}$$

$$= 2$$

Since these limits are different the limit of the function does not exist.

- Find the limits if they exist.

(a) $\displaystyle\lim_{(x,y)\to(1,2)} \frac{e^x\cos y - \ln x \sin y}{2x+y}$

First find the domain. If the limit point is in the domain then we may find the limit by plugging this limit point into the formula for the function.

(a) The domain is the set of (x,y) such that $2x+y \neq 0$. Since $(1,2)$ does not lie on this line it is in the domain and

$$\lim_{(x,y)\to(1,2)} \frac{e^x\cos y - \ln x \sin y}{2x+y}$$

$$= \frac{e^1\cos 2 - \ln 1 \sin 2}{2(1)+2}$$

$$= \frac{e\cos 2 - 0}{4} = e\cos 2.$$

(b) $\displaystyle\lim_{(x,y)\to(0,0)} x^2 - xy + e^{xy}$

Use the same hint as above.

(b) In this case the function is defined for all (x, y) and we may plug $(0,0)$ into the function.

$$\lim_{(x,y)\to(0,0)} x^2 - xy + e^{xy}$$
$$= (0)^2 - (0)(0) + e^0$$
$$= 1$$

(c) $\displaystyle\lim_{(x,y)\to(2,2)} \frac{x^2 - y^2}{x - y}\ln(x + y)$

Cancel the factor $x - y$ in the denominator and the numerator. Then plug in the values $(2, 2)$.

$$\lim_{(x,y)\to(2,2)} \frac{x^2 - y^2}{x - y}\ln(x + y)$$
$$= \lim_{(x,y)\to(2,2)} \frac{(x - y)(x + y)}{x - y}\ln(x + y)$$
$$= \lim_{(x,y)\to(2,2)} (x + y)\ln(x + y)$$
$$= (2 + 2)\ln(2 + 2)$$
$$= 4\ln 4$$
$$= 8\ln 2$$

SkillMaster 11.5.

• Explain why each function is continuous or discontinuous.
(a) The height of the ocean floor as measured from a satellite.

Consider what the ocean floor might look like.

(a) The height of the ocean floor is discontinuous because there may be cliffs with overhangs. As the satellite scans over the cliff there will be a sudden drop off.

(b) The barometric pressure in Death Valley as a function of location.

How does air pressure vary for nearby places?

(b) The barometric pressure is continuous because it cannot change instantaneously.

• Describe the largest set for which the function is continuous
(a) $f(x, y) = e^{x+y}\tan(xy)$

The functions are made up of basic continuous functions. They are continuous at each point where they are defined, that is on their domain.

(a) The exponential function is defined for all input values. The tangent function is defined except at points of the form $\pi/2 + n\pi$ for $n = 0, \pm 1, \pm 2, ...$ The domain is the set $D = \{(x, y)|xy \neq \pi/2 + n\pi, n$ an integer$\}$.

(b) $f(x, y) = \ln x^2 + x/(y+1)$

See the above hint.

(b) This is defined in all places where the input to the logarithm is positive and there is no division by zero. That is the domain $D = \{(x,y)|x \neq 0 \text{ and } y \neq -1\}$. D is the largest set upon which f is continuous.

(c) $f(x, y) = \dfrac{x + y + z}{z - x^2 + y^2}$

See the above hint.

(c) The domain of this function is the set of (x, y, z) for which division by zero does not occur. That is $D = \{(x,y,z)|z \neq x^2 + y^2\}$, that is f is continuous off the paraboloid $z = x^2 + y^2$.

• Find the largest set on which the function is continuous.
(a) $f(x, y, z) =$

$$\begin{cases} \dfrac{x^4 + y^4 - z^4}{x^4 + y^4 + z^4} & \text{if } (x,y,z) \\ & \neq (0,0,0) \\ 1 & \text{if } (x,y,z) \\ & = (0,0,0) \end{cases}$$

This functions will be continuous at all points where the first formula is defined. For it to be continuous at the exceptional point (the origin) two things must happen. First the limit must exist and second the limit must be equal to the value of the function at that point. In (b) the situation may be simplified by the substitution $u = x^4 + y^4$. Note that $(x, y) \to (0,0)$ if and only if $u \to 0$.

(a) Try limits along the x-axis and along the z-axis. (The limits along the x-axis and y-axis are the same.)

$$\lim_{(x,0,0)\to(0,0,0)} \frac{x^4 + y^4 - z^4}{x^4 + y^4 + z^4}$$
$$= \lim_{(x,0,0)\to(0,0,0)} \frac{x^4 + (0)^4 - (0)^4}{x^4 + (0)^4 + (0)^4}$$
$$= \lim_{(x,0,0)\to(0,0,0)} \frac{x^4}{x^4}$$
$$= 1$$

$$\lim_{(0,y,0)\to(0,0,0)} \frac{x^4 + y^4 - z^4}{x^4 + y^4 + z^4}$$
$$= \lim_{(0,0,z)\to(0,0,0)} \frac{(0)^4 + (0)^4 - z^4}{(0)^4 + (0)^4 + z^4}$$
$$= \lim_{(0,0,z)\to(0,0,0)} \frac{-z^4}{z^4}$$
$$= -1$$

Since these limits are not equal the function is not continuous at $(0,0,0)$. The largest set upon which the function is continuous is the set of (x, y, z) not equal to $(0,0,0)$.

(b) $f(x,y) =$

$$\begin{cases} \dfrac{\sin\left(\pi\left(x^4+y^4\right)\right)}{x^4+y^4} & \text{if } (x,y) \\ & \neq (0,0) \\[2ex] 1 & \text{if } (x,y) \\ & = (0,0) \end{cases}$$

See the above hint. Here, the situation may be simplified by the substitution $u = x^4 + y^4$. Note that $(x,y) \to (0,0)$ if and only if $u \to 0$.

(b) As suggested let $u = x^4 + y^4$ and compute the limit using l'Hospital's Rule.

$$\lim_{(x,y)\to(0,0)} \frac{\sin\left(\pi\left(x^4+y^4\right)\right)}{x^4+y^4}$$

$$= \lim_{u\to 0} \frac{\sin\left(\pi u\right)}{u} = \lim_{u\to 0} \frac{d(\sin\left(\pi u\right))/du}{d(u)/du}$$

$$= \lim_{u\to 0} \frac{\pi\cos(\pi u)}{1} = \pi\cos 0 = \pi$$

The limit exists but it is not equal to the value of the function so the function is not continuous at $(0,0)$. The largest set upon which the function is continuous is the set of all $(x,y) \neq (0,0)$.

(c) $f(x,y,z) = \dfrac{\ln x + \ln y + \ln z}{\sqrt{1-x-y-z}}$

The function is continuous on the set of (x,y,z) where it is defined.

(c) The logarithm is defined only for positive numbers so we need $x > 0, y > 0$, and $z > 0$. Also it is necessary for $1-x-y-z > 0$ or $x+y+z < 1$. This is the region in the first octant and cut off by the plane $x + y + z = 1$.

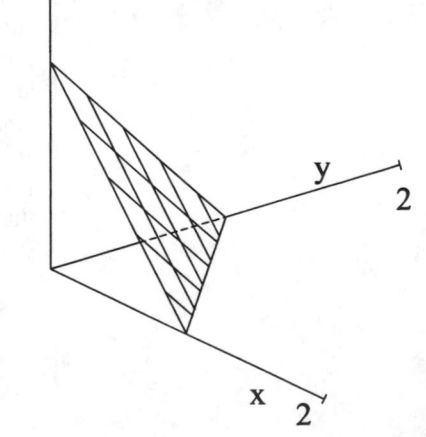

Section 11.3 – Partial Derivatives

Key Concepts:

- Partial Derivatives and their Interpretation as Slopes and Rates of Change.
- Higher Derivatives and Clairaut's Theorem
- Partial Differential Equations

Skills to Master:

- Compute and interpret partial derivatives.
- Use Clairaut's Theorem to compute higher partial derivatives.
- Verify if a given function satisfies a partial differential equation.

Discussion:

This section introduces the important concept of partial derivatives of functions of several variables. Partial derivatives can best be thought of as derivatives with respect to a particular variable, while all the other variables are treated as constant. Partial derivatives of functions of two variables are introduced first. Later, partial derivatives are discussed for functions of more than two variables.

This section also introduces higher order partial derivatives and partial differential equations. At first, this new material and terminology may seem to be somewhat overwhelming. By reading through the examples carefully and working through the problems, you should be able to master this material.

Key Concept: Partial Derivatives and their Interpretation as Slopes and Rates of Change.

For a function $z = f(x, y)$ of two variables, the partial derivative of f with respect to x is defined by

$$f_x(x, y) = \lim_{h \to 0} \frac{f(x + h, y) - f(x, y)}{h}.$$

Note that y is held constant in the limit and x varies.

The partial derivative of f with respect to y is defined by

$$f_y(x, y) = \lim_{h \to 0} \frac{f(x, y + h) - f(x, y)}{h}.$$

Note that x is held constant in the limit and that y varies.

The partial derivative can be interpreted as the rate of change of the function as x or y is held constant. The partial derivatives can also be interpreted as the slope of the tangent line to certain curves in the surface given by $z = f(x, y)$. Read *Example 2* carefully to gain a geometric understanding of this concept.

page 769.

You should master the various notation that is used for partial derivatives. For example, the partial derivative of $z = f(x, y)$ with respect to x can be written in the following ways.

$$f_x(x, y), \quad f_x, \quad \frac{\partial f}{\partial x}, \quad \frac{\partial}{\partial x} f(x, y), \quad \frac{\partial z}{\partial x}, \quad f_1, \quad D_1 f, \quad \text{and } D_x f.$$

One reason for the different notations is that partial derivatives arise in many applications of mathematics to other areas and each area has developed its preferred notation.

Key Concept: Higher Derivatives and Clairaut's Theorem

Since the partial derivative of a function can be viewed as a new function, it is possible to repeat the process of taking partial derivatives with the new function and obtain higher order partial derivatives. For example $f_{xy} = (f_x)_y$ represents the result of first taking the partial derivative with respect to x, and then taking the partial derivative with respect to y. Contrast this with $f_{yx} = (f_y)_x$ which represents the result of first taking the partial derivative with respect to y, and then taking the partial derivative with respect to x.

From this, it appears that the order in which partial derivatives are taken may lead to different results. While this can sometimes occur for complicated functions, Clairaut's Theorem shows that the order in which partial derivatives are taken does not matter as long as the first partial derivatives are continuous. More precisely,

> Suppose that f is defined on a disk D that contains the point (a, b).
>
> If both f_{xy} and f_{yx} are continuous on D, then $f_{xy}(a, b) = f_{yx}(a, b)$.

page 772.

The same result holds for higher order partial derivatives. Study *Example 6* for a specific case in which this holds.

Key Concept: Partial Differential Equations

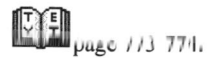page 773-774.

A partial differential equation is an equation that contains partial derivatives. These kinds of equations occur in many applications. They often arise where the behavior of the rate of change of a function with respect to certain variables is known to obey some physical law. Study *Examples 8 and 9* to see some specific instances of partial differential equations.

SkillMaster 11.6: Compute and interpret partial derivatives.

page 768-770.

page 769.

To compute partial derivatives, differentiate as you usually would with functions of one variable. Interpret all the variables except the one you are differentiating with respect to as constants. Study *Examples 1, 3, and 4* to see details on how to do this. You should interpret partial derivatives as slopes of tangent lines to certain curves in the surfaces. *Example 2* shows how to do this.

SkillMaster 11.7: Use Clairaut's Theorem to compute higher partial derivatives.

page 773.

Clairaut's Theorem states that the order in which partial derivatives are taken doesn't matter if the second order partial derivatives are continuous on a disk containing the point under consideration. This means that you can take partial derivatives in the order that is most convenient for calculation. Study *Example 7* to see how this works.

SkillMaster 11.8: Verify if a given function satisfies a partial differential equation.

page 773-774.

To verify whether a given function satisfies a particular partial differential equation, you need to substitute the function and its various derivatives into the equations. *Examples 8 and 9* give details on how to do this.

Worked Examples

For each of the following examples, first try to find the solution without looking at the middle or right columns. Cover the middle and right columns with a piece of paper. If you need a hint, uncover the middle column. If you need to see the worked solution, uncover the right column.

Example	Tip	Solution

SkillMaster 11.6.

• Consider the function
$f(x, y) =$

$$\begin{cases} xy\dfrac{x^2 - y^2}{x^2 + y^2} & \text{if } (x, y) \neq (0, 0) \\ \\ 0 & \text{if } (x, y) = (0, 0). \end{cases}$$

Use the definition to calculate $f_x(0, y)$ and $f_y(x, 0)$.

Tip: Recall the definition of $f_x(0, y)$ as $\lim\limits_{h \to 0} \dfrac{f(h, y) - f(0, y)}{h}$. The formula is simplified by the fact that $f(0, y) = 0$.

Solution:

$f_x(0, y)$

$= \lim\limits_{h \to 0} \dfrac{f(h, y) - f(0, y)}{h}$

$= \lim\limits_{h \to 0} \dfrac{f(h, y) - 0}{h}$

$= \lim\limits_{h \to 0} \dfrac{1}{h}(hy\dfrac{h^2 - y^2}{h^2 + y^2}) = \lim\limits_{h \to 0} y\dfrac{h^2 - y^2}{h^2 + y^2}$

$= y\dfrac{(0)^2 - y^2}{(0)^2 + y^2} = y(-1) = -y$

The computation of $f_y(x,0)$ is very similar. Again $f(x,0) = 0$, so the formula for the partial derivative simplifies to
$$f_x(x,0) = \lim_{k \to 0} \frac{f(x,k) - f(x,0)}{k}$$
$$= \lim_{k \to 0} \frac{1}{k} f(x,k).$$

$$f_y(x,0)$$
$$= \lim_{k \to 0} \frac{1}{k} f(x,k)$$
$$= \lim_{k \to 0} \frac{1}{k}\left(xk\frac{x^2 - k^2}{x^2 + k^2}\right)$$
$$= \lim_{k \to 0} \left(x\frac{x^2 - k^2}{x^2 + k^2}\right)$$
$$= \left(x\frac{x^2 - (0)^2}{x^2 + (0)^2}\right) = x\frac{x^2}{x^2} = x$$

- Calculate the first order partial derivatives of
$$z = f(x,y)$$
$$= \ln\left(\frac{xy}{\sin(x^2) + \cos(y^2)}\right).$$

First simplify the function using the laws of logarithms. Then take the partial derivatives. Remember that a partial derivative is computed in the same way as an ordinary derivative, with the exception that the other variables are treated as constants.

$$f(x,y)$$
$$= \ln\left(\frac{xy}{\sin(x^2) + \cos(y^2)}\right)$$
$$= \ln x + \ln y - \ln(\sin(x^2) + \cos(y^2))$$

$$f_x(x,y)$$
$$= \partial(\ln x)/\partial x + \partial(\ln y)/\partial x$$
$$\quad -\partial(\ln(\sin(x^2) + \cos(y^2)))/\partial x$$
$$= 1/x + 0$$
$$\quad -\frac{1}{\sin(x^2) + \cos(y^2)}\frac{\partial(\sin(x^2) + \cos(y^2))}{\partial x}$$
$$= 1/x - \frac{\cos(x^2)(2x)}{\sin(x^2) + \cos(y^2)}$$
$$= 1/x - \frac{2x\cos(x^2)}{\sin(x^2) + \cos(y^2)}$$

The computation of f_y is similar.

$$f_y(x,y)$$
$$= \partial(\ln x)/\partial y + \partial(\ln y)/\partial y$$
$$-\partial(\ln(\sin(x^2) + \cos(y^2)))/\partial y$$
$$= 0 + 1/y$$
$$-\frac{1}{\sin(x^2) + \cos(y^2)}\frac{\partial(\sin(x^2) + \cos(y^2))}{\partial y}$$
$$= 1/x - \frac{-\sin(y^2)(2y)}{\sin(x^2) + \cos(y^2)}$$
$$= 1/x + \frac{2y\sin(y^2)}{\sin(x^2) + \cos(y^2)}$$

• In spherical coordinates $\rho = \sqrt{x^2 + y^2 + z^2}$. Show $\partial\rho/\partial x = x/\rho$. Thus as $x > 0$ is increased, the rate of increase of ρ is less for larger ρ then for smaller ρ.

Try this two ways. First calculate the partial derivative using the rules of differentiation for functions. Remember that it is easier to compute the derivative of functions involving roots if they are expressed as exponents.

$$\partial\rho/\partial x$$
$$= \partial(\sqrt{x^2 + y^2 + z^2})/\partial x$$
$$= \partial((x^2 + y^2 + z^2)^{1/2})/\partial x$$
$$= (1/2)(x^2 + y^2 + z^2)^{-1/2}\frac{\partial(x^2 + y^2 + z^2)}{\partial x}$$
$$= \frac{2x + 0 + 0}{2(x^2 + y^2 + z^2)^{1/2}}$$
$$= \frac{x}{(x^2 + y^2 + z^2)^{1/2}}$$
$$= x/\rho$$

Second, rewrite $\rho = \sqrt{x^2 + y^2 + z^2}$ as $\rho^2 = x^2 + y^2 + z^2$ and use implicit differentiation.

$$\partial(\rho^2)/\partial x = \partial(x^2 + y^2 + z^2)/\partial x$$
$$(2\rho)\partial\rho/\partial x = 2x + 0 + 0$$
$$\rho\partial\rho/\partial x = x$$
$$\partial\rho/\partial x = x/\rho$$

- .Find the partial derivative with respect to x of the function $z = f(x^2 + y^2)$, where f is a function of one variable. Suppose that $f(2) = f'(2) = 1$. Consider the curve obtained by the intersection of the surface and the plane through $(1, 1, 0)$ and parallel to the xz-plane. Find the slope of the tangent line within this plane and tangent to the curve at $(1, 1, 1)$.

First notice that the slope of the tangent line is given by $\partial z/\partial x$ according to the geometric interpretation of partial derivatives. Second, using the chain rule and keeping y constant in the calculation we can compute this partial derivative.

$$
\begin{aligned}
& \partial z/\partial x \\
= \ & \partial (f(x^2 + y^2)/\partial x \\
= \ & f'(x^2 + y^2)\partial(x^2 + y^2)/\partial x \\
= \ & 2xf'(x^2 + y^2)
\end{aligned}
$$

The point to evaluate this at is $(x, y) = (1, 1)$.

$$
\begin{aligned}
& \partial z/\partial x \\
= \ & 2(1)f'(1^2 + 1^2) \\
= \ & 2f'(2) \\
= \ & 2
\end{aligned}
$$

The slope of the tangent line is 2.

- Consider the implicit function $e^{xyz} = x^2 + y^2 + z^2$. Find $\partial z/\partial x$ and $\partial z/\partial y$.

Use implicit differentiation to find an equation for $\partial z/\partial x$.

$$
\begin{aligned}
\partial(e^{xyz})/\partial x &= \frac{\partial(x^2 + y^2 + z^2)}{\partial x} \\
e^{xyz}\partial(xyz)/\partial x &= 2x + 0 + 2z\frac{\partial z}{\partial x} \\
ye^{xyz}\partial(xz)/\partial x &= 2x + 2z\frac{\partial z}{\partial x} \\
ye^{xyz}\left[z\frac{\partial(x)}{\partial x} + x\frac{\partial(z)}{\partial x}\right] &= 2x + 2z\frac{\partial z}{\partial x} \\
ye^{xyz}[z + x\partial(z)/\partial x] &= 2x + 2z\frac{\partial z}{\partial x} \\
yze^{xyz} + xye^{xyz}\partial z/\partial x &= 2x + 2z\frac{\partial z}{\partial x} \\
(xye^{xyz} - 2z)\partial z/\partial x &= 2x - yze^{xyz} \\
\partial z/\partial x &= \frac{2x - yze^{xyz}}{xye^{xyz} - 2z}
\end{aligned}
$$

SkillMaster 11.7.

• Consider the function
$f(x, y) =$

$$\begin{cases} xy\dfrac{x^2 - y^2}{x^2 + y^2} & \text{if } (x, y) \neq (0, 0) \\ \\ 0 & \text{if } (x, y) = (0, 0). \end{cases}$$

This function is continuous and the first order partials are also continuous. Find $f_{xy}(0, 0)$ and $f_{yx}(0, 0)$ and show that they are not equal. Why does this not contradict Clairaut's Theorem?

The definition of f_{xy} is $\partial f_x / \partial y$. So
$f_{xy}(0, 0) =$
$\lim\limits_{k \to 0} \frac{f_x(0,k) - f_x(0,0)}{k}$.
We have already computed
$f_x(0, y) = -y$.

$f_{xy}(0, 0)$

$$\begin{aligned} &= \lim_{k \to 0} \frac{f_x(0, k) - f_x(0, 0)}{k} \\ &= \lim_{k \to 0} \frac{-k - 0}{k} \\ &= \lim_{k \to 0} -1 \\ &= -1 \end{aligned}$$

The other computation is similar. The definition of f_{yx} is $\partial f_y / \partial x$. So
$f_{yx}(0, 0) =$
$\lim\limits_{h \to 0} \frac{f_y(h,0) - f_y(0,0)}{h}$.
We have already computed
$f_y(x, 0) = x$.

$f_{yx}(0, 0)$

$$\begin{aligned} &= \lim_{h \to 0} \frac{f_y(h, 0) - f_y(0, 0)}{h} \\ &= \lim_{h \to 0} \frac{h - 0}{h} \\ &= \lim_{h \to 0} 1 \\ &= 1 \end{aligned}$$

If the conclusion of a theorem is false then one of the hypotheses must be false. Clairaut's Theorem has the conclusion that the mixed partials are equal, $f_{xy} = f_{yx}$. In this case, the mixed partials are not equal since $1 \neq -1$.

Since the conclusion of Clairaut's Theorem fails, one of the hypotheses must be false. In this case, it turns out the partial derivatives f_{xy} and f_{yz} are not continuous near $(0, 0)$, although they are defined everywhere.

• Consider the polynomial
$w = f(x, y, z) =$
$Ax^3y^2 + Bxyz + Cz^2y^2 + 2.$
Suppose we are given:
(i) $f_{xyz} = 2$,
(ii) $f_{zzyy} = -1$,
and (iii) $f_{yxxxy} = 4$.
Calculate $A, B,$ and C. Find $f(3, 2, 1)$.

Formally compute the mixed partials whose values are given and observe what relationships the coefficients A, B, C have. Notice that Clairaut's Theorem applies here so mixed partials may be computed in any order.

f_x
$= \partial(Ax^3y^2)/\partial x + \partial(Bxyz)/\partial x$
$\quad + \partial(Cz^2y^2)/\partial x + \partial(2)/\partial x$
$= A3x^2y^2 + Byz + 0 + 0$
$= 3Ax^2y^2 + Byz$

f_{xy}
$= \partial f_x/\partial y$
$= \partial(3Ax^2y^2)/\partial y + \partial(Byz)/\partial y$
$= 3Ax^2(2y) + Bz$
$= 6Ax^2y + Bz$

f_{xyz}
$= \partial f_{xy}/\partial z$
$= \partial(6Ax^2y)/\partial z + \partial(Bz)/\partial z$
$= 0 + B$
$= B$

$$f_{xyz} = 2 = B$$

Thus, $B = 2$.

Continue and compute the other mixed partials in the problem.

f_z

$$= \ \partial(Ax^3y^2)/\partial z + \partial(Bxyz)/\partial z$$
$$+ \partial(Cz^2y^2)/\partial z + \partial(2)/\partial z$$
$$= \ 0 + Bxy + 2Czy^2 + 0$$
$$= \ Bxy + 2Czy^2$$

$$f_{zz} = \partial(f_z)/\partial z$$
$$= \ \partial(Bxy)/\partial z + \partial(2Czy^2)/\partial z$$
$$= \ 0 + 2Cy^2 = 2Cy^2$$

$$f_{zzy} = \partial f_{zz}/\partial y$$
$$= \ \partial(2Cy^2)/\partial y = 4Cy$$

Next,

$$f_{zzyy} = \partial f_{zzy}/\partial y$$
$$= \ \partial(4Cy)/\partial y = 4C = -1$$

Thus $C = -1/4$. Using the computation above:

$$f_{yxxxy} = (f_{xy})_{xxy}$$
$$= \ (6Ax^2y + Bz)_{xxy}$$
$$= \ (6A(2xy) + 0)_{xy}$$
$$= \ (12Ay)_y = 12A = 4$$

Thus, $A = 1/3$, and the equation is

$$w \ = \ f(x, y, z)$$
$$= \ (1/3)x^3y^2 + 2xyz + (-1/4)z^2y^2 + 2.$$

Finally,

$$f(3, 2, 1)$$
$$= \ (1/3)(3)^3(2)^2 + 2(3)(2)(1)$$
$$+ (-1/4)(1)^2(2)^2 + 2$$
$$= \ (9)(4) + 12 - 1 + 2$$
$$= \ 36 + 12 - 1 + 2 = 49.$$

595

SkillMaster 11.8.

• Show that $z = f(x, y)$ $= x^2 - y^2 + \tan^{-1}(y/x)$ is a harmonic function.

A function is harmonic if $f_{xx} + f_{yy} = 0$. Compute these partials. Recall that $d(\tan^{-1} u)/du = \dfrac{1}{1 + u^2}$.

f_x

$= \dfrac{\partial(x^2)}{\partial x} - \dfrac{\partial(y^2)}{\partial x} + \dfrac{\partial(\tan^{-1}(y/x))}{\partial x}$

$= 2x + 0 + \dfrac{1}{1 + (y/x)^2}\dfrac{\partial(y/x)}{\partial x}$

$= 2x + \dfrac{1}{1 + (y/x)^2}(-y/x^2)$

$= 2x - \dfrac{y}{x^2 + y^2}$

f_{xx}

$= 2 - (-\dfrac{y(2x)}{(x^2 + y^2)^2})$

$= 2 + \dfrac{2xy}{(x^2 + y^2)^2}$

f_y

$= \dfrac{\partial(x^2)}{\partial y} - \dfrac{\partial(y^2)}{\partial y} + \dfrac{\partial(\tan^{-1}(y/x))}{\partial y}$

$= 0 - 2y + \dfrac{1}{1 + (y/x)^2}(1/x)$

$= -2y + \dfrac{x}{x^2 + y^2}$

f_{yy}

$= \dfrac{\partial(-2y + \frac{x}{x^2+y^2})}{\partial y}$

$= -2 + \dfrac{x(-2y)}{(x^2 + y^2)^2}$

$= -2 - \dfrac{2xy}{(x^2 + y^2)^2}$

$= -f_{xx}$

Thus $f_{xx} + f_{yy} = 0$ and the function is harmonic.

• For what values of A, B does $u = \sin(Ax + Bt)$ satisfy the wave equation $\partial^2 u/\partial t^2 = a^2 \partial^2 u/\partial x^2$ for some $a \neq 0$.

A function satisfies the wave equation if
$$u_{tt} = a^2 u_{xx}.$$

$$
\begin{aligned}
u_{tt} &= \partial(u_t)/\partial t \\
&= \partial((\cos(Ax + Bt)B)/\partial t \\
&= -\sin(Ax + Bt)B^2 \\
&= -B^2 \sin(Ax + Bt)
\end{aligned}
$$

Similarly

$$
\begin{aligned}
u_{xx} &= \partial(u_x)/\partial x \\
&= \partial((\cos(Ax + Bt)A)/\partial x \\
&= -\sin(Ax + Bt)A^2 \\
&= -A^2 \sin(Ax + Bt).
\end{aligned}
$$

If $A \neq 0$ then
$$(-1/A^2)u_{xx} = \sin(Ax + Bt).$$
Substitute this into the equation for u_{tt}.

$$
\begin{aligned}
u_{tt} &= -B^2 \sin(Ax + Bt) \\
&= -B^2(-1/A^2)u_{xx}
\end{aligned}
$$

This reduces to
$$u_{tt} = \frac{B^2}{A^2}u_{xx}.$$

Thus $u = \sin(Ax + Bt)$ satisfies the wave equation with $a^2 \neq 0$ if and only if $A \neq 0$ and in this case $a = B/A$.

597

• Show that
$w = \ln(x^2 + y^2 + z^2)$
satisfies the partial differential equation
$x\dfrac{\partial w}{\partial x} + y\dfrac{\partial w}{\partial y} + z\dfrac{\partial w}{\partial z} = 2.$

First compute the partial
$\partial w/\partial x =$
$$\dfrac{\partial(\ln(x^2 + y^2 + z^2)}{\partial x}$$
$$= \dfrac{1}{x^2 + y^2 + z^2}(2x)$$
$$= \dfrac{2x}{x^2 + y^2 + z^2}$$
By symmetry the other first order partials must have the same form.
$\partial w/\partial y$
$$= \dfrac{2y}{x^2 + y^2 + z^2}$$
and
$\partial w/\partial z$
$$= \dfrac{2z}{x^2 + y^2 + z^2}.$$
Substitute the formulas for the partial derivatives into the equation to verify it.

$$x\dfrac{\partial w}{\partial x} + y\dfrac{\partial w}{\partial y} + z\dfrac{\partial w}{\partial z}$$
$$= x\dfrac{2x}{x^2 + y^2 + z^2}$$
$$+ y\dfrac{2y}{x^2 + y^2 + z^2}$$
$$+ z\dfrac{2z}{x^2 + y^2 + z^2}$$
$$= \dfrac{2x^2 + 2y^2 + 2z^2}{x^2 + y^2 + z^2}$$
$$= 2.$$

Section 11.4 –Tangent Planes and Linear Approximation

Key Concepts:

- The Tangent Plane to a Surface
- The Linear Approximation to a Function
- Differentiable Functions
- Differentials and the Estimation of Error
- Tangent Planes to Parametric Surfaces

Skills to Master:

- Compute the tangent plane to a surface given by a function of two variables.
- Determine if a function is differentiable.
- Use linearization to approximate values of differentiable functions.
- Use differentials to estimate the error in an approximation.
- Compute tangent planes to parametric surfaces.

Discussion:

This section introduces the concepts of the tangent plane to a surface at a point and the linear approximation to a function of two variables. Just as the tangent line at a point was a good approximation to a function of a single variable near that point, the tangent plane is a good approximation to a function of two variables. Make sure that you study the pictures that go with the examples in this section. The pictures should give you a geometric understanding of tangent planes and linear approximations.

Key Concept: The Tangent Plane to a Surface

page 779.

The tangent plane to a surface S given by $z = f(x,y)$ at a point $P(x_0, y_0, z_0 = f(x_0, y_0))$ can be thought of as the plane through P containing all tangent lines to curves on S passing through P. See *Figure 1* for an illustration of this. To make sure that the surface is smooth enough to have a tangent plane, we impose the condition that f have continuous partial derivatives at P. With this assumption, the tangent plane to S at P is the plane with equation

$$z - z_0 = f_x(x_0, y_0)(x - x_0) + f_y(x_0, y_0)(y - y_0) \text{ or}$$
$$z = f_x(x_0, y_0)(x - x_0) + f_y(x_0, y_0)(y - y_0) + z_0.$$

Note that can be written in the familiar form
$$z = Ax + By + C$$
where $A = f_x(x_0, y_0)$, $B = f_y(x_0, y_0)$, and $C = z_0 - x_0 f_x(x_0, y_0) - y_0 f_y(x_0, y_0)$.

page 779.

Study *Example 1* to see an illustration of finding the tangent plane.

Key Concept: The Linear Approximation to a Function

The linear function whose graph is the tangent plane to $z = f(x,y)$ at the point $(a, b, f(a,b))$ is called the linearization of the function $z = f(x,y)$ at (a,b). This linear function has already been derived in the discussion of the tangent plane, and is given by
$$L(x,y) = f(a,b) + f_x(a,b)(x - a) + f_y(a,b)(y - b).$$
Compare this formula with the formula for the tangent plane in the previous Key Concept.

The linear approximation or tangent surface approximation to f at (a,b) is the approximation
$$f(x,y) \approx f(a,b) + f_x(a,b)(x - a) + f_y(a,b)(y - b).$$
Make sure that you understand the terminology in this section.

Key Concept: Differentiable Functions

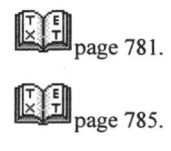
page 781.

Without extra conditions on the function, the linear approximation may not be a good approximation to the function as illustrated in the discussion accompanying *Figure 4* . This leads to the concept of a differentiable function.   A function $z = (f(x, y)$ is differentiable at (a, b) if
$$\Delta z = f(a + \Delta x, b + \Delta y) - f(a, b)$$
can be expressed in the form
$$\Delta z = f_x(a, b)\Delta x + f_y(a, b)\Delta y + \epsilon_1 \Delta x + \epsilon_2 \Delta y \text{ where}$$
$$\epsilon_1 \text{ and } \epsilon_2 \text{ go to } 0 \text{ as } (\Delta x, \Delta y) \text{ goes to } (0, 0).$$
This may seem like a strange condition, but it is what is needed for the tangent surface approximation to be a good approximation.

page 781.

To see that many functions are differentiable, read and understand *Theorem 8*

If the partial derivatives f_x and f_y exist near (a, b) and are continuous at (a, b) then f is differentiable at (a, b).

Key Concept: Differentials and the Estimation of Error

For functions of two variables $z = f(x, y)$, the differentials dx and dy are defined to be independent variables that can be given any values and dz is defined by
$$dz = f_x(x, y)dx + f_y(x, y)dy = \frac{\partial z}{\partial x}dx + \frac{\partial z}{\partial y}dy.$$
Note that dz depends on x, y, dx, and dy. If we take $dx = x - a$ and $dy = y - b$, then
$$dz = f_x(a, b)(x - a) + f_y(a, b)(y - b).$$
page 781.

page 785.

Note that this is the same expression that occurs in the formula for *linear approximation*. The differential dz can be used as an estimate of the maximum error that occurs in $z = f(x, y)$ if x and y are only known to an accuracy of dx and dy. Study *Example 5* to see how this works.

Key Concept: Tangent Planes to Parametric Surfaces

page 734.

Review the concept of parametric surfaces from *Section 10.5* if you need to. The tangent plane to a parametric surface S given by

$$\mathbf{r}(u,v) = x(u,v)\mathbf{i} + y(u,v)\mathbf{j} + \mathbf{z}(u,v)\mathbf{k}$$

at a point P_0 with position vector $\mathbf{r}(u_0, v_0)$ is the plane that contains the vectors $\mathbf{r}_u$ and $\mathbf{r}_v$ and has $\mathbf{r}_u \times \mathbf{r}_v$ for a normal vector.

Here, $\mathbf{r}_u$ is the "partial derivative" of $\mathbf{r}$ with respect to u,

$$\mathbf{r}_u = \frac{\partial x}{\partial u}(u_0, v_0)\mathbf{i} + \frac{\partial y}{\partial u}(u_0, v_0)\mathbf{j} + \frac{\partial z}{\partial u}(u_0, v_0)\mathbf{k}$$

and $\mathbf{r}_v$ is the "partial derivative" of $\mathbf{r}$ with respect to v,

$$\mathbf{r}_v = \frac{\partial x}{\partial v}(u_0, v_0)\mathbf{i} + \frac{\partial y}{\partial v}(u_0, v_0)\mathbf{j} + \frac{\partial z}{\partial v}(u_0, v_0)\mathbf{k}.$$

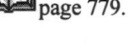
page 786-787.

Study *Figures 9 and 10* and *Example 7* to see a geometric picture and a specific example of this

SkillMaster 11.9: Compute the tangent plane to a surface given by a function of two variables.

To compute the tangent plane to a surface S given by $z = f(x,y)$ at a point (x_0, y_0, z_0), use the formula

$$z = f_x(x_0, y_0)(x - x_0) + f_y(x_0, y_0)(y - y_0) + z_0.$$

First compute the partial derivatives $f_x(x_0, y_0)$ and $f_y(x_0, y_0)$ at the point (x_0, y_0). Then substitute these values and the value for $z_0 = f(x_0, y_0)$ into the formula. *See*

page 779. *Example 1* for a specific instance of this.

SkillMaster 11.10: Determine if a function is differentiable.

To determine if a function $z = f(x,y)$ is differentiable, you need to determine if

there are quantities ϵ_1 and ϵ_2 that go to 0 as $(\Delta x, \Delta y)$ goes to $(0,0)$ so that
$$\Delta z = f_x(a,b)\Delta x + f_y(a,b)\Delta y + \epsilon_1 \Delta x + \epsilon_2 \Delta y.$$
If the function has continuous partial derivatives, it follows immediately that it is differentiable.

The Worked Example for SkillMaster 11.10 below shows how to do this in general, even if it is not known that the function has continuous partial derivatives.

SkillMaster 11.11: Use linearization to approximate values of differentiable functions.

To use linearization to approximate the value of a differentiable function $f(x,y)$ at a point x, y near (a, b), use the formula
$$f(x,y) \approx f(a,b) + f_x(a,b)(x-a) + f_y(a,b)(y-b).$$
First compute the partial derivatives, and then substitute those values into the formula. Study *Example 2* to see how this works.

page 782.

SkillMaster 11.12: Compute differentials and use them to estimate the error in an approximation.

To compute the differential dz, use the formula
$$dz = f_x(x,y)dx + f_y(x,y)dy = \frac{\partial z}{\partial x}dx + \frac{\partial z}{\partial y}dy.$$
To use differentials to estimate the error in an approximation, let dx and dy be the possible errors in the values of x and y, and then use the formula above to get dz, an estimate in the error for z. Study *Example 5* to see how this works.

page 785.

SkillMaster 11.13: Compute tangent planes to parametric surfaces.

To compute the tangent plane to a parametric surface
$$\mathbf{r}(u,v) = x(u,v)\mathbf{i} + y(u,v)\mathbf{j} + \mathbf{z}(u,v)\mathbf{k},$$

page 787. compute r_u, r_v, and $r_u \times r_v$. See the definitions of these in the Key Concept above. Study *Example 7* to see how this works.

Worked Examples

For each of the following examples, first try to find the solution without looking at the middle or right columns. Cover the middle and right columns with a piece of paper. If you need a hint, uncover the middle column. If you need to see the worked solution, uncover the right column.

Example	Tip	Solution

SkillMaster 11.9.

• Find the equation of the tangent plane to the function
$z = f(x, y)$
$= \sin x + \sin y + \sin(2x + 3y)$ at the point
$(x_0, y_0, z_0) = (0, 0, 0)$.

First find the partial derivatives.

$f_x(x, y)$

$$= \frac{\partial(\sin x)}{\partial x} + \frac{\partial(\sin y)}{\partial x}$$
$$+ \frac{\partial(\sin(2x + 3y))}{\partial x}$$
$$= \cos x + 0 + \cos(2x + 3y)(2)$$
$$= \cos x + 2\cos(2x + 3y)$$

$f_x(0, 0)$

$$= \cos 0 + 2\cos(2(0) + 3(0))$$
$$- 1 + 2(1) = 3.$$

$f_y(x, y)$

$$= \frac{\partial(\sin x)}{\partial y} + \frac{\partial(\sin y)}{\partial y} + \frac{\partial(\sin(2x + 3y)}{\partial y}$$
$$= 0 + \cos y + \cos(2x + 3y)(3)$$
$$= \cos y + 3\cos(2x + 3y)$$

$f_y(0, 0)$

$$= \cos 0 + 3\cos(2(0) + 3(0))$$
$$= 1 + 3 = 4$$

The tangent plane has the form
$$z = z_0$$
$$+ f_x(x_0, y_0)(x - x_0)$$
$$+ f_y(x_0, y_0)(y - y_0).$$
Plug in the appropriate values for (x_0, y_0, z_0).

z
$$= 0 + f_x(0,0)(x - 0) + f_y(0,0)(y - 0)$$
$$= f_x(0,0)(x) + f_y(0,0)(y) = 3x + 4y$$

Note how the tangent plane slices through the surface.

SkillMaster 11.10.

• Show that the function
$$z = f(x, y)$$
$$= \frac{xy}{x^2 + y^2 + 1}$$ is differentiable at the point $(0, 0)$.

Recall that a function is differentiable at a point if both the partial derivative are exist and are continuous near the point. Compute the partial derivatives and check that they are continuous. Use the quotient rule to compute f_x.

$f_x(x, y)$
$$= \frac{(x^2 + y^2 + 1)\partial(xy)/\partial x}{(x^2 + y^2 + 1)^2}$$
$$- \frac{xy\partial(x^2 + y^2 + 1)/\partial x}{(x^2 + y^2 + 1)^2}$$
$$= \frac{(x^2 + y^2 + 1)y - xy(2)x}{(x^2 + y^2 + 1)^2}$$
$$= \frac{y(y^2 - x^2 + 1)}{(x^2 + y^2 + 1)^2}$$

Use symmetry. Since $f(x, y)$ is symmetric, that is it remains the same when x and y are interchanged, $f_y(x, y)$ is the expression obtained by interchanging x and y in the expression for $f_x(x, y)$.

$$f_x(x, y)$$
$$= \frac{x(x^2 - y^2 + 1)}{(y^2 + x^2 + 1)^2}$$

Both $f_x(x, y)$ and $f_y(x, y)$ are continuous (because the denominator is always positive), so $f(x, y)$ is differentiable.

- Show that the function $z = f(x, y)$ is differentiable at the point $(0, 0)$ where
$f(x, y) =$

$$\begin{cases} x^2 & \text{if } x \geq 0 \\ xy & \text{if } x < 0 \end{cases}.$$

Note that the function is continuous because both pieces of the definition agree on the boundary, i.e. $\lim_{h \to 0} f(h, y) = 0$. First compute the partial derivatives.

By definition,

$$f_x(0, y) = \lim_{h \to 0} \frac{f(h, y) - f(0, 0)}{h}$$

$$= \begin{cases} \lim_{h \to 0} \dfrac{h^2}{h} = 0 & \text{if } h \geq 0 \\ \lim_{h \to 0} \dfrac{h \cdot 0}{h} = 0 & \text{if } h < 0 \end{cases}$$

Since both of these limits are 0, we have $f_x(0, 0) = 0$.

$$f_y(0, 0) = \lim_{h \to 0} \frac{f(0, h) - f(0, 0)}{h}$$

$$= \begin{cases} \lim_{h \to 0} \dfrac{0^2}{h} & \text{if } h \geq 0 \\ \lim_{h \to 0} \dfrac{0 \cdot h}{h} & \text{if } h < 0 \end{cases}$$

Since both of these limits are 0, we have $f_y(0, 0) = 0$

607

The definition says that a function is differentiable at $(0,0)$ if there are functions ε_1 and ε_2 of Δx and Δy that go to 0 as $(\Delta x, \Delta y) \to (0,0)$, so that $\Delta z = f_x(0,0)\Delta x + f_y(0,0)\Delta y + \varepsilon_1 \Delta x + \varepsilon_2 \Delta y$. Since $f(0,0) = 0$ it is follows that $\Delta z = f(\Delta x, \Delta y) - f(0,0) = f(\Delta x, \Delta y)$. Since $f_x(0,0) = f_y(0,0) = 0$ showing differentiability amounts to finding ε_1 and ε_2 so that $f(\Delta x, \Delta y) = \varepsilon_1 \Delta x + \varepsilon_2 \Delta y$.

Case 1. Suppose $\Delta x \geq 0$ then
$$
\begin{aligned}
f(\Delta x, \Delta y) &= (\Delta x)^2 \\
&= (\Delta x)\Delta x + (0)\Delta y.
\end{aligned}
$$

Case 2. Suppose that $\Delta x < 0$ then
$$
\begin{aligned}
f(\Delta x, \Delta y) &= (\Delta x)(\Delta y) \\
&= (\Delta y)\Delta x + (0)\Delta y.
\end{aligned}
$$

This shows how to define ε_1 and ε_2. Set $\varepsilon_2 = 0$ and set $\varepsilon_1 = \Delta x$ if $\Delta x \geq 0$ and set $\varepsilon_1 = \Delta y$ otherwise (i.e. if $\Delta x < 0$). Then
$$
f(\Delta x, \Delta y) = \varepsilon_1 \Delta x + \varepsilon_2 \Delta y
$$
Since ε_1 and ε_2 both go to 0 as $(\Delta x, \Delta y) \to (0,0)$, this shows the function is differentiable at $(0,0)$ using the definition. Notice that we could not simply apply Theorem 8 since the partial derivatives are not continuous.

SkillMaster 11.11.

• Let $z = f(x, y) = (1 + 2\sqrt{x} - \sqrt{y})^{1/3}$. Show that f is differentiable at $(9, 36)$ and find the linearization of f at $(9, 36)$.

The easiest way to check if a function which is expressed as a single formula is differentiable at a point is to check that its partial derivatives are continuous near the point.

$f_x(x, y)$
$$= \partial((1 + 2\sqrt{x} - \sqrt{y})^{1/3})/\partial x$$
$$= (1/3)(1 + 2\sqrt{x} - \sqrt{y})^{-2/3}$$
$$\cdot (2x^{-1/2})(1/2)$$
$$= (1/3\sqrt{x})(1 + 2\sqrt{x} - \sqrt{y})^{-2/3}$$

$f_x(9, 36)$
$$= (1/3\sqrt{9})(1 + 2\sqrt{9} - \sqrt{36})^{-2/3}$$
$$= (1/9)(1 + 6 - 6)^{-2/3}$$
$$= 1/9.$$

Similarly

$f_y(x, y)$
$$= \partial((1 + 2\sqrt{x} - \sqrt{y})^{1/3})/\partial y$$
$$= (1/3)(1 + 2\sqrt{x} - \sqrt{y})^{-2/3}$$
$$\cdot (y^{-1/2})(1/2)$$
$$= (1/6\sqrt{y})(1 + 2\sqrt{x} - \sqrt{y})^{-2/3}$$

$f_y(9, 36)$
$$= (1/6\sqrt{36})(1 + 2\sqrt{9} - \sqrt{36})^{-2/3}$$
$$= (1/36)(1 + 6 - 6)^{-2/3}$$
$$= 1/36.$$

Both of these functions are well-defined and continuous so the function is differentiable.

The linearization of $f(x, y)$ at $(9, 36)$ is
$L(x, y) = f(9, 36) + f_x(9, 36)(x - 9) + f_y(9, 36)(y - 36)$
To fill in the formula for the linearization compute

$$f(9, 36)$$
$$= (1 + 2\sqrt{9} - \sqrt{36})^{1/3}$$
$$= (1 + 2(3) - 6)^{1/3}$$
$$= 1^{1/3}$$
$$= 1.$$

Thus
$$L(x, y)$$
$$= 1 + (1/9)(x - 9) + (1/36)(y - 36)$$
$$x/9 + y/36 - 1.$$

· Use the linearization in the previous problem at $(9, 36)$ to estimate $f(8.8, 36.6)$.

The linearization is an approximation to the function near the given point. The values 9 and 8.8 are close as are the values 36 and 36.6. Thus $f(8.8, 36.6) \approx L(8.8, 36.6)$.

$$f(8.8, 36.6)$$
$$\approx L(8.8, 36.6)$$
$$= 8.8/9 + 36.6/36 - 1$$
$$\approx 0.9944$$

The correct value to four decimal places is 0.9594

· Use the linearization of a suitable three variable equation to approximate

$$\sqrt{(1.03) + (3.96)^2 + (2.02)^3}.$$

Try using the function
$$f(x, y, z) = \sqrt{x + y^2 + z^3}$$
at the point $(1, 4, 2)$.
We are lucky that
$$f(1, 4, 2) = \sqrt{1 + 4^2 + 2^3} = \sqrt{25} = 5.$$
The linearization is
$$L(x, y, z) = 5 + f_x(1, 4, 2)(x - 1) + f_y(1, 4, 2)(y - 4) + f_z(1, 4, 2)(z - 2).$$
Compute the partial derivatives to fill in this formula. The first partial is
$$f_x(x, y, z) = (1/2)(x + y^2 + z^3)^{-1/2}$$
$$f_x(1, 4, 2) = (1/2)(1 + 2^2 + 3^3)^{-1/2} = 1/10.$$

To find the approximation, plug in the values $(1.03, 3.96, 2.02)$ for (x, y, z) in L.

Continuing from the hint,
$$f_y(x, y, z)$$
$$= (1/2)(x + y^2 + z^3)^{-1/2}(2y)$$
$$= y(x + y^2 + z^3)^{-1/2}$$
$$f_x(1, 4, 2)$$
$$= 4(1 + 4^2 + 2^3)^{-1/2}$$
$$= 4/5$$
$$f_z(x, y, z)$$
$$= (1/2)(x + y^2 + z^3)^{-1/2}(3z^2)$$
$$= (3z^2/2)(x + y^2 + z^3)^{-1/2}$$
$$f_z(1, 4, 2)$$
$$= (12/2)(1 + 4^2 + 2^3)^{-1/2}$$
$$= 6/5$$
Now we can fill in the linearization of f.
$$L(x, y, z)$$
$$= 5 + (1/10)(x - 1)$$
$$+ (4/5)(y - 4) + (6/5)(z - 2)$$
$$= x/10 + 4y/5 + 6z/5 - 3/10$$

$$\sqrt{(1.03) + (3.96)^2 + (2.02)^3}$$
$$= f(1.03, 3.96, 2.02)$$
$$\approx L(1.03, 3.96, 2.02)$$
$$= 1.03/10 + 4(3.96)/5 + 6(2.02)/5 - 3/10$$
$$= 5.395.$$

SkillMaster 11.12.

• Find the differential of the function $z = f(x, y) = e^x \cos y$

The formula for the differential is
$$dz = \frac{\partial z}{\partial x} dx + \frac{\partial z}{\partial y} dy.$$

$$dz = [\partial(e^x \cos y)/\partial x] dx$$
$$+ [\partial(e^x \cos y)/\partial y] dy$$
$$= e^x(\cos y) dx - e^x(\sin y) dy$$

611

· The total resistance of two resistors in parallel is given by $R = \dfrac{R_1 R_2}{R_1 + R_2}$ where R_1 and R_2 are the resistances of the resistors. Suppose that R_1 is measured to be 200Ω with a maximum error of 3% and that R_2 is measured to be 400Ω with a maximum error of 5%. Estimate the maximum error of R.

First compute the differential dR in terms of dR_1 and dR_2.

$$
\begin{aligned}
dR &= \frac{\partial R}{\partial R_1} dR_1 + \frac{\partial R}{\partial R_2} dR_2 \\
&= \left(\frac{R_2}{R_1 + R_2}\right)^2 dR_1 \\
&+ \left(\frac{R_1}{R_1 + R_2}\right)^2 dR_2
\end{aligned}
$$

Evaluate this differential at the point $R_1 = 200, R_2 = 400$.

$$
\begin{aligned}
dR &= \left(\frac{400}{200 + 400}\right)^2 dR_1 \\
&+ \left(\frac{200}{200 + 400}\right)^2 dR_2 \\
dR &= (4/9)dR_1 + (1/9)dR_2
\end{aligned}
$$

The maximum error for R_1 is 3% or 6Ω and the maximum error for R_2 is 5% or 20Ω. To estimate the maximum error for R substitute 6 for dR_1 and 20 for dR_2.

The estimated maximum error is

$$
\begin{aligned}
dR &= (4/9)dR_1 + (1/9)dR_2 \\
&= (4/9)6 + (1/9)(20) \\
&= 8/3 + 20/9 \\
&= 48/9 \\
&= 16/3 \\
&\approx 5.33.
\end{aligned}
$$

SkillMaster 11.13.

• Find the equation of the tangent plane to the helioid $x = u\cos v$, $y = u\sin v$, $z = 2v$ at the point given by $(u, v) = (2, \pi/4)$.

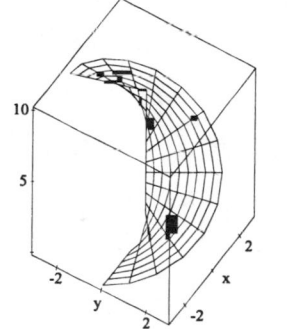

The plane passes through the point on the surface $x = 2\cos\pi/4$, $y = 2\sin\pi/4$, $z = 2(\pi/4)$ or $(\sqrt{2}, \sqrt{2}, \pi/2)$. To find the equation of the tangent plane find a normal vector to this plane. Two vectors in the tangent plane are given by the first partial derivatives with respect to u and to v.

$$< \partial x/\partial u, \partial y/\partial u, \partial z/\partial u >$$
$$= < \cos v, \sin v, 0 >$$
A tangent vector at the point is found by plugging in $u = 2, v = \pi/4$.
$$< \cos\pi/4, \sin\pi/4, 0 >$$
$$= <1/\sqrt{2}, 1/\sqrt{2}, 0> = (1/\sqrt{2}) < 1, 1, 0 > .$$
Since scalar multiples are irrelevant to the direction, $< 1, 1, 0 >$ is a tangent vector.
Similarly, another tangent vector is
$$< \partial x/\partial v, \partial y/\partial v, \partial z/\partial v >$$
$$= < -u\sin v, u\cos v, 2 >$$
Evaluate at the point $u = 2, v = \pi/4$:
$$< -2\sin\pi/4, 2\cos\pi/4, 2 >$$
$$= < -\sqrt{2}, \sqrt{2}, 2 > .$$

To find a normal vector, is the same as finding a vector orthogonal to both $< 1, 1, 0 >$ and $< -\sqrt{2}, \sqrt{2}, 2 >$. This may be done by finding the cross product of the two tangent vectors.

A normal vector is
$$\begin{vmatrix} \mathbf{i} & \mathbf{j} & \mathbf{k} \\ 1 & 1 & 0 \\ -\sqrt{2} & \sqrt{2} & 2 \end{vmatrix}$$
$$= [(1)(2) - (0)(\sqrt{2})]\mathbf{i}$$
$$\quad -[(1)(2) - (0)(-\sqrt{2})]\mathbf{j}$$
$$\quad +[(1)(\sqrt{2}) - (1)(-\sqrt{2})]\mathbf{k}$$
$$= 2\mathbf{i} - 2\mathbf{j} + 2\sqrt{2}\mathbf{k} = < 2, -2, 2\sqrt{2} >$$
$$= 2 < 1, -1, \sqrt{2} > .$$
Thus a normal vector is $< 1, -1, \sqrt{2} > .$

The equation of a plane through the point (x_0, y_0, z_0) with normal vector $< a, b, c >$ is $a(x - x_0) + b(y - y_0) + c(z - z_0) = 0.$

The tangent plane is
$$(x - \sqrt{2}) - (y - \sqrt{2}) + \sqrt{2}(z - \pi/2) = 0$$
$$\text{or } x - y + \sqrt{2}z - \pi/\sqrt{2} = 0.$$

Section 11.5 – The Chain Rule

Key Concepts:

- The Chain Rule in Specific Cases
- Tree Diagrams and the General Form of the Chain Rule
- Implicit Functions and Implicit Differentiation

Skills to Master:

- Compute derivatives using the chain rule.
- Use implicit differentiation to compute derivatives.

Discussion:

This section introduces the chain rule for functions of several variables. In the case of a function of n variables, $u = f(x_1, x_2, \cdots x_n)$, a variable u is a function of the n variables

$$x_1, x_2, \cdots x_n.$$

If, in turn, each of the variables x_i is a function of m other variables,

$$t_1, t_2, \cdots t_m,$$

then the original function $u = f(x_1, x_2, \cdots x_n)$ can also be viewed as a function of the m variables t_j. This section discusses how to compute the various partial derivatives

$$\frac{\partial u}{\partial t_j}$$

in terms of other partial derivatives. Pay careful attention to the specific cases discussed.

Key Concept: The Chain Rule in Specific Cases

Two special cases of the chain rule are discussed in this section. For the first special case,

$z = f(x, y)$ is a differentiable function of x, and y, and

$x = g(t)$ and $y = h(t)$ in turn are differentiable functions of t.

Then

$$\frac{dz}{dt} = \frac{\partial f}{\partial x}\frac{dx}{dt} + \frac{\partial f}{\partial y}\frac{dy}{dt}.$$

Note that partial derivative notation is used for f since f is a function of two variables, and ordinary derivative notation is used for x and y since they are functions of the single variable t.

For the second special case,

$z = f(x, y)$ is a differentiable function of x, and y, and

$x = g(s, t)$ and $y = h(s, tt)$ in turn are differetntiable functions of s and t.

Then

$$\frac{\partial z}{\partial s} = \frac{\partial f}{\partial x}\frac{\partial x}{\partial s} + \frac{\partial f}{\partial y}\frac{\partial y}{\partial s} \text{ and}$$

$$\frac{\partial z}{\partial t} = \frac{\partial f}{\partial x}\frac{\partial x}{\partial t} + \frac{\partial f}{\partial y}\frac{\partial y}{\partial t}.$$

Note that partial derivative notation is used throughout this case. *Study Examples 1, 2, and 3* to gain a better understanding of these special cases.

page 790-791.

Key Concept: The General Form of the Chain Rule

For the general form of the chain rule, we assume that

u is a differentiable function of the n variables $x_1, x_2, \cdots x_n$ and that

each x_j is a differentiable function of the m variables $t_1, t_2, \cdots t_m$.

Then for each i from 1 to m,

$$\frac{\partial u}{\partial t_i} = \frac{\partial u}{\partial x_1}\frac{\partial x_1}{\partial t_i} + \frac{\partial u}{\partial x_2}\frac{\partial x_2}{\partial t_i} + \cdots \frac{\partial u}{\partial x_n}\frac{\partial x_n}{\partial t_i}.$$

Note that the partial derivative of u needs to be taken with respect to each variable

page 792-793.

x_j and the result multiplied by $\frac{\partial x_j}{\partial t_i}$. Study *Examples 4 through 7* to see how this works.

Key Concept: Implicit Functions and Implicit Differentiation

If $F(x, y) = 0$ defines y implicitly as a function of x, we can use the forms of the chain rule developed in this section to derive a formula for $\frac{dy}{dx}$:

$$\frac{dy}{dx} = -\frac{\dfrac{\partial F}{\partial x}}{\dfrac{\partial F}{\partial y}} = -\frac{F_x}{F_y}.$$

Similarly, if z is given implicitly as a function of x and y by a formula $F(, x, y, z) = 0$, we can use the forms of the chain rule developed in this section to derive formulas for $\frac{\partial z}{\partial x}$ and $\frac{\partial z}{\partial y}$:

$$\frac{\partial z}{\partial x} = -\frac{\dfrac{\partial F}{\partial x}}{\dfrac{\partial F}{\partial z}} \quad \text{and} \quad \frac{\partial z}{\partial y} = -\frac{\dfrac{\partial F}{\partial y}}{\dfrac{\partial F}{\partial z}}.$$

page 794-795.

Study *Examples 8 and 9* to see how this can be applied.

SkillMaster 11.14: Compute derivatives using the chain rule

page 792-793.

To compute derivatives using the various forms of the chain rule developed in this section, first take the partial derivatives and derivatives needed and then substitute these values into the formulas. You should not need to memorize the formulas if you understand the process for deriving them. Study the tree diagrams in *Figures 2, 3, and 4* to see how to remember the chain rule formulas.

SkillMaster 11.15: Use implicit differentiation to compute derivatives.

In using implicit differentiation, you will be given a formula like $F(x, y) = 0$ or $F(x, y, z) = 0$ and you will be asked to find various derivatives or partial derivatives. Use the formulas listed in the Key Concept above to find these derivatives. Make sure that you understand how the chain rule was used in deriving these formulas.

Worked Examples

For each of the following examples, first try to find the solution without looking at the middle or right columns. Cover the middle and right columns with a piece of paper. If you need a hint, uncover the middle column. If you need to see the worked solution, uncover the right column.

Example	Tip	Solution

SkillMaster 11.14.

· Calculate dz/du by both the chain rule and by direct substitution, then differentiation.
$z = x \cos y$, $x = 2e^u - 1$, $y = 2e^{-u} + 1$.

Use Formula 2 on page 780 of the text.
$$\frac{dz}{dt} = \frac{\partial f}{\partial x}\frac{dx}{dt} + \frac{\partial f}{\partial y}\frac{dy}{dt}.$$

$\partial f/\partial x$
$$= \partial(x \cos y)/\partial x$$
$$= \cos y$$
$\partial f/\partial y$
$$= \partial(x \cos y)/\partial y$$
$$= -x \sin y$$
$\partial x/\partial u$
$$= \partial(2e^u - 1)/\partial u$$
$$= 2e^u$$
$\partial y/\partial u$
$$= \partial(2e^{-u} + 1)/\partial u$$
$$= -2e^{-u}$$

dz/du
$$= [\cos y][2e^u]$$
$$+ [-x \sin y]\left[-2e^{-u}\right]$$
$$= 2e^u \cos y + 2e^{-u}x \sin y$$
$$= 2e^u \cos(2e^{-u} + 1)$$
$$+ 2e^{-u}(2e^u - 1)\sin(2e^{-u} + 1)$$
$$= 2e^u \cos(2e^{-u} + 1)$$
$$+ 4\sin(2e^{-u} + 1)$$
$$- 2e^{-u}\sin(2e^{-u} + 1)$$

As a check, substitute the expressions for x and y into the formula for z and take partial derivatives.

z
$= x \cos y$
$=$
$= 2e^u \cos(2e^{-u} + 1)$
$\quad - \cos(2e^{-u} + 1)$

dz/du
$= 2e^u d(\cos(2e^{-u} + 1))/du$
$\quad + \cos(2e^{-u} + 1)d(2e^u)/du$
$\quad + \sin(2e^{-u} + 1)(-2e^{-u})$
$= 2e^u(-\sin(2e^{-u} + 1))(-2e^{-u})$
$\quad + 2e^u \cos(2e^{-u} + 1)$
$\quad - 2e^{-u} \sin(2e^{-u} + 1)$
$= 4\sin(2e^{-u} + 1)$
$\quad + 2e^u \cos(2e^{-u} + 1)$
$\quad - 2e^{-u} \sin(2e^{-u} + 1)$

Calculate $\partial z/\partial t$ where
$z =$
$x^2 \ln(1 + y^2)$ and $x = te^s$, $y = se^t$.

Use Formula 3 page 791 in the text to find these partial derivatives.
$\partial z/\partial t =$
$(\partial z/\partial x)(\partial x/\partial t) +$
$(\partial z/\partial y)(\partial y/\partial t)$

$\partial z/\partial x$
$\quad = 2x \ln(1 + y^2)$
$\partial z/\partial y$
$\quad = \dfrac{2x^2 y}{1 + y^2}$
$\partial x/\partial t$
$\quad = e^s$
$\partial y/\partial t$
$\quad = se^t$
$\partial z/\partial t$
$\quad = (\partial z/\partial x)(\partial x/\partial t)$
$\quad\quad + (\partial z/\partial y)(\partial y/\partial t)$
$\quad = 2xe^s \ln(1 + y^2)$
$\quad\quad + \dfrac{2x^2 y se^t}{1 + y^2}$

619

· Use the chain rule to calculate $\partial z/\partial x$ and $\partial z/\partial y$ where $z = \ln(\cos^2 x + \sin^2 y) \cdot \sin^{-1}(\sqrt{1 - x^2 - y^2})$

Use the chain rule by first setting $u = \cos^2 x + \sin^2 y$ and $v = \sqrt{1 - x^2 - y^2}$ so that $z = \ln u \sin^{-1} v$. The chain rule says $\partial z/\partial x = (\partial z/\partial u)(\partial u/\partial x) + (\partial z/\partial v)(\partial v/\partial x)$ and $\partial z/\partial y = (\partial z/\partial u)(\partial u/\partial y) + (\partial z/\partial v)(\partial v/\partial y)$.

$\partial z/\partial u$

$$= \frac{\sin^{-1}(\sqrt{1 - x^2 - y^2})}{\cos^2 x + \sin^2 y}$$

$\partial z/\partial v$

$$= \frac{\ln(\cos^2 x + \sin^2 y)}{\sqrt{x^2 + y^2}}$$

$\partial u/\partial x$

$$= -2 \cos x \sin x$$
$$= -\sin(2x)$$

$\partial u/\partial y$

$$= 2 \sin y \cos y$$
$$= \sin(2y)$$

$\partial v/\partial x$

$$= \frac{-x}{\sqrt{1 - x^2 - y^2}}$$

$\partial v/\partial y$

$$= \frac{-y}{\sqrt{1 - x^2 - y^2}}$$

Now put these pieces together inside the Chain Rule.

$\partial z/\partial x$

$$= \quad (\partial z/\partial u)(\partial u/\partial x)$$
$$+(\partial z/\partial v)(\partial v/\partial x)$$

$$= \quad \frac{-\sin(x)\sin^{-1}(\sqrt{1-x^2-y^2})}{\cos^2 x + \sin^2 y}$$
$$+\frac{-x\ln(\cos^2 x + \sin^2 y)}{\sqrt{x^2+y^2}\sqrt{1-x^2-y^2}}$$

$\partial z/\partial y$

$$= \quad (\partial z/\partial u)(\partial u/\partial y)$$
$$+(\partial z/\partial v)(\partial v/\partial y)$$

$$= \quad \frac{\sin(2y)\sin^{-1}(\sqrt{1-x^2-y^2})}{\cos^2 x + \sin^2 y}$$
$$+\frac{-y\ln(\cos^2 x + \sin^2 y)}{\sqrt{x^2+y^2}\sqrt{1-x^2-y^2}}$$

· Find $\partial z/\partial t$ at $s = 1, t = 0$ where $z = f(x^2 + y^2)$, $x = s\cos t$, $y = h(s,t)$ where $h(1,0) = -1, f'(0) = 3, f'(2) = 5, \partial h(1,0)/\partial s = -1$, and $\partial h(1,0)/\partial t = 2$.

The Chain Rule gives $\partial z/\partial t = (\partial z/\partial x)(\partial x/\partial t) + (\partial z/\partial y)(\partial y/\partial t)$. It will save time and effort to plug in the values for x, y, s, t into the expressions for each partial derivative as soon as the partial derivative is computed (but not before). First compute the values of x and y then compute the partials.

$$x = (1)\cos 0 = 1$$
$$y = h(1,0) = -1$$

$\partial z/\partial x$

$$= \quad f'(x^2 + y^2)(2x)$$
$$= \quad f'(2)(2)$$
$$= \quad 10$$

$\partial z/\partial y$

$$= \quad f'(x^2 + y^2)(2y)$$
$$= \quad f'(2)(-2)$$
$$= \quad -10$$

$\partial x/\partial t$

$$= \quad -s\sin t$$
$$= \quad -\sin 0$$
$$= \quad 0$$

$\partial y/\partial t$

$$= \quad h_t(1,0)$$
$$= \quad 2$$

621

Now put these together into the Chain Rule.

$$\partial z/\partial t$$
$$= (\partial z/\partial x)(\partial x/\partial t)$$
$$\quad +(\partial z/\partial y)(\partial y/\partial t)$$
$$= (10)(0) + (-10)(2)$$
$$= -20.$$

· Draw the tree diagram and use it to write out the Chain Rule for the functions. $w = f(x, y, z)$, $x = x(r, t, s)$, $y = y(r, t)$, $z = z(r, s)$.

In the tree diagram an edge goes from the dependent variable w to the intermediate variables x, y, and z that w is a function of. The branches go from the intermediate variables to the variables that they are functions of.

· Let $z = \dfrac{x - y}{x + y}$, where $x = u^2 + v^2 + w^2$ and $y = uvw$. Draw the tree diagram and find the partial derivatives $\partial z/\partial u$, $\partial z/\partial v$, $\partial z/\partial w$ at the point $(u, v, w) = (1, -1, 1)$.

First find the values of x and y when $(u, v, w) = (1, -1, 1)$.

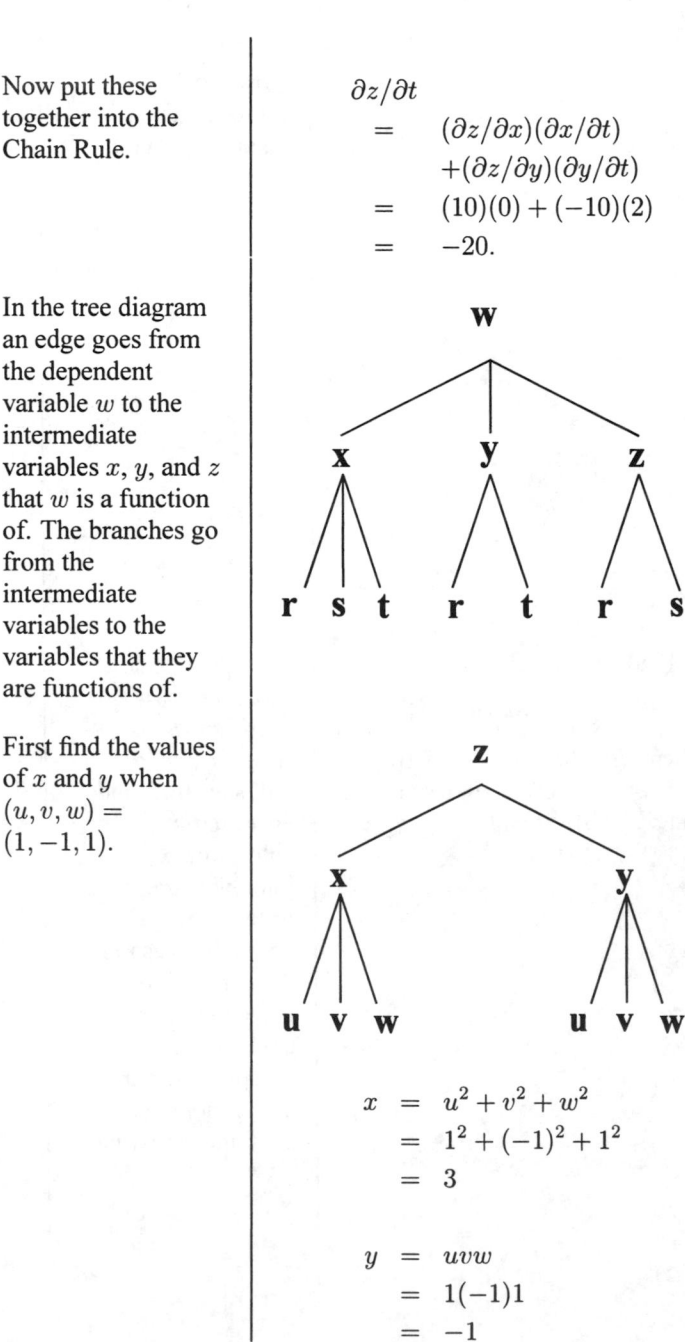

$$x = u^2 + v^2 + w^2$$
$$= 1^2 + (-1)^2 + 1^2$$
$$= 3$$

$$y = uvw$$
$$= 1(-1)1$$
$$= -1$$

$\partial z / \partial u =$
$(\partial z / \partial x)(\partial x / \partial u) +$
$(\partial z / \partial y)(\partial y / \partial u)$
with similar
formulas for $\partial z / \partial v$
and $\partial z / \partial w$. Find the
partials that make up
these formulas and
evaluate them at the
point.

$\partial z / \partial x$

$$= \frac{(x + y) - (x - y)}{(x + y)^2}$$

$$= \frac{2}{(x + y)^2}$$

$$= \frac{2(-1)}{(3 - 1)^2}$$

$$= -2/4$$

$$= -1/2$$

Similarly

$\partial z / \partial y$

$$= \frac{-2x}{(x + y)^2}$$

$$= \frac{-2(3)}{(3 - 1)^2}$$

$$= -6/4$$

$$= -3/2$$

$\partial x / \partial u$

$$= 2u = 2$$

$\partial x / \partial v$

$$= 2v = -2$$

$\partial x / \partial w$

$$= 2w = 2$$

$\partial y / \partial u$

$$= vw = -1$$

$\partial y / \partial v$

$$= uw = 1$$

$\partial y / \partial w$

$$= uv = -1$$

Now put these together into the Chain Rule.

$\partial z/\partial u$

$$= (\partial z/\partial x)(\partial x/\partial u)$$
$$+(\partial z/\partial y)(\partial y/\partial u)$$
$$= (-1/2)(2) + (-3/2)(-1)$$
$$= 1/2.$$

$\partial z/\partial v$

$$= (\partial z/\partial x)(\partial x/\partial v)$$
$$+(\partial z/\partial y)(\partial y/\partial v)$$
$$= (1/2)(-2) + (-3/2)(1)$$
$$= -1/2.$$

$\partial z/\partial w$

$$= (\partial z/\partial x)(\partial x/\partial w)$$
$$+(\partial z/\partial y)(\partial y/\partial w)$$
$$= (-1/2)(2) + (-3/2)(-1)$$
$$= 1/2.$$

· A triangle has two sides of length 10 cm and 15 cm which form an angle of $\pi/3$ radians. These sides are increasing at rates of 4 cm/s and 3 cm/s respectively while the angle is decreasing at a rate of $-1/\sqrt{3}$ radians per second. At this instant, what rate is the third side changing. Is it increasing or decreasing?

The Law of Cosines says that $s^2 = x^2 + y^2 - 2xy\cos(\theta)$ where x, y are the other sides, and θ is the angle between them. Initially, take $x = 10$, $y = 15$, and $\theta = \pi/3$. The other given information is that $dx/dt = 4$ cm/s, $dy/dt = 3$ cm/s, and $d\theta/dt = -1/\sqrt{3}$ radians per second. Notice that the positive increase of the sides will tend to make the third side larger while the decrease in the angle between these sides will tend to make the third side smaller.

Substitute the values of x, y, and θ to find the value of s.

$$
\begin{aligned}
s &= \sqrt{x^2 + y^2 - 2xy\cos(\theta)} \\
&= \sqrt{10^2 + 15^2 - 2(10)(15)(1/2)} \\
&= \sqrt{175} \\
&= 5\sqrt{7} \\
&\approx 13.23
\end{aligned}
$$

Now use the Chain Rule.

$$ds/dt$$
$$= (ds/dx)(dx/dt)$$
$$+(ds/dy)(dy/dt)$$
$$+(ds/d\theta)(d\theta/dt)$$

$$ds/dx$$
$$= \frac{x - y\cos\theta}{\sqrt{x^2 + y^2 - 2xy\cos\theta}}$$
$$= \frac{x - y\cos\theta}{s} = \frac{(10) - (15)(0.5)}{5\sqrt{7}}$$
$$= 1/2\sqrt{7}$$

$$ds/dy$$
$$= \frac{y - x\cos\theta}{\sqrt{x^2 + y^2 - 2xy\cos\theta}}$$
$$= \frac{y - x\cos\theta}{s} = \frac{(15) - (10)(0.5)}{5\sqrt{7}}$$
$$= 2/\sqrt{7}$$

$$ds/d\theta$$
$$= \frac{xy\sin\theta}{\sqrt{x^2 + y^2 - 2xy\cos\theta}}$$
$$= \frac{xy\sin\theta}{s} = \frac{(10)(15)(\sqrt{3}/2)}{10\sqrt{7}}$$
$$= 15\sqrt{3}/\sqrt{7}$$

Now substitute these values into the Chain Rule Formula to get the rate of change ds/dt.

$$ds/dt$$
$$= (ds/dx)(dx/dt)$$
$$+(ds/dy)(dy/dt)$$
$$+(ds/d\theta)(d\theta/dt)$$
$$= (1/2\sqrt{7})(4)$$
$$+(2/\sqrt{7})(3)$$
$$+(15\sqrt{3}/\sqrt{7})(-1/\sqrt{3})$$
$$= -7/\sqrt{7}$$
$$= -\sqrt{7} \text{ cm/s}$$
$$\approx -2.65 \text{ cm/s}$$

· The portion of a tree usable for lumber may be viewed as a right circular cylinder. If the tree's height increases as a rate of 24 in/year and the radius increases at a rate of 3 in/year. At what rate is the tree's volume growing when it has a height of 240 in and a radius of 30 in?

The formula for the volume of a cylinder is $V = \pi r^2 h$ where r is the radius of the base and h is the height. The problem asks for dV/dt the rate of change of the volume with respect to time. We are given $dr/dt = 3$ and $dh/dt = 24$. At the initial time $r = 30$ in and $h = 240$ in, thus $V = \pi(30)^2(240) = \pi 216,000$ in^3. Use the Chain Rule.

dV/dt

$$
\begin{aligned}
&= (\partial V/\partial r)(dr/dt) \\
&\quad + (\partial V/\partial h)(dh/dt) \\
&= \pi(2r)h(3) + \pi r^2(24) \\
&= 6\pi h + 24\pi r^2 \\
&= 6\pi(240) + 24\pi(30)^2 \\
&= 23,040\pi
\end{aligned}
$$

SkillMaster 11.15.

· Find the derivative dy/dx of the family of Cassinian ovals, $(x^2 + y^2)^2 - 2x^2 + 2y^2 - C = 0$ where C is a fixed constant.

Formula 6 page 794 gives
$dy/dx = -F_x/F_y$

F_x

$$
\begin{aligned}
&= 2(x^2 + y^2)(2x) - 4x \\
&= 4x(x^2 + y^2) - 4x
\end{aligned}
$$

F_y

$$
\begin{aligned}
&= 2(x^2 + y^2)(2y) + 4y \\
&= 4y(x^2 + y^2) + 4y
\end{aligned}
$$

dy/dx

$$
\begin{aligned}
&= -\frac{4x(x^2 + y^2) - 4x}{4y(x^2 + y^2) + 4y} \\
&= -\frac{x(x^2 + y^2) - x}{y(x^2 + y^2) + y}
\end{aligned}
$$

· For the previous problem, for what values of (x, y) does the formula that you found fail?

The formula fails when the conditions for the Implicit Function Theorem fails. These conditions are that F_x and F_y be continuous and that $F_y \neq 0$. The partial derivatives are clearly continuous.

$$
\begin{aligned}
0 &= F_y \\
&= 4y(x^2 + y^2) + 4y \\
&= 4y(x^2 + y^2 + 1)
\end{aligned}
$$

This is zero if and only if $y = 0$ since the other factors are always positive. Thus the formula holds for all $y \neq 0$.

· Find $\partial z / \partial x$ and $\partial z / \partial y$ where $z^4 + x^2 z^3 + y^2 + xy - 2 = 0$. Where is the formula valid?

Formula 7 (on page 795) says $\partial z / \partial x = -F_x / F_z$ and $\partial z / \partial y = -F_y / F_z$. This is valid at all points where $F_z \neq 0$.

$$
\begin{aligned}
F_x & \\
&= 2xz^3 + y \\
F_y & \\
&= 2y + x \\
F_z & \\
&= 4z^3 + 3x^2 z^2 \\
\partial z / \partial x & \\
&= -F_x / F_z \\
&= -\frac{2xz^3 + y}{4z^3 + 3x^2 z^2} \\
\partial z / \partial y & \\
&= -F_y / F_z \\
&= -\frac{2y + x}{4z^3 + 3x^2 z^2}
\end{aligned}
$$

These formulas are valid at all points where $F_z \neq 0$.

$$
\begin{aligned}
F_z & \\
&= 4z^3 + 3x^2 z^2 \\
&= z^2(4z + 3x^2)
\end{aligned}
$$

The formula fails on the $z-$axis ($z = 0$) and on the parabolic cylinder $z = -(3/4)x^2$.

628

Section 11.6 –Directional Derivatives and the Gradient Vector

---

Key Concepts:

- Directional Derivatives
- The Gradient Vector
- Tangent Planes to Level Surfaces

Skills to Master:

- Compute directional derivatives.
- Find and apply gradient vectors.
- Find tangent planes and normal lines to level surfaces.

Discussion:

This section introduces directional derivatives and gradient vectors and uses gradient vectors to find tangent planes to level surfaces. You have already seen a specific type of directional derivative, namely partial derivatives with respect to x or y. The partial derivative with respect to x is just the directional derivative in the direction of the unit vector $\mathbf{i}$ and the partial derivative with respect to y is just the directional derivative with respect to the unit vector $\mathbf{j}$. If you keep this in mind when reading this section, the concepts will be easier to understand. Pay careful attention to all the examples.

Key Concept: Directional Derivatives

The directional derivative of a function $z = f(x, y)$ at (x_0, y_0) in the direction of a unit vector $\mathbf{u} = \langle a, b \rangle$ is given by

$$\begin{aligned} D_{\mathbf{u}}f(x_0, y_0) &= \lim_{h \to 0} \frac{f(x_0 + ha, y_0 + hb) - f(x_0, y_0)}{h} \\ &= \lim_{h \to 0} \frac{f((x_0, y_0) + h\mathbf{u}) - f(x_0, y_0)}{h}. \end{aligned}$$

If $f(x, y, z)$ is a function of three variables, then the directional derivative of f at (x_0, y_0, z_0) in the direction of a unit vector $\mathbf{u} = \langle a, b, c \rangle$ is given by

$$\begin{aligned} D_{\mathbf{u}}f(x_0, y_0, z_0) &= \lim_{h \to 0} \frac{f(x_0 + ha, y_0 + hb, z_0 + hc) - f(x_0, y_0, z_0)}{h} \\ &= \lim_{h \to 0} \frac{f((x_0, y_0, z_0) + h\mathbf{u}) - f(x_0, y_0, z_0)}{h}. \end{aligned}$$

page 801. Study *Figure 5* to gain a geometric understanding of this concept.

Both of these formulas can be rewritten in the more compact notation

$$D_{\mathbf{u}}f(\mathbf{x}_0) = \lim_{h \to 0} \frac{f(\mathbf{x}_0 + h\mathbf{u}) - f(\mathbf{x}_0)}{h}$$

where the vector $\mathbf{x}_0$ represents either $\langle x_0, y_0 \rangle$ or $\langle x_0, y_0, z_0 \rangle$.

Key Concept: The Gradient Vector

If f is a function of (x, y), then the gradient of f is the vector function

$$\nabla f(x, y) = \langle f_x(x, y), f_y(x, y) \rangle = \frac{\partial f}{\partial x}\mathbf{i} + \frac{\partial f}{\partial y}\mathbf{j}.$$

If f is a function of (x, y, z), then the gradient of f is the vector function

$$\nabla f(x, y, z) = \langle f_x(x, y, z), f_y(x, y, z), f_z(x, y, z) \rangle = \frac{\partial f}{\partial x}\mathbf{i} + \frac{\partial f}{\partial y}\mathbf{j} + \frac{\partial f}{\partial z}\mathbf{k}.$$

One significance of the gradient vector is that if f is differentiable, then f has a directional derivative in the direction of any unit vector $\mathbf{u}$ and

$$D_{\mathbf{u}}f = \nabla f \cdot \mathbf{u}.$$

page 802. That is, directional derivatives are given by taking the dot product of the gradient vector with the direction. Study *Figure 6* to see a geometric picture of the gradient in two dimensions.

Key Concept: Tangent Planes to Level Surfaces

If S is a level surface given by $F(x, y, z) = k$, the tangent plane to the surface at $P(x_0, y_0, z_0)$ is the plane passing through (x_0, y_0, z_0) with normal vector $\nabla F(x_0, y_0, z_0)$. So the tangent plane has equation

$$F_x(x_0, y_0, z_0)(x - x_0) + F_y(x_0, y_0, z_0)(y - y_0) + F_z(x_0, y_0, z_0)(z - z_0) = 0, \text{ or}$$

$$\nabla F(\mathbf{x}_0) \cdot \mathbf{x}_0 = 0$$

page 806-807.

where $\mathbf{x}_0 = (x_0, y_0, z_0)$. Study *Figures 9 and 10* to see geometric pictures of this.

SkillMaster 11.16: Compute directional derivatives.

To compute directional derivatives, either use the definitions presented in the Key Concept above, or in the case of differentiable functions, use

$$D_{\mathbf{u}}f = \nabla f \cdot \mathbf{u}.$$

Recall that if f has continuous partial derivatives, it is differentiable. Study *Examples 2, 4 and 5* to see how this is done.

page 801-803.

SkillMaster 11.17: Find and apply gradient vectors.

To find gradient vectors, use the definitions presented in the Key Concept above. Applications of gradient vectors include computing directional derivatives and finding the maximum value of the directional derivatives. Since the directional derivative in the direction $\mathbf{u}$ is given by

$$D_{\mathbf{u}}f = \nabla f \cdot \mathbf{u}.$$

for differentiable functions, and since ∇f at a point is fixed, the maximum value of this dot product occurs when $\mathbf{u}$ has the same direction as the gradient. This maximum value is $|\nabla f|$. Study *Figures 7 and 8* and *Examples 6 and 7* to see how this works.

page 804-805.

SkillMaster 11.18: Find tangent planes to level surfaces.

To find the tangent plane to a level surface given by $F(x, y, z) = k$ at $\mathbf{x}_0 = (x_0, y_0, z_0)$ use the formula

$$\nabla F(\mathbf{x}_0) \cdot \mathbf{x}_0 = 0.$$

page 806. Study *Example 8* to see how this works.

Worked Examples

For each of the following examples, first try to find the solution without looking at the middle or right columns. Cover the middle and right columns with a piece of paper. If you need a hint, uncover the middle column. If you need to see the worked solution, uncover the right column.

Example	Tip	Solution
SkillMaster 11.16.		
· Consider the function $f(x,y) = e^x(x+y)$. Find the directional derivative at $(0,0)$ from the direction of $\theta = \pi/6$.	The directional derivative of $f(x,y)$ in the direction θ is $f_x(0,0)\cos\theta + f_y(0,0)\sin\theta$.	$f_x(x,y)$ $\quad = \quad e^x(x+y) + e^x(1)$ $\quad = \quad (1+x+y)e^x$ $f_y(x,y)$ $\quad = \quad e^x$
	It is important to calculate partial (and ordinary) derivatives at a point using the formula with the variables. As soon as the derivatives have been computed, it saves time to substitute in the values for the variables.	$f_x(0,0)$ $\quad = \quad (1+0+0)e^0$ $\quad = \quad 1$ $f_y(0,0)$ $\quad = \quad e^0$ $\quad = \quad 1$ $D_\mathbf{u}f(0,0)$ $\quad = \quad f_x(0,0)\cos\theta$ $\qquad +f_y(0,0)\sin\theta$ $\quad = \quad \cos(\pi/6) + \sin(\pi/6)$ $\quad - \quad \sqrt{3}/2 + 1/2$ $\quad = \quad (\sqrt{3}+1)/2$
· Find the directional derivative of $f(x,y) = \pi/2 + \tan^{-1}(y/x)$ at the point $(1,-2)$ in the direction of $\mathbf{v} = 3\mathbf{i} + 4\mathbf{j}$.	First find a unit vector in the direction of $\mathbf{v}$.	$\mathbf{u} \quad = \quad \dfrac{\mathbf{v}}{|\mathbf{v}|}$ $\quad = \quad \dfrac{3\mathbf{i}+4\mathbf{j}}{\sqrt{3^2+4^2}}$ $\quad = \quad (3/5)\mathbf{i} + (4/5)\mathbf{j}$

For $\mathbf{u} = (a, b)$ the directional derivative is $D_{\mathbf{u}} f = f_x a + f_y b$

$$f_x(x, y)$$
$$= \frac{1}{1 + (y/x)^2}(-y/x^2)$$
$$= \frac{-y}{x^2 + y^2}$$
$$f_y(x, y)$$
$$= \frac{1}{1 + (y/x)^2}(1/x)$$
$$= \frac{x}{x^2 + y^2}$$

Immediately substitute the values $x = 1, y = -2$.

$$f_x(1, -2)$$
$$= \frac{-(-2)}{1^2 + (-2)^2}$$
$$= 2/5.$$
$$f_y(1, -2)$$
$$= \frac{1}{1^2 + (-2)^2}$$
$$= 1/5$$

Finally the directional derivative is

$$D_{\mathbf{u}} f(1, -2)$$
$$= (2/5)(3/5) + (1/5)(4/5)$$
$$= 10/25$$
$$= 2/5.$$

· Find the directional derivative of $f(x,y,z) = \sqrt{10 - x^2 - y^2 - z^2}$ at the point $(1,1,-2)$ in the direction of $\mathbf{v} = 3\mathbf{i} + 4\mathbf{j} - 12\mathbf{k}$.

Let
$$w = f(x,y,z)$$
$$= \sqrt{10 - x^2 - y^2 - z^2}.$$
Using the Chain Rule
$$\partial w/\partial x$$
$$= (-x)/(10 - x^2 - y^2 - z^2)^{1/2}$$
$$= -x/w.$$
By symmetry,
$$\partial w/\partial y = -y/w$$
and
$$\partial w/\partial z = -z/w.$$

A vector with direction $\mathbf{v}$ is
$$\mathbf{u} = \mathbf{v}/|\mathbf{v}|.$$

If $\mathbf{u} = (a,b,c)$ then
$$D_{\mathbf{u}}f = f_x a + f_y b + f_z c.$$

Evaluate the partial derivatives at the points,

$$w$$
$$= \sqrt{10 - 1^2 - 1^2 - (-2)^2}$$
$$= \sqrt{4}$$
$$= 2$$
$$\partial w/\partial x = -1/2$$
$$\partial w/\partial y = -1/2$$
$$\partial w/\partial z = -(-2)/2$$
$$= 1$$

$$\mathbf{u} = \mathbf{v}/|\mathbf{v}|$$
$$= (3\mathbf{i} + 4\mathbf{j} - 12\mathbf{k})/\sqrt{3^2 + 4^2 + 12^2}$$
$$= (3\mathbf{i} + 4\mathbf{j} - 12\mathbf{k})/13$$
$$= (3/13)\mathbf{i} + (4/13)\mathbf{j} - (12/13)\mathbf{k}$$

$$D_{\mathbf{u}}f(1,1,-2)$$
$$= (-1/2)(3/13)$$
$$+(-1/2)(4/13)$$
$$+(1)(12/13)$$
$$= 17/26$$

SkillMaster 11.17.

· Find the gradient of $f(x, y) = \frac{x^2 - y^2}{4} + \tan^{-1}(xy)$. Find the rate of change of f at the point $(\sqrt{2}, \sqrt{2}/2)$ in the direction of steepest increase.

First compute the gradient, then evaluate the gradient at $(\sqrt{2}, \sqrt{2}/2)$.

$$f_x(x, y)$$
$$= \quad x/2 + \frac{y}{1 + (xy)^2}$$
$$f_x(\sqrt{2}, \sqrt{2}/2)$$
$$= \quad \sqrt{2}/2 + \frac{\sqrt{2}/2}{1 + (\sqrt{2}\sqrt{2}/2)}$$
$$= \quad \sqrt{2}/2 + \sqrt{2}/4$$
$$= \quad 3\sqrt{2}/4$$
$$f_y(x, y)$$
$$= \quad -y/2 + \frac{x}{1 + (xy)^2}$$
$$f_y(\sqrt{2}, \sqrt{2}/2)$$
$$= \quad -\sqrt{2}/4 + \frac{\sqrt{2}}{1 + 1^2}$$
$$= \quad \sqrt{2}/4$$
$$\nabla f(\sqrt{2}, \sqrt{2}/2)$$
$$= \quad <3\sqrt{2}/4, \sqrt{2}/4>$$

Use Theorem 8 (page 804) the rate of change in the direction of maximum increase is $|\nabla f|$.

$$\left|\nabla f(\sqrt{2}, \sqrt{2}/2)\right|$$
$$= \quad \left|<3\sqrt{2}/4, \sqrt{2}/4>\right|$$
$$= \quad (\sqrt{2}/4)\left|<3, 1>\right|$$
$$= \quad (\sqrt{2}/4)\sqrt{10}$$
$$= \quad \sqrt{5}/2.$$

· Suppose that $w = f(\rho)$ and $\rho = \sqrt{x^2 + y^2 + z^2}$. Show $\nabla f(\rho) = \frac{f'(\rho)}{\rho} < x, y, z >$ and find $|\nabla f(\rho)|$.

Using the Chain Rule,
$\partial w/\partial x = f'(\rho)\partial\rho/\partial x$.

$\partial\rho/\partial x$
$$= \frac{x}{\sqrt{x^2 + y^2 + z^2}}$$
$$= \frac{x}{\rho}$$

By symmetry
$$\partial\rho/\partial y = \frac{y}{\rho},$$
$$\partial\rho/\partial z = \frac{z}{\rho}$$

$\nabla f(\rho)$
$$= < \partial w/\partial x, \partial w/\partial y, \partial w/\partial z >$$
$$= f'(\rho) < \partial\rho/\partial x, \partial\rho/\partial y, \partial\rho/\partial z >$$
$$= f'(\rho) < \frac{x}{\rho}, \frac{y}{\rho}, \frac{z}{\rho} >$$
$$= \frac{f'(\rho)}{\rho} < x, y, z >$$

$|\nabla f(\rho)|$
$$= \left| \frac{f'(\rho)}{\rho} < x, y, z > \right|$$
$$= \frac{|f'(\rho)|}{\rho} |< x, y, z >|$$
$$= \frac{|f'(\rho)|}{\rho} \sqrt{x^2 + y^2 + z^2}$$
$$= \frac{|f'(\rho)|}{\rho} \rho$$
$$= |f'(\rho)|$$

· Use the previous problem to find the gradient of $W = f(x, y, z) = gmM/\sqrt{x^2 + y^2 + z^2}$ where g, m, M are constants. Find the rate of increase of f in the direction of maximum increase and find a vector in this direction at the point $(1/\sqrt{3}, 1/\sqrt{3}, 1/\sqrt{3})$.

The rate of increase of f in the direction of maximum increase is $|\nabla f|$.

$\rho = \sqrt{1/3 + 1/3 + 1/3}$
$= \sqrt{1} = 1$.

$$W$$
$$= f(x, y, z)$$
$$= gmM/\rho$$
$$\left| \nabla f(1/\sqrt{3}, 1/\sqrt{3}, 1/\sqrt{3}) \right|$$
$$= \frac{f(\rho)}{\rho} \left| < 1/\sqrt{3}, 1/\sqrt{3}, 1/\sqrt{3} > \right|$$
$$= \frac{gmM}{\rho^2} \left| < 1/\sqrt{3}, 1/\sqrt{3}, 1/\sqrt{3} > \right|$$
$$= \frac{gmM}{\rho^2}(1) = gmM$$

is the rate of maximum increase at the point.

· Allegra is the captain of a space ship who finds herself on the hot side of Mercury and her space suit is starting to melt. The temperature is locally given by $T(x, y, z) = K(e^{-x} + e^{-2y} + e^{-3z})$. In which direction should she start moving in order to reduce her temperature as quickly as possible?

The direction needed to travel is the direction of maximum decrease or $-\nabla T$.

$$T_x = K(-e^{-x})$$
$$= -Ke^{-x}$$
$$T_y = -2Ke^{-2y}$$
$$T_z = -3Ke^{-3z}$$
$$-\nabla T$$
$$= K < e^{-x}, 2e^{-2}, 3e^3 >$$

She should initially travel in the same direction as the vector $< e^{-x}, 2e^{-2}, 3e^3 >$.

SkillMaster 11.18.

· Find the equation of the tangent plane and the normal line through the point $(5, -2, 3)$ on the surface given by $x^2 + xyz + -2y^2 + z^3 = 14$.

The gradient of $F(x, y, z)$
$=$
$x^2 + xyz + -2y^2 + z^3$
evaluated at $(5, -2, 3)$ is a vector that is normal to the surface.
The tangent plane satisfies
$\nabla F(5, -2, 3)$
$\cdot (\mathbf{r} - <5, -2, 3>$
$) = 0.$

∇F
$=\ <2x + yz, xz - 4y, xy + 3z^2>$
$\nabla F(5, -2, 3)$
$=\ <2(5) - (2)3, 5(3) + 4(2), 5(-2) + 3(3)^2 >$
$=\ <4, 23, 17>$
The tangent plane equation is
0
$=\ <4, 23, 17> \cdot <x - 5, y + 2, z - 3>$
$=\ 4(x - 5) + 23(y + 2) + 17(z - 3)$
$=\ 4x + 23y + 17z - 25.$
So
$$4x + 23y + 17z = 25.$$
is the equation of the tangent plane.

The equation of the normal line is
$<4, 23, 17> t$
$+ <5, -2, 3>$
in parametric form.

In symmetric form the equation of the normal line is
$$\frac{x - 5}{4} = \frac{y + 2}{23} = \frac{z\ \ 3}{17}.$$

639

· Find the acute angle between the two surfaces $2xe^z+xy^2+z^3e^x = 3$ and $x^3+y^3-xy+\sin(x+y+z) = 1$ at the point $(1, -1, 0)$.

The acute angle between the surfaces is interpreted to mean the acute angle between the tangent planes. The angle between tangent planes is the same as the acute angle between the normal vectors. These normal vectors are easy to compute because they are the three variable gradients of the functions.

The first normal vector is the gradient of
$$F(x,y,z)$$
$$= 2xe^z + xy^2 + z^3e^x$$
or
$$\nabla F$$
$$= <2e^z + y^2 + z^3e^x, 2xy, 2xe^z + 3z^2e^x>$$
$$\nabla F(1,-1,0)$$
$$= <2+1+0, -2, 2+0>$$
$$= <3, -2, 2>$$
is the normal vector.

The second normal vector is the gradient of
$$G(x,y,z)$$
$$= x^3 + y^3 - xy$$
$$+ \sin(x+y+z)$$
$$\nabla G$$
$$= <3x^2 - y + \cos(x+y+z),$$
$$3y^2 - x + \cos(x+y+z),$$
$$\cos(x+y+z) >$$
$$\nabla G(1,-1,0)$$
$$= <3+1+1, 3-1+1, 1>$$
$$= <5, 3, 1>.$$

The angle between two vectors is given by the dot product rule,
$\cos(\theta) =$
$\mathbf{u} \cdot \mathbf{v}/ |\mathbf{u}|\,|\mathbf{v}|$.

The cosine of the angle between the two surfaces is
$$\frac{<3,-2,2>\cdot<5,3,1>}{|<3,-2,2>||<5,3,1>|}$$
$$= \frac{15 - 6 + 2}{\sqrt{17}\sqrt{35}}$$
$$= \frac{11}{\sqrt{17}\sqrt{35}} = \cos\theta.$$
The angle is
$$\theta$$
$$= \cos^{-1}(\frac{11}{\sqrt{17}\sqrt{35}}) \approx 63.19°$$

Section 11.7 – Maximum and Minimum Values

Key Concepts:

- Local Maximum and Minimum Values of a Function
- Second Derivative Tests for Extreme Values
- Absolute Maximum and Minimum Values

Skills to Master:

- Find local maximum and minimum values.
- Compute the absolute maximum and minimum values of a function.

Discussion:

In this section, you will see how to find maximum and minimum values for functions of two variables. Just as derivatives were important in finding maximum and minimum values of functions of one variable, partial derivatives will be important here. Make sure that you understand how the results in this section are obtained. Do not just memorize the formulas. Understanding these results will allow you to apply the techniques to functions of more than two variables if needed.

Key Concept: Local Maximum and Minimum Values of a Function

A function of two variables, $f(x, y)$ has a *local maximum at* (a, b) if
$$f(x, y) \leq f(a, b) \text{ when } (x, y) \text{ is near } (a, b).$$

A function of two variables, $f(x, y)$ has a *local minimum at* (a, b) if
$$f(x, y) \geq f(a, b) \text{ when } (x, y) \text{ is near } (a, b).$$

641

page 811.

In the above cases, the number $f(a, b)$ is called the *local maximum value* or *local minimum value*. Make sure that you use the correct terminology. Study *Figure 1* to see a geometric picture of local maximum and minimum values.

Key Concept: Second Derivative Tests for Extreme Values

A function of two variables, $f(x, y)$ has a critical point at (a, b) if
$$f_x(a, b) = f_y(a, b) = 0$$
or if one of these partial derivatives doesn't exist. If the second partial derivatives of f are continuous on a disc with center (a, b), and if (a, b) is a critical point of f, we let
$$D = D(a, b) = f_{xx}(a, b) f_{yy}(a, b) - [f_{xy}(a, b)]^2 .$$
Note that $f_{xy} = f_{yx}$ because of the assumptions about f. The second derivative test states:

(a) If $D > 0$ and $f_{xx}(a, b) > 0$, then $f(a, b)$ is a local minimum value

(b) If $D > 0$ and $f_{xx}(a, b) < 0$, then $f(a, b)$ is a local maximum value

(c) If $D < 0$, then $f(a, b)$ is not a local maximum or local minimum value.

page 812-813.

In case (c), (a, b) is called a saddle point. Study *Figures 3 and 4* and *Example 3* to gain a better understanding of the second derivative test.

Key Concept: Absolute Maximum and Minimum Values

A function of two variables, $f(x, y)$ has an *absolute maximum or minimum at* (a, b) if
$$f(x, y) \leq f(a, b) \text{ when } (x, y) \text{ is near } (a, b)$$
$$\text{or } f(x, y) \geq f(a, b) \text{ when } (x, y) \text{ is near } (a, b) \text{ respectively.}$$
The extreme value theorem for functions of two variables states that

If f is continuous on a closed bounded set in R^2, then

f attains an absolute maximum and absolute minimum value at some points in D.

SkillMaster 11.19: Find local maximum and minimum values.

To find local maximum and minimum values of functions of two variables, examine the critical points of f and use the second derivative test if possible. Recall that (a, b) is a critical point if

$$f_x(a, b) = f_y(a, b) = 0$$

or if one of these partial derivatives doesn't exist. The reason that you just have to check critical points is because if f has a local maximum or minimum value at (a, b), and if the first order partial derivatives of f exist there, then $f_x(a, b) = f_y(a, b) = 0$. Study *Examples 1, 2, and 3* to see how this works.

page 811-812.

SkillMaster 11.20: Compute the absolute maximum and minimum values of a function.

To find the absolute maximum and minimum values of a continuous function $f(x, y)$ on a closed and bounded set D,

(1) Find the values of f at the critical points of f in D.

(2) Find the extreme values of f on the boundary of D.

(3) The largest of the values from (1) and (2) is the absolute maximum. The smallest of the values is the absolute minimum.

page 817.

Study *Example 7* to gain a better understanding of how this works.

Worked Examples

For each of the following examples, first try to find the solution without looking at the middle or right columns. Cover the middle and right columns with a piece of paper. If you need a hint, uncover the middle column. If you need to see the worked solution, uncover the right column.

Example	Tip	Solution
SkillMaster 11.19.		

· Given the function $f(x, y) = x^3 - y^2 + 2y - 9x$, find all the critical points of f.
Classify the critical points of f using the second derivative test.

Recall that the critical points of f are those points for which $\nabla f = 0$.

$$f_x$$
$$= 3x^2 - 9$$
$$f_y$$
$$= -2y + 2$$
$$\nabla f = 0$$
if and only if
$$3x^2 - 9 = 0$$
$$x^2 - 3 = 0$$
$$x = \pm\sqrt{3}$$
and
$$-2y + 2 = 0$$
$$-y + 1 = 0$$
$$y = 1.$$
The critical points are $(\sqrt{3}, 1)$ and $(-\sqrt{3}, 1)$.

· Use the second derivative test to classify the critical points in the previous problem.

To use the second derivative test at a critical point (a, b), first compute $D = f_{xx}f_{yy} - [f_{xy}]^2$ at that point.
If $D > 0$ and $f_{xx}(a, b) > 0$ then (a, b) is a local minimum.
If $D > 0$ and $f_{xx}(a, b) < 0$ then (a, b) is a local maximum.
If $D < 0$ then (a, b) is neither a local minimum nor local maximum.
Hint: sometimes when there is more than one critical point it is easier to compute D as a function of (x, y) and then plug in the critical points to apply the second derivative test.

$$f_{xx} = 6x$$
$$f_{yy} = -2$$
$$f_{xy} = 0$$

$$f_{xx}f_{yy} - [f_{xy}]^2$$
$$= (6x)(-2) - 0^2$$
$$= -12x$$

At $(\sqrt{3}, 1)$
$$D = -12\sqrt{3} < 0$$
so $(\sqrt{3}, 1)$ is neither a maximum nor a minimum but is a saddle point.

At $(-\sqrt{3}, 1)$
$$D = 12\sqrt{3} > 0,$$
$$f_{xx} = -6\sqrt{3} < 0$$
so $(\sqrt{3}, 1)$ is a maximum.

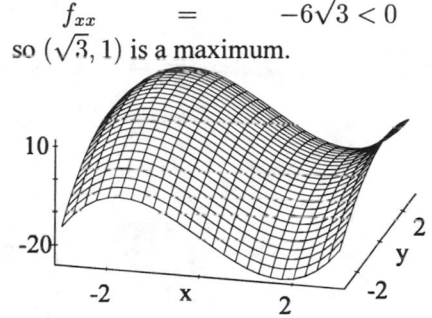

· Given the function $f(x,y) = 4xy - x^4 - y^4$, find all the critical points of f.

Use the same hint as in the preceding problem.

$$f_x = 4y - 4x^3$$
$$f_y = 4x - 4y^3$$

So

$$y = x^3$$
$$x = y^3 = (x^3)^3 = x^9$$
$$0 = x^9 - x$$
$$= x(x^8 - 1)$$

The solutions are $x = 0, 1, -1$ with corresponding y values equal to x. The critical points are $(0,0)$, $(1,1)$, $(-1,-1)$.

Classify the critical points of f using the second derivative test.

Again compute the second partials and compute
$D = f_{xx}f_{yy} - [f_{xy}]^2$.

$$f_{xx} = -12x^2$$
$$f_{yy} = -12y^2$$
$$f_{xy} = 4$$
$$D = (-12x^2)(-12y^2) - (4)(4)$$
$$= 144x^2y^2 - 16$$

Plug in the critical points to apply the second derivative test.

$$D(0,0) = 144(0) - 16$$
$$= -16 < 0$$

so $(0,0)$ is a saddle point.

$$D(1,1) = 144(1) - 16$$
$$= 128 > 0$$
$$f_{xx}(1,1) = -12(1) < 0$$

so $(1,1)$ is a maximum. Similarly $(-1,-1)$ is a maximum.

· Suppose that a function has the form $f(x, y) = F(ax + by + c)$ where F is differentiable and a, b, and c are constants.

Show that the second derivative test fails at each critical point.

The second derivative test fails if $D = f_{xx}f_{yy} - [f_{xy}]^2 = 0$. Use the chain rule to formally find the first and second partial derivatives. To start $f_x(x, y) = aF'(ax + by + c)$ and $f_y(x, y) = bF'(ax + by + c)$.

$$f_{xx}(x, y) = a^2 F''(ax + by + c)$$

$$f_{yy}(x, y) = b^2 F''(ax + by + c)$$

$$f_{xy}(x, y) = abF''(ax + by + c)$$

$$D = (a^2 b^2 - (ab)^2)F''(ax + by + c)$$
$$= 0$$

So the second derivative test fails.

· If it rains on the surface whose height is $f(x,y) = \dfrac{1}{x} + \dfrac{1}{y} + xy$ where will a puddle begin to form?

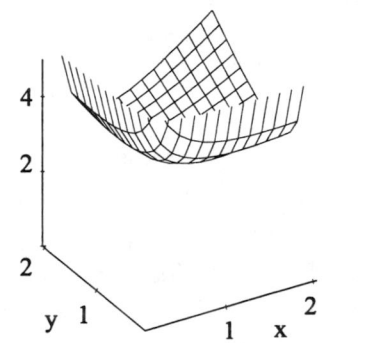

A puddle will form at any point which is a local minimum. Look for critical points where $D > 0$ and $f_{xx} > 0$.

$$f_x$$
$$= -x^{-2} + y$$
$$f_y$$
$$= -y^{-2} + x$$

Setting these to zero, the critical points must satisfy

$$y = 1/x^2$$
$$x = 1/y^2$$

Substituting $1/y^2$ for x in the first equation

$$y = 1/x^2$$
$$= 1/(1/y^2)^2$$
$$= y^4.$$

So

$$y^4 - y = 0$$
$$y(y-1)(y^2 + y + 1) = 0.$$

The last factor has no roots (use the quadratic formula) so possible solutions are $y = 0$, $y = 1$. The domain does not include $y = 0$ so the only critical point is $(1,1)$.

$$f_{xx} = 2x^{-3}$$
$$f_{yy} = 2y^{-3}$$
$$f_{xy} = 1$$

$$D(1,1)$$
$$= (2)(2) - (1)(1)$$
$$= 3 > 0$$

Also

$$f_{xx}(1,1) = 2(1^{-3})$$
$$= 2 > 0$$

so the second derivative test implies $(1,1)$ is a local minimum and water will begin to form a puddle there.

Find the maxima, minima, and saddle points in the contour plot.

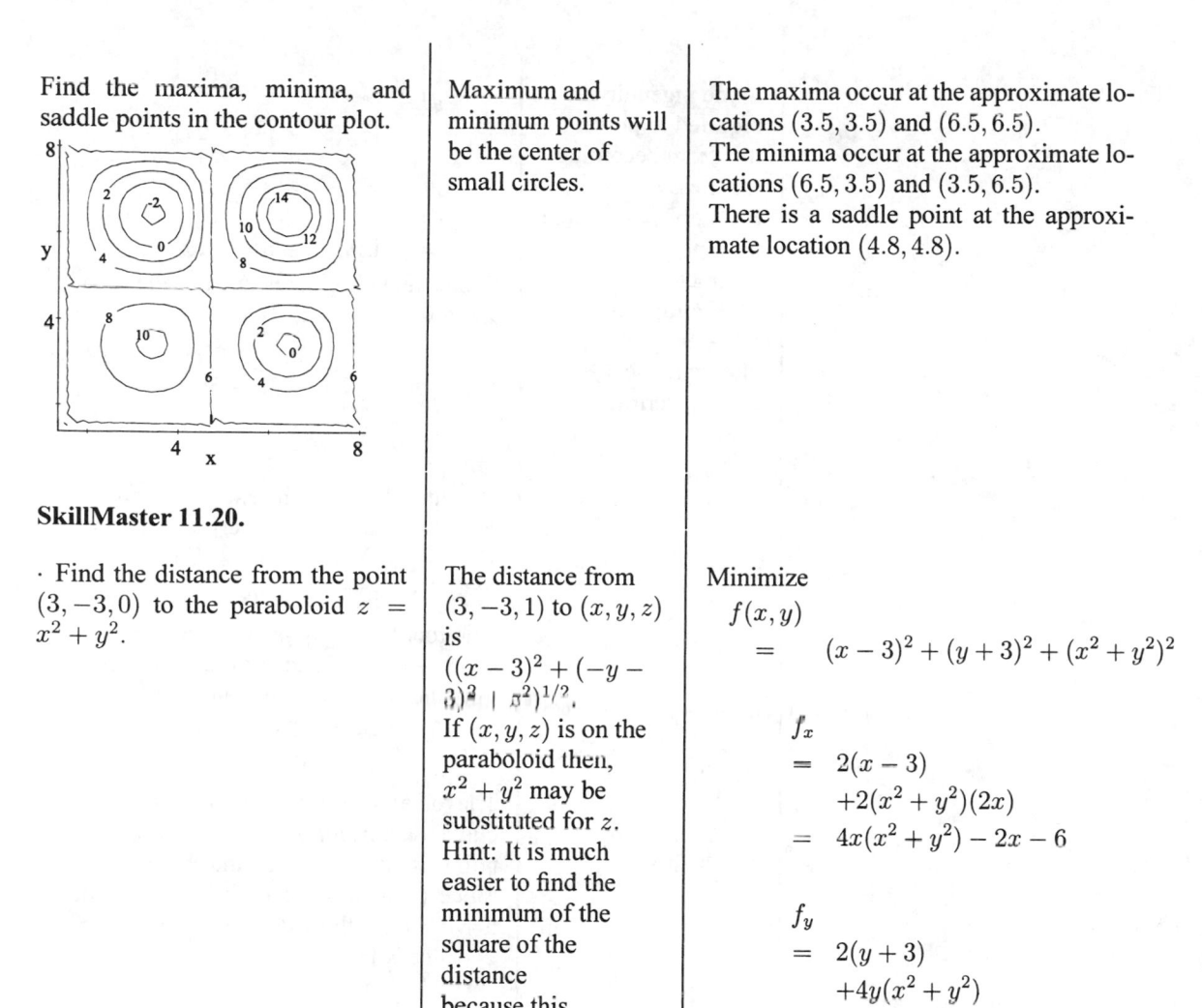

| Maximum and minimum points will be the center of small circles. | The maxima occur at the approximate locations $(3.5, 3.5)$ and $(6.5, 6.5)$. The minima occur at the approximate locations $(6.5, 3.5)$ and $(3.5, 6.5)$. There is a saddle point at the approximate location $(4.8, 4.8)$. |

SkillMaster 11.20.

· Find the distance from the point $(3, -3, 0)$ to the paraboloid $z = x^2 + y^2$.

The distance from $(3, -3, 1)$ to (x, y, z) is $((x-3)^2 + (-y - 3)^2 + z^2)^{1/2}$. If (x, y, z) is on the paraboloid then, $x^2 + y^2$ may be substituted for z. Hint: It is much easier to find the minimum of the square of the distance because this eliminates the square root.

Minimize
$$f(x, y)$$
$$= (x-3)^2 + (y+3)^2 + (x^2 + y^2)^2$$

$$f_x$$
$$= 2(x-3)$$
$$\quad + 2(x^2 + y^2)(2x)$$
$$= 4x(x^2 + y^2) - 2x - 6$$

$$f_y$$
$$= 2(y+3)$$
$$\quad + 4y(x^2 + y^2)$$
$$= 4y(x^2 + y^2) - 2y + 6$$

Some ingenuity is required to prevent this from becoming too messy.
Multiply f_x by y and multiply f_y by x and subtract.
This computes $yf_x - xf_y$ to eliminate the $x^2 + y^2$ terms.

$$yf_x$$
$$= 4xy(x^2 + y^2) - 2xy - 6y$$

$$xf_y$$
$$= 4xy(x^2 + y^2) - 2xy + 6x$$

Subtracting eliminates most of the terms and gives

$$0 = yf_x - xf_y$$
$$= -6y - 6x$$

$$y = -x$$

Substitute this into the expression for f_x, set equal to 0 to find the critical points.

$$8x^3 - 2x - 6 = 0$$
$$4x^3 - x - 3 = 0$$

One root found by inspection is $x = 1$, so $(x - 1)$ is a factor of the polynomial. Find the other factor by long division.

$$0 = 4x^3 - 3x - 1$$
$$= (x - 1)(4x^2 + 4x + 3)$$

The other factor has no root (use the quadratic formula) so the only critical point is at $x = 1$ and $y = -1$. Since $(1, -1)$ is the only critical point, it MUST be the minimum point. The z-value is $1^2 + (-1)^2 = 2$.

Remember the problem asks to find the minimum distance. You still have to find the distance between $(3, -3, 0)$ and $(1, -1, 2)$.

The distance is
$$\sqrt{(3 - 1)^2 + (-3 - (-1))^2 + (0 - 2)^2}$$
$$= \sqrt{4 + 4 + 4}$$
$$= 2\sqrt{3}.$$

· Find the highest points on the Martian crater given by $f(x,y) = 10(x^2 + y^2)e^{-x^2-y^2}$ on the disk $D = \{(x,y)|x^2 + y^2 \leq 2\}$

Notice that x and y always appear in the expression for f as expressions in $x^2 + y^2$. This means the level curves of f are circles with the origin as center. It is tempting to use the substitution $r^2 = x^2 + y^2$. This would work but it is simpler to use the substitution $u = x^2 + y^2$ so $f(x,y) = g(u) = 10ue^{-u}$.

Don't forget that to find the absolute maximum it is necessary to also check the values of y at the endpoints as well as the critical points.

Minimize
$$g(u) = 10ue^{-u}, \; 0 \leq u \leq 2$$
$$g'(u) = 10e^{-u} - 10ue^{u}$$
$$= 10e^{-u}(1 - u)$$
All the factors are positive except the last.
$$g'(u) = 0$$
if and only if
$$1 - u = 0$$
$$u = 1$$
So $u = 1$ is the critical point.

$$g(1) = 10e^{-1} \approx 3.68$$
$$g(0) = 0$$
$$g(2) = 20e^{-2} \approx 2.71$$
The largest value is $10e^{-1}$ which occurs at $u = 1$. The absolute maxima of f occur on the unit circle.

· A triangular plate occupies a region D shown. The temperature on the plate is given by $T = x^2 + xy + 2y^2 - 3x + 2y + 4$. Find the hottest and coolest points on this plate.

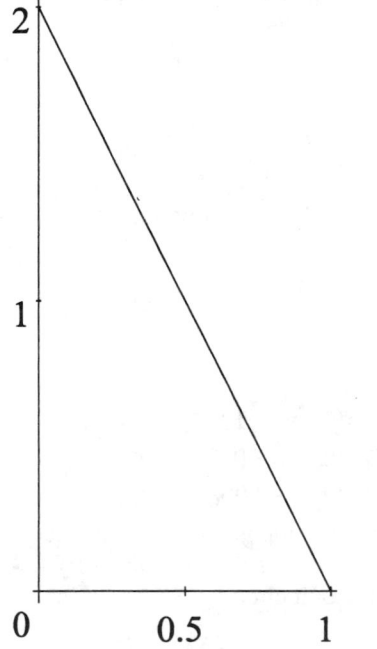

The critical points in the interior of D have to be checked, but so do the points on the boundary.

Set the gradient equal to 0.
$$0 = T_x$$
$$= 2x + y - 3$$

$$0 = T_y$$
$$= x + 4y + 2$$

$$y = -2x + 3$$
$$0 = x + 4(-2x + 3) + 2$$
$$= x - 8x + 12 + 2$$
$$= -7x + 14$$

$$7x = 14$$
$$x = 2$$

Any point with $x = 2$ is outside the region D so there are no critical points on the region. The absolute maximum and the absolute minimum have to be on the boundary of D.

To find the absolute extrema on the boundary, first remember that the T values of the corners have to be checked. To find interior points on segments of the boundary, represent as a one variable problem. For example, the segment from $(0,0)$ to $(0,2)$ may be represented as $(0,y)$ for $0 \leq y \leq 2$.

For this segment along the y–axis

$$g_1(y)$$
$$= \quad T(0,y)$$
$$= \quad 2y^2 + 2y + 4$$

Set

$$0 \quad = \quad g_1'(y)$$
$$= \quad 4y + 2$$
$$y \quad = \quad -1/2$$

This is not on the segment so it is not a candidate for an extreme point. It is not necessary to check its value.

Represent the segment along the x-axis by $(x, 0)$ for $0 \leq x \leq 1$. Here

$$
\begin{aligned}
g_2(x) & \\
&= T(x, 0) \\
&= x^2 - 3x + 4.
\end{aligned}
$$

Set

$$
\begin{aligned}
0 &= g_2'(x) \\
&= 2x - 3 \\
x &= 3/2
\end{aligned}
$$

This is also not on the segment so it is not necessary to check its value.

The equation for the hypotenuse of the triangle is

$$
y = -2x + 2, \ 0 \leq x \leq 2.
$$

Set

$$
\begin{aligned}
g_3(x) & \\
&= T(x, -2x + 2) \\
&= x^2 + x(-2x + 2) \\
&\quad + 2(-2x + 2)^2 - 3x \\
&\quad + 2(-2x + 2) + 4 \\
&= 7x^2 - 21x + 16
\end{aligned}
$$

Set

$$
\begin{aligned}
0 &= g_3'(x) \\
&= 14x - 21 \\
x &= 3/2,
\end{aligned}
$$

also outside the domain D.

The absolute maximum and the absolute minim must occur at one (or more) of the corners. Check these values.

$$
\begin{aligned}
T(0, 0) &= 4 \\
T(0, 2) &= 16 \\
T(1, 0) &= 2
\end{aligned}
$$

The hottest point is at $(0, 2)$ and the coolest point is at $(1, 0)$.

Section 11.8 – Lagrange Multipliers

Key Concepts:

- Lagrange Multipliers and Extreme Values with Constraints
- Extreme Values Subject to Two Constraints

Skills to Master:

- Use the method of Lagrange multipliers to find extreme values subject to constraints.

Discussion:

In the previous section, the method for finding absolute maximum and minimum values of a function on a set D involved finding the maximum and minimum values of the function on the boundary of D. This section introduces a technique, called the method of Lagrange multipliers, to deal with this kind of problem and with other problems that involve finding the maximum or minimum value for a function that is subject to certain constraints. Study all the examples carefully so that you become proficient at this method.

Key Concept: Lagrange Multipliers and Extreme Values with Constraints

---

To find the maximum and minimum values of a differentiable function $f(x, y, z)$ subject to the constraint that $g(x, y, z) = 0$,

(a) Find all values x, y, z, and λ such that
$$\nabla f(x,y,z) = \lambda \nabla g(x,y,z) \text{ and}$$
$$g(x,y,z) = k$$

(b) Evaluate f at each point (x,y,z) from part (a). The largest value is the maximum. The smallest value is the minimum.

page 824-825.

Study *Examples 1 and 2* and *Figure 2* to gain a better understanding of this method.

Key Concept: Extreme Values Subject to Two Constraints

To find the maximum and minimum values of a differentiable function $f(x,y,z)$ subject to the constraints that $g(x,y,z) = k$ and $h(x,y,z) = c$, find all values x, y, z, μ and λ such that
$$\nabla f(x,y,z) = \lambda \nabla g(x,y,z) + \mu \nabla h(x,y,z),$$
$$g(x,y,z) = k \text{ and } h(x,y,z) = c.$$
Then proceed as in the previous Key concept. Note that there are five equations to work with here.

Study Figure 5.

SkillMaster 11.21: Use the method of Lagrange multipliers to find extreme values subject to constraints.

Worked Examples

For each of the following examples, first try to find the solution without looking at the middle or right columns. Cover the middle and right columns with a piece of paper. If you need a hint, uncover the middle column. If you need to see the worked solution, uncover the right column.

Example	Tip	Solution

SkillMaster 11.21.

· Maximize the function $3xy + 1$ on the ellipse $\dfrac{x^2}{a^2} + \dfrac{y^2}{b^2} = 1$ where a and b are positive constants.

Tip
Use Lagrange multipliers. Set $g(x,y) = \dfrac{x^2}{a^2} + \dfrac{y^2}{b^2}$. Minimize $f(x,y) = 3xy + 1$ subject to the constraint $g(x,y) = 1$. Search for points at which ∇f is parallel to ∇g, $\nabla f = \lambda \nabla g$

Solution

$$< 3y, 3x >= \lambda < 2x/a^2, 2y/b^2 >$$

Case 1. $\lambda = 0$. Then $x = y = 0$. This is not a solution since
$$g(0,0) = 0 \neq 1.$$
We are left with Case 2. $\lambda \neq 0, x \neq 0, y \neq 0$.
Solve for λ.
$$3ya^2/2x = \lambda = 3xb^2/2y$$

$$\frac{ya^2}{x} = \frac{xb^2}{y}$$
$$y^2 = (b/a)^2 x^2$$

Plug this expression for y^2 into the constraint

$$1 = g(x,y) = \frac{x^2}{a^2} + \frac{y^2}{b^2}$$
$$= \frac{x^2}{a^2} + \frac{(b/a)^2 x^2}{b^2}$$
$$= \frac{x^2}{a^2} + \frac{x^2}{a^2} = \frac{2x^2}{a^2}$$
$$x = \pm a/\sqrt{2}.$$

Resubstituting gives similarly
$$y = \pm b/\sqrt{2}.$$

There are four candidates, $(a/\sqrt{2}, b/\sqrt{2})$, $(-a/\sqrt{2}, -b/\sqrt{2})$, $(-a/\sqrt{2}, b/\sqrt{2})$, and $(a/\sqrt{2}, -b/\sqrt{2})$. Plug these into f.

$$
\begin{aligned}
f(a/\sqrt{2}, b/\sqrt{2}) &= 3ab2 + 1 \\
f(-a/\sqrt{2}, -b/\sqrt{2}) &= 3ab2 + 1 \\
f(-a/\sqrt{2}, b/\sqrt{2}) &= -3ab2 + 1 \\
f(a/\sqrt{2}, -b/\sqrt{2}) &= -3ab2 + 1
\end{aligned}
$$

The maximizing point are
$$(a/\sqrt{2}, b/\sqrt{2})$$
and
$$(-a/\sqrt{2}, -b/\sqrt{2}).$$
(The other two are minimizing points.)

· Show that the sum of the sines of the angles of a triangle is maximized when the triangle is equilateral.

Give names to the angles of the triangle, say x, y, and z.
The function to maximize is
$f(x, y, z) = \sin x + \sin y + \sin z$.
The constraint is
$g(x, y, z) = x + y + z = \pi$,
$x > 0, y > 0$, and $z > 0$.

Solve
$$\nabla f = \lambda \nabla g$$
$$< \cos x, \cos y, \cos z >= \lambda < 1, 1, 1 >.$$
Thus
$$\lambda = \cos x = \cos y = \cos z.$$
All the angles are between 0 and π. The cosine function takes on unique values (in other words, $[0, \pi]$ is the domain of the inverse cosine). Thus
$$x = y = z$$
The constraint condition says
$$x + y + z = \pi.$$
Substituting
$$
\begin{aligned}
x + x + x &= \pi \\
3x &= \pi \\
x &= \pi/3.
\end{aligned}
$$
Thus every angle is $\pi/3$ or $60°$ and the triangle is equilateral.

· Find the minimum of $f(x, y, z) = 2x^2 + y^2 + 3z^2$ for (x, y, z) constrained to be on the line that is the intersection of the planes $x + y = 4$ and $x - y + z = 6$.

You could solve for the line that is the intersection of the two planes and use Lagrange multipliers. Instead, solve the Lagrange multiplier problem with two constraints. Set
$$g(x, y, z) = x + y$$
and
$$h(x, y, z) = x - y + z.$$
The problem becomes: minimize
$$f(x, y, z) = 2x^2 + y^2 + 3z^2$$
subject to the constraints
$$g(x, y, z) = 4 \text{ and}$$
$$h(x, y, z) = 6.$$

Solve
$$\nabla f = \lambda \nabla g + \mu \nabla h.$$

∇f
$$= \ <4x, 2y, 6z>$$
$$= \ \lambda <1, 1, 0> + \mu <1, -1, 1>$$

$$4x = \lambda + \mu$$
$$2y = \lambda - \mu$$
$$6z = \mu$$

Now solve for λ and μ in terms of x, y, z. Substitute $6z = \mu$ into the first two equations:
$$4x - 6z = \lambda$$
$$2y + 6z = \lambda$$
Equate the two expressions for λ:
$$4x - 6z = 2y + 6z$$

$$4x - 2y - 12z = 0$$
$$2x - y - 6z = 0.$$
Combine this equation with the two constraint equations and solve for x, y, z.
$$2x - y - 6z = 0$$
$$x + y + 0 = 4$$
$$x - y + z = 6$$
The middle equation is
$$x = -y + 4.$$
Substitute this for x in the other two equations.
$$2(-y + 4) - y - 6z = 0$$
$$(-y + 4) - y + z = 6$$
or
$$3y + 6z = 8$$
$$2y - z = -2$$
$$z = 2y + 2.$$

Substitute this into the first equation to solve for y.

$$3y + 6(2y + 2) = 8$$
$$15y = -4$$
$$y = -4/15$$

Use this to get the x and y values.

$$z = 2(-4/15) + 2$$
$$= 1\frac{7}{15}$$

$$x = -(-4/15) + 4$$
$$= 4\frac{4}{15}.$$

This gives the point $(\frac{64}{15}, -\frac{4}{15}, \frac{22}{15})$ at which the minimum occurs.

The minimum value is

$$f(\frac{64}{15}, -\frac{4}{15}, \frac{22}{15})$$
$$= 2(\frac{64}{15})^2 + (-\frac{4}{15})^2 + 3(\frac{22}{15})^2$$
$$= \frac{2(4096) + 16 + 484}{225}$$
$$= \frac{8692}{225}$$
$$\approx 38.63$$

SkillMasters for Chapter 11

SkillMaster 11.1:   Find the domain and range of functions of two variables.

SkillMaster 11.2:   Use different descriptions of functions.

SkillMaster 11.3:   Describe the level surfaces of functions of three variables.

SkillMaster 11.4:   Determine if a function $f(x, y)$ has a limit at (a, b).

SkillMaster 11.5:   Determine the points of continuity of a function $f(x, y)$.

SkillMaster 11.6:   Compute and interpret partial derivatives.

SkillMaster 11.7:   Use Clairaut's Theorem to compute higher partial derivatives.

SkillMaster 11.8:   Verify if a given function satisfies a partial differential equation.

SkillMaster 11.9:   Compute the tangent plane to a surface given by a function of two variables.

SkillMaster 11.10:   Determine if a function is differentiable.

SkillMaster 11.11:   Use linearization to approximate values of differentiable functions.

SkillMaster 11.12:   Compute differentials and use them to estimate the error in an approximation.

SkillMaster 11.13:   Compute tangent planes to parametric surfaces.

SkillMaster 11.14:   Compute derivatives using the chain rule

SkillMaster 11.15:   Use implicit differentiation to compute derivatives.

SkillMaster 11.16:   Compute directional derivatives.

SkillMaster 11.17:   Find and apply gradient vectors.

SkillMaster 11.18:   Find tangent planes to level surfaces.

SkillMaster 11.19:   Find local maximum and minimum values.

SkillMaster 11.20:   Compute the absolute maximum and minimum values of a function.

SkillMaster 11.21:   Use the method of Lagrange multipliers to find extreme values subject to constraints.

Chapter 12 - Multiple Integrals

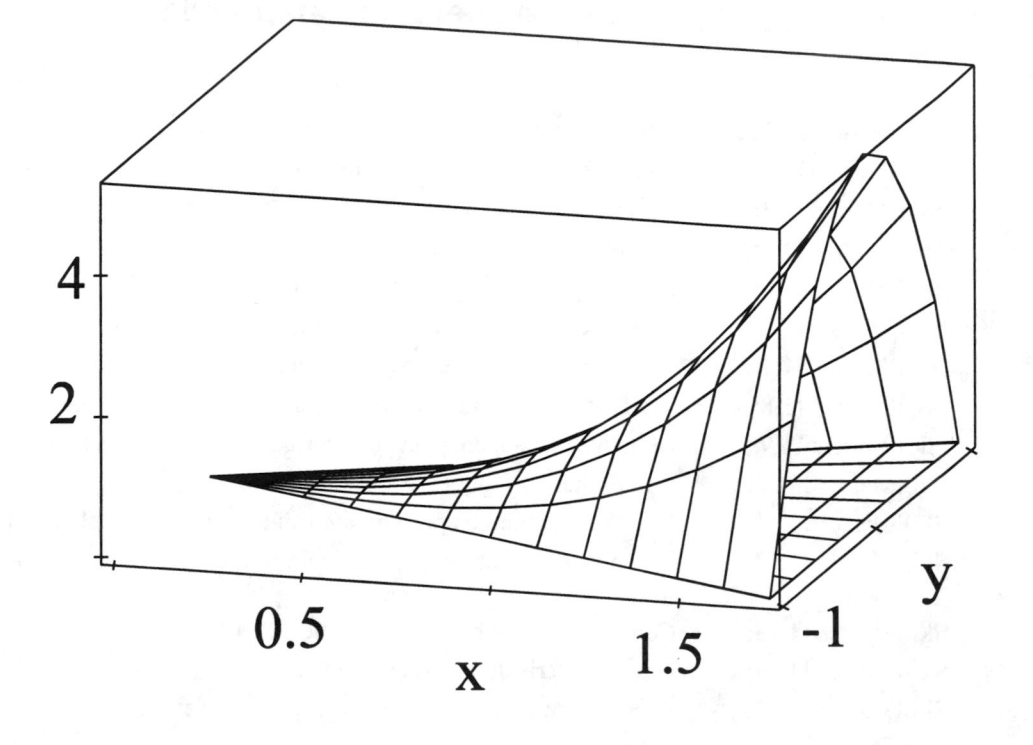

Section 12.1 – Double Integrals over Rectangles

Key Concepts:

- Volume, Double Integrals, and Double Riemann Sums
- Average Value of a Function
- Properties of Double Integrals

Skills to Master:

- Use double Riemann sums to approximate double integrals.
- Evaluate double integrals by computing volumes.
- Use the properties of double integrals to estimate or evaluate double integrals.

Discussion:

page 841-843.

This section introduces double integrals of functions $f(x, y)$ over rectangles in the plane. Just as single integrals were initially discussed as a means for computing the area under a curve, double integrals are initially discussed as a means for computing the volume under surfaces. Study *Figures 4, 5, and 8* to get a geometric picture of how double integrals are used to compute volume.

As you study multiple integrals, you will be able to use much of what you learned about single integrals.

Key Concept: Volume, Double Integrals, and Double Riemann Sums

If $f(x, y)$ is a function defined on a rectangle
$$R = [a, b] \times [c, d] = \{(x, y) \mid a \le x \le b, c \le y \le d\},$$

663

a double Riemann Sum of f over R is a sum of the form

$$\sum_{i=1}^{m} \sum_{j=1}^{n} f(x_{ij}^{*}, y_{ij}^{*}) \Delta A.$$

page 841.

Here x_{ij}^{*} is in the interval $[a + (i-1)\frac{(b-a)}{m}, a + i\frac{(b-a)}{m}]$, y_{ij}^{*} is in the interval $[c + (j-1)\frac{(d-c)}{c}, c + j\frac{(d-c)}{n}]$ and where $\Delta A = \frac{(b-a)(d-c)}{mn}$. Study *Figure 3* to get a better understanding of this.

If $f(x, y) \geq 0$, the double Riemann Sum is meant to be an approximation of the volume of the solid bounded above by the surface formed by $f(x, y)$ and bounded below by the rectangle R in the xy plane.

The double integral of f over the rectangle R is

$$\iint\limits_{R} f(x, y)\, dA = \lim_{m,n \to \infty} \sum_{i=1}^{m} \sum_{j=1}^{n} f(x_{ij}^{*}, y_{ij}^{*}) \Delta A$$

provided this limit exists. The double integral always exists for continuous functions. We define the volume to be equal to the double integral.

page 844.

Study the *Midpoint Rule* to see one method of *approximating* double integrals.

Key Concept: Average Value of a Function

In a manner similar to the way the average value was defined for functions of one variable, the average value of $f(x, y)$ over a rectangle R is defined to be

$$f_{ave} = \frac{1}{A(R)} \iint\limits_{R} f(x, y)\, dA$$

where $A(R)$ is the area of the rectangle. If $f(x, y) \geq 0$, the volume of the rectangular box with base R and height f_{ave} is the same as the volume of the solid that lies under the graph of f. See *Figure 11* for a geometric representation of this.

page 845.

Key Concept: Properties of Double Integrals

Double integrals have properties similar to those of single integrals.

- $$\iint\limits_{R} f\left[(x,y) + g(x,y)\right]\,dA = \iint\limits_{R} f(x,y)\,dA + \iint\limits_{R} g(x,y)\,dA$$

- $$\iint\limits_{R} c \cdot f(x,y)\,dA = c \iint\limits_{R} f(x,y)\,dA$$

- If $f(x,y) \geq g(x,y)$ for all (x,y) in R, then $\displaystyle\iint\limits_{R} f(x,y)\,dA \geq \iint\limits_{R} g(x,y)\,dA.$

SkillMaster 12.1: Use double Riemann sums to approximate double integrals.

---

page 843-844.

To use double Riemann sums to approximate double integrals, partition the rectangle R into smaller rectangles and choose points in each subrectangle to evaluate the function at. Study *Examples 1 and 3* to see how this process is carried out.

SkillMaster 12.2: Evaluate double integrals by computing volumes.

---

page 843.

If you know the volume under a surface given be $f(x,y)$ you can use your knowledge of this volume to compute the exact value of certain double integrals. Study Example 2 to see how this is done.

SkillMaster 12.3: Use the properties of double integrals to estimate or evaluate double integrals.

The properties of double integrals can be used to break up complicated double integrals into simpler ones that are easier to evaluate or approximate. See the worked examples for SkillMaster 12.3 below.

Worked Examples

For each of the following examples, first try to find the solution without looking at the middle or right columns. Cover the middle and right columns with a piece of paper. If you need a hint, uncover the middle column. If you need to see the worked solution, uncover the right column.

Example	Tip	Solution

SkillMaster12.1.

· Estimate the volume that lies below the surface $z = e^{2x-y}$ and above the rectangular region $R = [1,3] \times [1,4]$ Use a Riemann sum with $m = 4$ and $n = 3$, and take the sample points to be the upper right hand corner of each subrectangle.

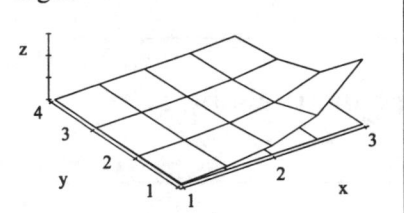

Subdivide the rectangle $[1,3] \times [1,4]$ by drawing vertical lines through $x = 1.5, 2,$ and 2.5 (the lines through $x = 1$ and $x = 3$) are already part of the boundary. Next draw horizontal lines through $y = 2$ and 3. There are 12 points at which to evaluate the function. The area of each subrectangle is $\Delta A = (2/4)(3/3) = 1/2$.

The approximation is ΔA times the sum

$$f(1.5,2) + f(2,2) + f(2.5,2) + f(3,2)$$
$$+f(1.5,3) + f(2,3) + f(2.5,3) + f(3,3)$$
$$+f(1.5,4) + f(2,4) + f(2.5,4) + f(3,4)$$

$$\begin{aligned} &= e^1 + e^2 + e^3 + e^4 \\ &\quad + e^0 + e^1 + e^2 + e^3 \\ &\quad + e^{-1} + e^0 + e^1 + e^2 \\ &= 127.46. \end{aligned}$$

The estimate of the integral is
$$(127.46)\Delta A$$
$$= (127.46)/2$$
$$= 63.73.$$

· Use the Midpoint Rule with $n = m = 2$ to find an approximation to the volume of the solid that lies above the rectangular region $R = [0, 2] \times [0, 2]$ and below the function $f(x, y) = 2 - x$.

Besides the boundary of the square region R draw a vertical line through $x = 1$ and an horizontal line through $y = 1$ to make the subrectangles. The midpoints of these subrectangles are $(0.5, 0.5)$, $(1.5, 0.5)$, $(0.5, 1.5)$, and $(1.5, 1.5)$. Also each subrectangle has area 1, so $\Delta A = 1$.

$$
\begin{aligned}
\int\int_R & 2 - x \, dA \\
\approx \ & \Delta A[(f(0.5, 0.5) + f(1.5, 0.5) \\
& + f(0.5, 1.5) + f(1.5, 1.5)] \\
= \ & [(2 - 0.5) + (2 - 1.5) \\
& + (2 - 0.5) + (2 - 1.5)] \\
= \ & 1.5 + 0.5 + 1.5 + 0.5 \\
= \ & 4
\end{aligned}
$$

· Use the Midpoint Rule with $n = m = 2$ to estimate the average temperature on a rectangular plate coordinatized as $R = [0, 0.5] \times [0, 0.5]$ with the temperature at a point given by $T(x, y) = \sin x \sin y$.

Complete the subrectangles by adding a vertical line through $x = 0.25$ and a horizontal line through $y = 0.25$. The midpoints of these subrectangles are $(0.125, 0.125)$, $(0.125, 0.375)$, $(0.375, 0.125)$, and $(0.375, 0.375)$. The area of each rectangle is $\Delta A = (0.25)(0.25) = 0.0625$.

$$
\begin{aligned}
\int\int_R & \sin x \sin y \, dA \\
\approx \ & \Delta A[\sin 0.125 \sin 0.125 \\
& + \sin 0.125 \sin 0.375 \\
& + \sin 0.375 \sin 0.125 \\
& + \sin 0.375 \sin 0.375] \\
\\
= \ & 0.0625[\sin 0.125 \\
& + \sin 0.375]^2 \\
= \ & 0.0625[0.1247 \\
& + 0.6249]^2 \\
= \ & 0.0139
\end{aligned}
$$

· Estimate the value of $\int \int f(x,y)\,dA$ using the Midpoint Rule with $m = n = 3$.
The region of integration is $[0,3] \times [0,3]$. The contour lines of the function f on this region are indicated below.

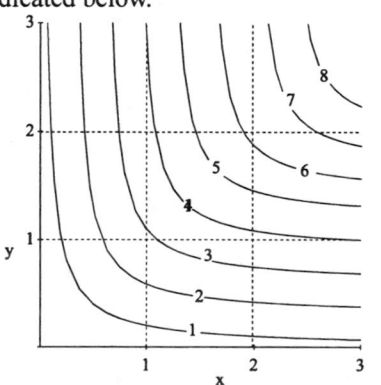

Estimate the value of the function at the midpoint of each rectangle.

Using $m = n = 3$, the region $[0,3] \times [0,3]$ is divided into 9 squares. Using the contour values of the function as a guide, the values of the function on the midpoints of the bottom three squares are about

1.2, 2.1, and 2.5.

The values of the function on the midpoints of the middle three squares are about

2.2, 4.4, and 5.5.

The values of the function on the midpoints of the top three squares are about

2.5, 5.5, and 7.8.

The area of each square is 1, so the Midpoint rule gives an approximation to $\int \int f(x,y)\,dA$ of

$$\Delta A(1.2 + 2.1 + 2.5 + 2.2 + 4.4 + 5.5 + 2.5 + 5.5 + 7.8)$$
$$= 1 \cdot 33.7 = 33.7.$$

SkillMaster 12.2.

· Find the integral $\int \int_R 2 - x\,dA$ where $R = [0,2] \times [0,2]$ by finding the volume of the solid region under the graph of $z = 2 - x$ and above the region R in the xy plane.

The function $z = 2 - x$ describes a plane through the points $(0,0,2)$, $(0,2,2)$, $(2,0,0)$, and $(2,2,0)$. The shape is a wedge which may be obtained by slicing the cube with side length 2 through one diagonal.

The volume of the wedge is 1/2 the volume of the cube of side length 2 or $(1/2)2^3 = 4$.

$$\int \int_R 2 - x\,dA = 4$$

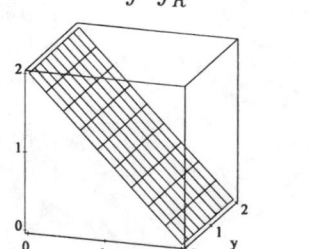

· Find the integral $\int\int_R y^2\, dA$ where $R = [0,3] \times [0,2]$ by finding the volume of the solid region under the graph of $z = y^2$ and above the region R in the xy plane.

The solid is a right cylinder with cross section given by $z = y^2$ for $0 \le y \le 2$. The volume of such a cylinder is the length times the area of the cross section.

The area of the cross section is

$$\int_0^2 y^2\, dy$$

$$= \left.\frac{y^3}{3}\right|_0^2$$

$$= \frac{2^3}{3} - \frac{0^3}{3}$$

$$= 8/3.$$

The volume is this area times the length or

$$(8/3)(3)$$

$$= 8$$

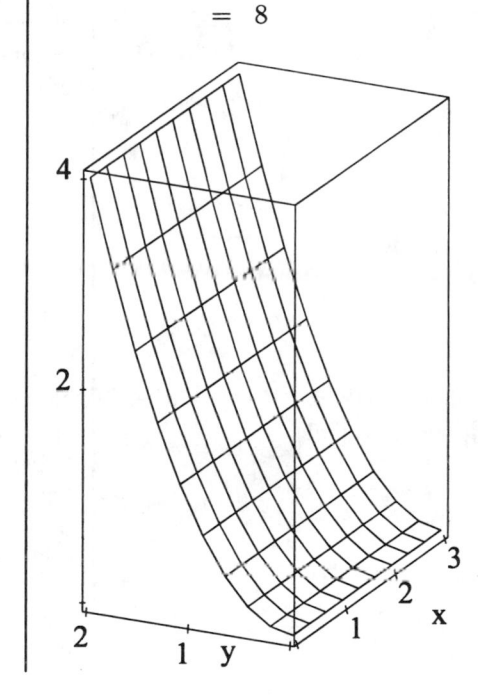

SkillMaster 12.3.

· Suppose that $\iint_R f(x,y)\,dA = 4$ and $\iint_R g(x,y)\,dA = 3$. Find $\iint_R 2f(x,y) - 3g(x,y)\,dA$.

Use the Properties of Double Integrals.

$$\iint_R 2f(x,y) - 3g(x,y)\,dA$$
$$= \iint_R 2f(x,y)\,dA - \iint_R 3g(x,y)\,dA$$
$$= 2\iint_R f(x,y)\,dA - 3\iint_R g(x,y)\,dA$$
$$= 2(4) - 3(3)$$
$$= -1$$

· Find $\iint_R 4 - y^2\,dA$ where $R = [0,3] \times [0,2]$.

Use the properties to express the double integral as the sum of two integrals.

$$\iint_R 4 - y^2\,dA$$
$$= \iint_R 4\,dA - \iint_R y^2\,dA$$
$$= \iint_R 4\,dA - 8$$
$$= 4(3)(2) - 8$$
$$= 24 - 8 = 6.$$

(The second integral has already been found to be equal to 8 in a worked out example above.)

· Show $\iint_R e^{-x-y}\,dA \le 1$ where $R = [0,1] \times [0,1]$.

Notice that e^{-u} gets smaller as u gets larger. This means that the largest value of e^{-x-y} for $0 \le x \le 1$ and $0 \le y \le 1$ occurs when $x = y = 0$ so that on this rectangle $e^{-x-y} \le e^0 \le 1$. Use Property 9 (page 847) of double integrals.

$$\iint_R e^{-x-y}\,dA$$
$$\le \iint_R 1\,dA$$
$$= \text{volume of a } 1 \times 1 \times 1 \text{ cube} = 1$$

· Let $f(x,y)$ be any continuous function and $R = [0,1] \times [0,2]$. Show that $\int \int_R \sin(f(x,y))\, dA \leq 2$.

Notice that regardless of what u is $|\sin u| \leq 1$ so for any x,y

$|\sin(f(x,y))| \leq 1$.

$$\int \int_R \sin(f(x,y))\, dA$$

$$\leq \int \int_R 1\, dA$$

$$= \text{volume of a } 1 \times 2 \times 1 \text{ cube}$$

$$= 2.$$

Section 12.2 – Iterated Integrals

Key Concepts:

- Iterated Integrals
- Fubini's Theorem

Skills to Master:

- Evaluate double integrals using iterated integrals.

Discussion:

This section gives a method for computing double integrals by computing certain single integrals that you already know. This allows you to use all the integration techniques that you already know. Study the examples in this section carefully.

Key Concept: Iterated Integrals

An iterated integral for a function $f(x, y)$ over a region $R = [a, b] \times [c, d]$ is an integral of the form

$$\int_a^b \left[\int_c^d f(x, y)\, dy \right]\, dx$$

where

$$\int_c^d f(x, y)\, dy$$

is evaluated by treating x as a constant. Note that to evaluate an iterated integral, you evaluate two ordinary integrals. Study *Example 1* to see how iterated integrals are evaluated.

page 850.

672

Key Concept: Fubini's Theorem

Fubini's theorem states that if f is continuous on the rectangle $R = [a, b] \times [c, d]$, then

$$\iint\limits_{R} f(x, y)\, dA = \int_a^b \left[\int_c^d f(x, y)\, dy \right]\, dx = \int_c^d \left[\int_a^b f(x, y)\, dy \right]\, dx.$$

page 851.

Note that in using Fubini's Theorem, you can either integrate with respect to y first, or with respect to x. Study *Figures 1 and 2* and the explanation that goes with these figures to get a better understanding of why Fubini's Theorem is true.

SkillMaster 12.4: Evaluate double integrals using iterated   integrals.

page 852-853.

To evaluate double integrals as iterated integrals, use Fubini's theorem to write the double integral as an iterated integral. Sometimes it may be easier to integrate first with respect to one variable that it is to integrate first with respect to the other variable. Study *Examples 3 and 4* to see how to do this. If the function $f(x, y)$ can be written as a product $g(x)h(y)$ of functions of a single variable, you can evaluate the integral

$$\iint\limits_{R} f(x, y)\, dA$$

as the product

$$\int_a^b g(x)\, dx \cdot \int_c^d h(y)\, dy$$

page 853.

See *Example 5* for a specific example of this.

Worked Examples

For each of the following examples, first try to find the solution without looking at the middle or right columns. Cover the middle and right columns with a piece of paper. If you need a hint, uncover the middle column. If you need to see the worked solution, uncover the right column.

Example	Tip	Solution

SkillMaster 9.4.

· Calculate the integral.

$$\int_0^1 \int_1^4 \frac{\sqrt{y}}{1+x^2}\, dy\, dx$$

Tip: Write the integral as a product of two single integrals.

Solution:

$$\int_0^1 \int_1^4 \frac{\sqrt{y}}{1+x^2}\, dy\, dx$$

$$= \left(\int_0^1 \frac{1}{1+x^2}\, dx \right) \left(\int_1^4 y^{1/2}\, dy \right)$$

$$= \left(\tan^{-1}(x)\Big|_0^1 \right) \left(\frac{2}{3} y^{3/2}\Big|_1^4 \right)$$

$$= (\pi/4)\frac{2}{3}\left((4^{3/2}) - (1^{3/2}) \right)$$

$$= (\pi/4)(14/3)$$

$$= 7\pi/6.$$

· Evaluate the iterated integral.

$$\int_{-1}^1 \int_0^2 (x^3 y^2 - 3x^2 y^2)\, dy\, dx$$

Tip: First use linearity of the iterated integral.

Solution:

$$= \int_{-1}^1 \int_0^2 x^3 y^2\, dy\, dx$$

$$-3 \int_{-1}^1 \int_0^2 x^2 y^2\, dy\, dx$$

Tip: Now express these as products of single integrals.

Solution:

$$= \left(\int_{-1}^1 x^3\, dx \right) \left(\int_0^2 y^2\, dy \right)$$

$$-3 \left(\int_{-1}^1 x^2\, dx \right) \left(\int_0^2 y^2\, dy \right)$$

The first integral with respect to x is equal to zero because x^3 is an odd function and the interval of integration is symmetric. The second integral with respect to x is equal to twice the integral from 0 to 2 by symmetry.

$$= (0)\left(\int_0^2 y^2\,dy\right)$$

$$-3\left(2\int_0^1 x^2\,dx\right)\left(\int_0^2 y^2\,dy\right)$$

$$= -6\left(\left.\frac{x^3}{3}\right|_0^1\right)\left(\left.\frac{y^3}{3}\right|_0^2\right)$$

$$= -6(1/3)(8/3)$$

$$= -16/3.$$

· Evaluate the double integral.

$$\int\int_R \frac{dA}{\sqrt{x+y}}$$

where $R = [4,9] \times [0,4]$

First express the double integral as an iterated integral. Then use the substitution $u = x + y$.

$$\int\int_R \frac{dA}{\sqrt{x+y}}$$

$$= \int_4^9 \int_0^4 \frac{1}{\sqrt{x+y}}\, dy\, dx$$

Use the substitution

$$u = x + y$$
$$du = dy$$
$$u = x \text{ when } y = 0$$
$$u = 4 + x \text{ when } y = 4$$

$$\int_4^9 \int_0^4 \frac{1}{\sqrt{x+y}}\, dy\, dx$$

$$= \int_4^9 \int_x^{4+x} \frac{1}{\sqrt{u}}\, du\, dx$$

$$= \int_4^9 \left(2u^{1/2}\Big|_x^{4+x}\right) dx$$

$$= 2\int_4^9 (4+x)^{1/2} - x^{1/2}\, dx$$

$$= 2\frac{2}{3}(4+x)^{3/2} - 2\frac{2}{3}x^{3/2}\Big|_4^9$$

$$= \frac{4}{3}13^{3/2} - \frac{4}{3}9^{3/2}$$

$$\quad -\frac{4}{3}8^{3/2} + \frac{4}{3}4^{3/2}$$

$$= \frac{4}{3}(13\sqrt{13} - 27 - 16\sqrt{2} + 8)$$

$$\approx 6.9930$$

· Evaluate the double integral.

$$\int\int_R xy\sec^2(x^2y)\,dA$$

where $R = [0,1] \times [0,\pi/4]$

This is a case in which the order of integration is important. Integrating with respect to y first is very difficult. (Try it!) Instead integrate with respect to x first.

$$\int\int_R xy\sec(x^2y)\,dA$$

$$= \int_0^{\pi/4}\left(\int_0^1 xy\sec(x^2y)\,dx\right)dy$$

Use the substitution

$$u = x^2y$$
$$du = 2xy\,dx$$
$$u = 0 \text{ when } x = 0$$
$$u = y \text{ when } x = 1$$

$$\int_0^{\pi/4}\left(\int_0^1 xy\sec^2(x^2y)\,dx\right)dy$$

$$= \int_0^{\pi/4}(1/2)\int_0^y\sec^2(u)\,du\,dy$$

$$= \frac{1}{2}\int_0^{\pi/4}\tan(u)|_0^y\,dy$$

$$= \frac{1}{2}\int_0^{\pi/4}\tan(y)\,dy$$

$$= \frac{1}{2}\ln(\sec y)|_0^{\pi/4}$$

$$= \frac{1}{2}(\ln(\sec(\pi/4)) - \ln(\sec(0)))$$

$$= \frac{1}{2}\ln(\sqrt{2})$$

$$= \frac{\ln 2}{4} \approx 0.1733$$

· Find the average value of the "monkey saddle" function
$$x^3 - 3xy^2$$
on the rectangle
$$R = [0, \sqrt{3}] \times [-1, 1]$$

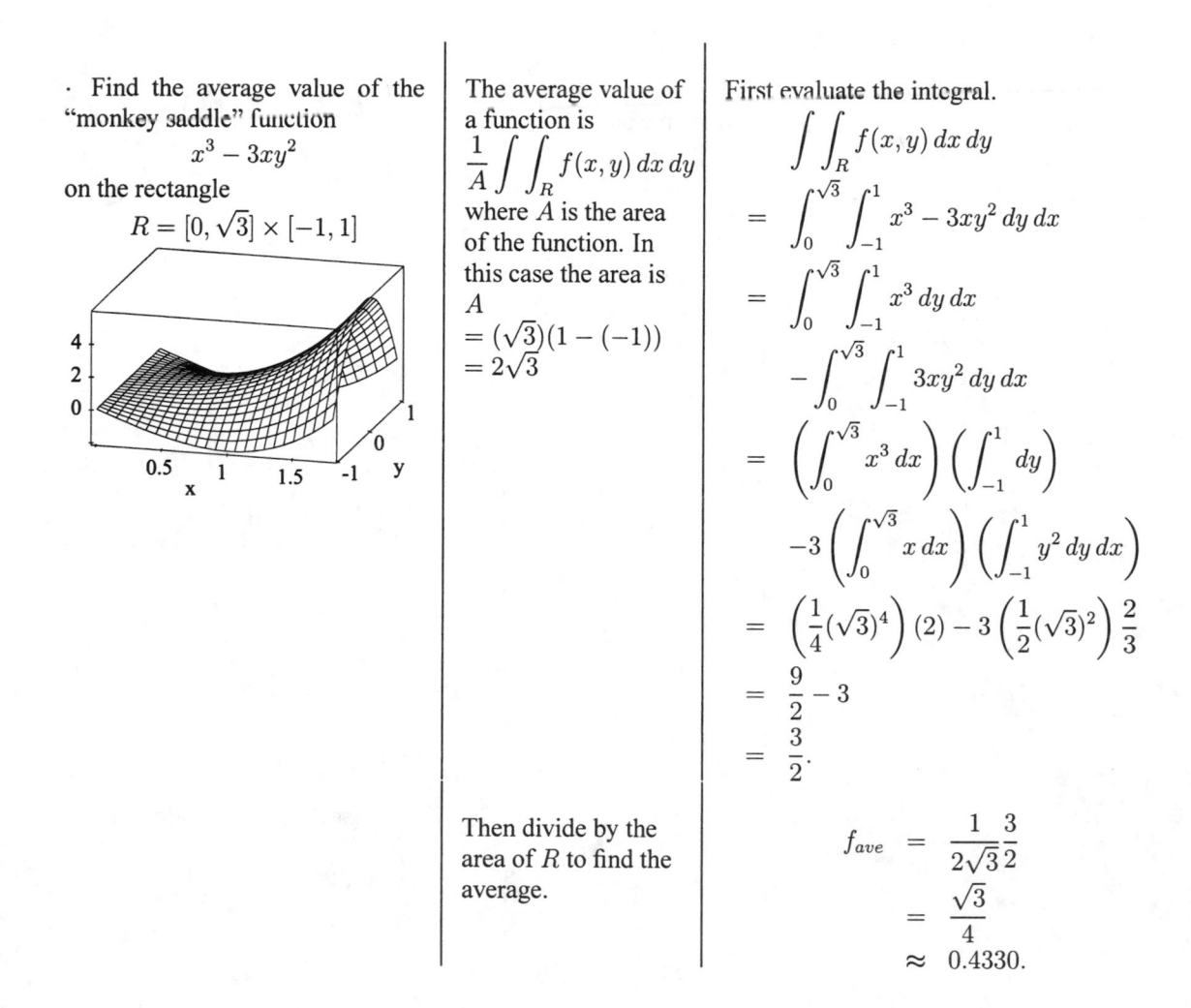

The average value of a function is
$$\frac{1}{A} \int \int_R f(x, y) \, dx \, dy$$
where A is the area of the function. In this case the area is
$$A$$
$$= (\sqrt{3})(1 - (-1))$$
$$= 2\sqrt{3}$$

Then divide by the area of R to find the average.

First evaluate the integral.
$$\int \int_R f(x, y) \, dx \, dy$$
$$= \int_0^{\sqrt{3}} \int_{-1}^1 x^3 - 3xy^2 \, dy \, dx$$
$$= \int_0^{\sqrt{3}} \int_{-1}^1 x^3 \, dy \, dx$$
$$- \int_0^{\sqrt{3}} \int_{-1}^1 3xy^2 \, dy \, dx$$
$$= \left(\int_0^{\sqrt{3}} x^3 \, dx \right) \left(\int_{-1}^1 dy \right)$$
$$- 3 \left(\int_0^{\sqrt{3}} x \, dx \right) \left(\int_{-1}^1 y^2 \, dy \, dx \right)$$
$$= \left(\frac{1}{4}(\sqrt{3})^4 \right) (2) - 3 \left(\frac{1}{2}(\sqrt{3})^2 \right) \frac{2}{3}$$
$$= \frac{9}{2} - 3$$
$$= \frac{3}{2}.$$

$$f_{ave} = \frac{1}{2\sqrt{3}} \frac{3}{2}$$
$$= \frac{\sqrt{3}}{4}$$
$$\approx 0.4330.$$

· Find the volume of the solid lying under the surface $z + x^2 + y^2 = 5$ and above the rectangle $R = [0,1] \times [0,2]$

First sketch the region so that the correct integral may be set up. In particular, it must be checked that the surface lies completely over the rectangle,

$$z = 5 - x^2 - y^2$$

is an inverted paraboloid with vertex at $(0,0,5)$. The intersection with the x, y–plane is a circle of radius $\sqrt{5}$ and center $(0,0)$,

$$x^2 + y^2 = 5.$$

The diagonal corner from the origin $(1,2)$ lies on this circle since $1^2 + 2^2 = 5$.

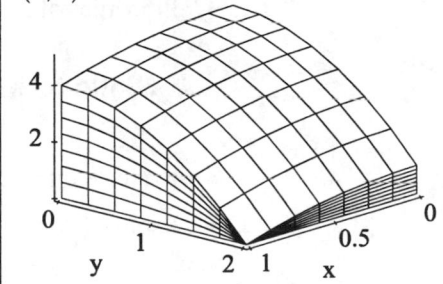

Now that it is established that the surface lies above the rectangle, it is clear that the volume is $\int\int_R 5 - x^2 - y^2 \, dA$.

$$\int\int_R 5 - x^2 - y^2 \, dA$$

$$= 5\int\int_R dA - \int\int_R x^2 \, dA - \int\int_R y^2 \, dA$$

$$= 5\int_0^1 \int_0^2 dy \, dx - \int_0^1 \int_0^2 x^2 \, dy \, dx$$

$$- \int_0^1 \int_0^2 y^2 \, dy \, dx$$

$$= 5(1-0)(2-0) - (2-0)\left(\int_0^1 x^2 \, dx\right)$$

$$- (1-0)\left(\int_0^2 y^2 \, dy\right)$$

$$= 10 - 2\left(\int_0^1 x^2 \, dx\right) - \left(\int_0^2 y^2 \, dy\right)$$

$$= 10 - \frac{2}{3} - \frac{2^3}{3}$$

$$= \frac{20}{3}.$$

Section 12.3 – Double Integrals over General Regions

Key Concepts:

- Double Integrals over General Regions
- Properties of Double Integrals

Skills to Master:

- Evaluate Integrals over General Regions.
- Estimate double integrals using the properties of double integrals.

Discussion:

The previous two sections considered double integrals over rectangular regions. In this section, you will learn how to work with and evaluate double integrals over more general regions in the plane. Make sure that you understand the geometric reasoning that goes along with the discussion in this section.

Key Concept: Double Integrals over General Regions

If D is a bounded region in the xy plane, and $f(x, y)$ is defined on D, we define the double integral of f over D

$$\iint\limits_{D} f(x, y)dA$$

by defining a new function $F(x, y)$ on a rectangle R containing D and setting

$$\iint\limits_{D} f(x, y)dA = \iint\limits_{R} F(x, y)dA.$$

page 855.

The new function is defined to be equal to f on D and is defined to be equal to 0 elsewhere. Study *Figures 1-4* to gain a geometric understanding of this definition.

680

Double integrals over more general regions D are most easily evaluated if D can be expressed as the region between the graphs of two functions of x (a type I region) or as the region between two functions of y (a type II regions). A type I region is a region of the form

$$D = \{(x, y) \mid a \leq x \leq b, g_1(x) \leq y \leq g_2(x)\}.$$

A type II region is a region of the form

$$D = \{(x, y) \mid c \leq y \leq d, h_1(y) \leq x \leq h_2(y)\}.$$

page 856.

Study *Figures 5, 6 and 7* for examples of these regions. SkillMaster 12.5 below discusses how to evaluate double integrals over type I or type II regions.

Key Concept: Properties of Double Integrals

page 847.

The three properties of double integrals over rectangles discussed in Section 12.1 carry over to double integrals over more general regions. Some new properties are the following:

- $$\iint_D f(x, y)dA = \iint_{D_1} f(x, y)dA + \iint_D f(x, y)dA \text{ if } D = D_1 \cup D_2 \text{ where } D_1$$

  and D_2 don't overlap except possible on their boundaries.

- $$\iint_D 1 dA \text{ is the area of } D.$$

- If $m \leq f(x, y) \leq M$ for all (x, y) in D, then

  $$m \cdot A(D) \leq \iint_D f(x, y)dA \leq M \cdot A(D)$$

page 861.

where $A(D)$ is the area of D. The first property above can sometimes be used to break up a region that is not of type I or type II into smaller regions that are of type I or type II. See *Figure 18* for a specific case where this occurs.

SkillMaster 12.5: Evaluate Integrals over General Regions.

To evaluate integrals over more general regions, try to express the region as a type I or a type II region. If the region is of type I and f is continuous, then

$$\iint\limits_D f(x,y)dA = \int_a^b \left[\int_{g_1(x)}^{g_2(x)} f(x,y)dy \right] dx.$$

Note that this is like the iterated integrals that you encountered earlier except that the limits of the inner integral are functions of x.

If the region is of type II and f is continuous, then

$$\iint\limits_D f(x,y)dA = \int_c^d \left[\int_{h_1(x)}^{h_2(x)} f(x,y)dx \right] dy.$$

Note that this is like the iterated integrals that you encountered earlier except that the limits of the inner integral are functions of y.

page 857-859.

Study *Examples 1-5* to see how this evaluation is carried out.

SkillMaster 12.6: Estimate double integrals using the properties of double integrals.

page 861.

If you can bound a function $f(x,y)$ between constant values m and M, then you can use the properties of double integrals to approximate the integral. Study *Example 6* to see how this works.

Worked Examples

For each of the following examples, first try to find the solution without looking at the middle or right columns. Cover the middle and right columns with a piece of paper. If you need a hint, uncover the middle column. If you need to see the worked solution, uncover the right column.

Example	Tip	Solution

SkillMaster 12.5.

· Evaluate the iterated integral. $\int_{-2}^{1}\int_{y-1}^{1-y^2} xy\,dx\,dy$. The region of integration is shown below.

Tip: First, integrate with respect to y so that x is treated as a constant. Then there will be a single integral with respect to y.

Solution:

$$\int_{-2}^{1}\int_{y-1}^{1-y^2} xy\,dx\,dy$$

$$= \int_{-2}^{1} y \int_{y-1}^{1-y^2} x\,dx\,dy$$

$$= \int_{-2}^{1} y \left(\frac{x^2}{2}\Big|_{y-1}^{1-y^2}\right) dy$$

$$= \frac{1}{2}\int_{-2}^{1} y((1-y^2)^2 - (y-1)^2)\,dy$$

$$= \frac{1}{2}\int_{-2}^{1} y(1 - 2y^2 + y^4 - y^2 + 2y - 1)\,dy$$

$$= \frac{1}{2}\int_{-2}^{1} (-3y^3 + y^5 + 2y^2)\,dy$$

$$= \frac{1}{2}\left(\frac{-3y^4}{4} + \frac{y^6}{6} + \frac{2y^3}{3}\right)$$

$$= (-\frac{3}{8} + \frac{1}{12} + \frac{2}{6}) - (-\frac{48}{8} + \frac{64}{12} + \frac{16}{6})$$

$$= \frac{47}{24}$$

· Evaluate the iterated integral. $\int_0^1 \int_1^{e^x} \frac{1}{y} \, dy \, dx$. The region of integration is shown below.

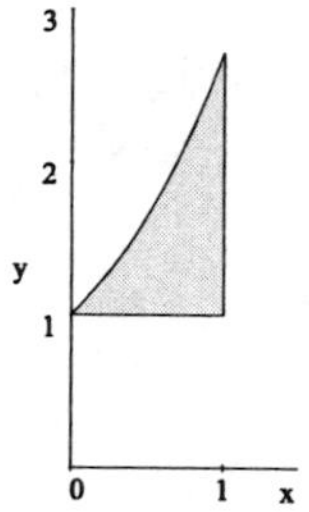

First evaluate the integral in the middle, keeping x constant, then integrate with respect to x.

$$\int_0^1 \int_1^{e^x} \frac{1}{y} \, dy \, dx$$

$$= \int_0^1 \left(\int_1^{e^x} \frac{1}{y} \, dy \right) dx$$

$$= \int_0^1 \ln y \Big|_1^{e^x} \, dx$$

$$= \int_0^1 \ln(e^x) - \ln 1 \, dx$$

$$= \int_0^1 x \, dx$$

$$= \frac{1}{2}$$

· Calculate the volume of the solid below the monkey saddle
$$z = x^3 - 3xy^2$$
and above the triangular region D in the $xy-$ plane bounded by the lines
$$y = \frac{1}{\sqrt{3}}x,$$
$$y = -\frac{1}{\sqrt{3}}x, \text{ and}$$
$$x = \sqrt{3}.$$

Look for the points of intersection of the monkey saddle and the $x, y-$plane.
$$0$$
$$= x^3 - 3xy^2$$
$$= x(x^2 - 3y^2)$$
$$= x(x - \sqrt{3}y) \cdot$$
$$(x + \sqrt{3}y).$$
Thus z is positive on the region D. D is most easily defined as a Type I region.
$$D = \{(x, y)|$$
$$0 \le x \le \sqrt{3},$$
$$-\sqrt{3} \le y \le \sqrt{3}\}$$

Volume

$$= \iint_D x^3 - 3xy^2 \, dA$$

$$= \int_0^{\sqrt{3}} \int_{-x/\sqrt{3}}^{x/\sqrt{3}} (x^3 - 3xy^2) \, dy \, dx$$

$$= 2 \int_0^{\sqrt{3}} \int_0^{x/\sqrt{3}} (x^3 - 3xy^2) \, dy \, dx$$

because the function (of y) is even over a symmetric interval.

$$= 2 \int_0^{\sqrt{3}} x^3 y - 3xy^3/3 \Big|_{-0}^{x/\sqrt{3}} \, dx$$

$$= 2 \int_0^{\sqrt{3}} x^3(x/\sqrt{3}) - 3x(x^3)/9\sqrt{3} \, dx$$

$$= 2 \int_0^{\sqrt{3}} \frac{x^4}{\sqrt{3}} - \frac{x^4}{3\sqrt{3}} \, dx$$

$$= \frac{4}{3\sqrt{3}} \int_0^{\sqrt{3}} x^4 \, dx$$

$$= \frac{4}{3\sqrt{3}} \frac{x^5}{5} \Big|_0^{\sqrt{3}}$$

$$= \frac{12}{5}$$

· Find the volume of the solid that lies above the x, y−plane, below the plane $z = x$, and inside the cylinder $x^2 + y^2 = 9$.

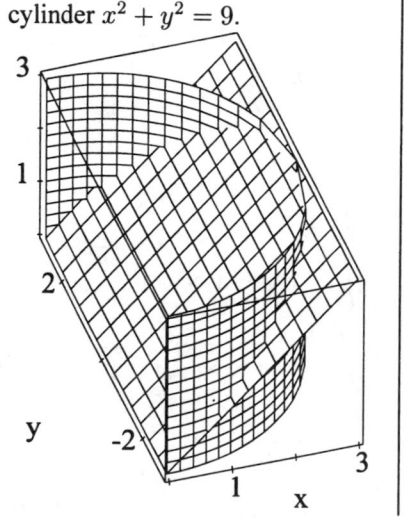

The wedge-shaped region is sketched above. The region D is the half disk which is the intersection of the disk $x^2 + y^2 \leq 9$ and the half plane $x \geq 0$. This is described as a Type II region $D = \{-3 \leq y \leq 3,\ 0 \leq x \leq \sqrt{9 - y^2}\}$.

The volume is

$$\int \int_D x \, dA$$

$$= \int_{-3}^{3} \int_{0}^{\sqrt{9-y^2}} x \, dx \, dy$$

$$= \int_{-3}^{3} \frac{x^2}{2} \Big|_{0}^{\sqrt{9-y^2}} dy$$

$$= \frac{1}{2} \int_{-3}^{3} 9 - y^2 \, dy$$

$$= \frac{1}{2} 2 \int_{0}^{3} 9 - y^2 \, dy$$

$$= 9y - y^3/3 \Big|_{0}^{3}$$

$$= 9(3) - (3)^3/3$$

$$= 18.$$

· Calculate the integral

$$\int_0^{\pi/4} \int_x^{\pi/4} \frac{\sin y}{y} \, dy \, dx$$

The integral is impossible to calculate as is. The order of integration must be reversed.

The region D is a triangle bounded by the lines

$$x = 0$$
$$y = \pi/4$$
$$y = x$$

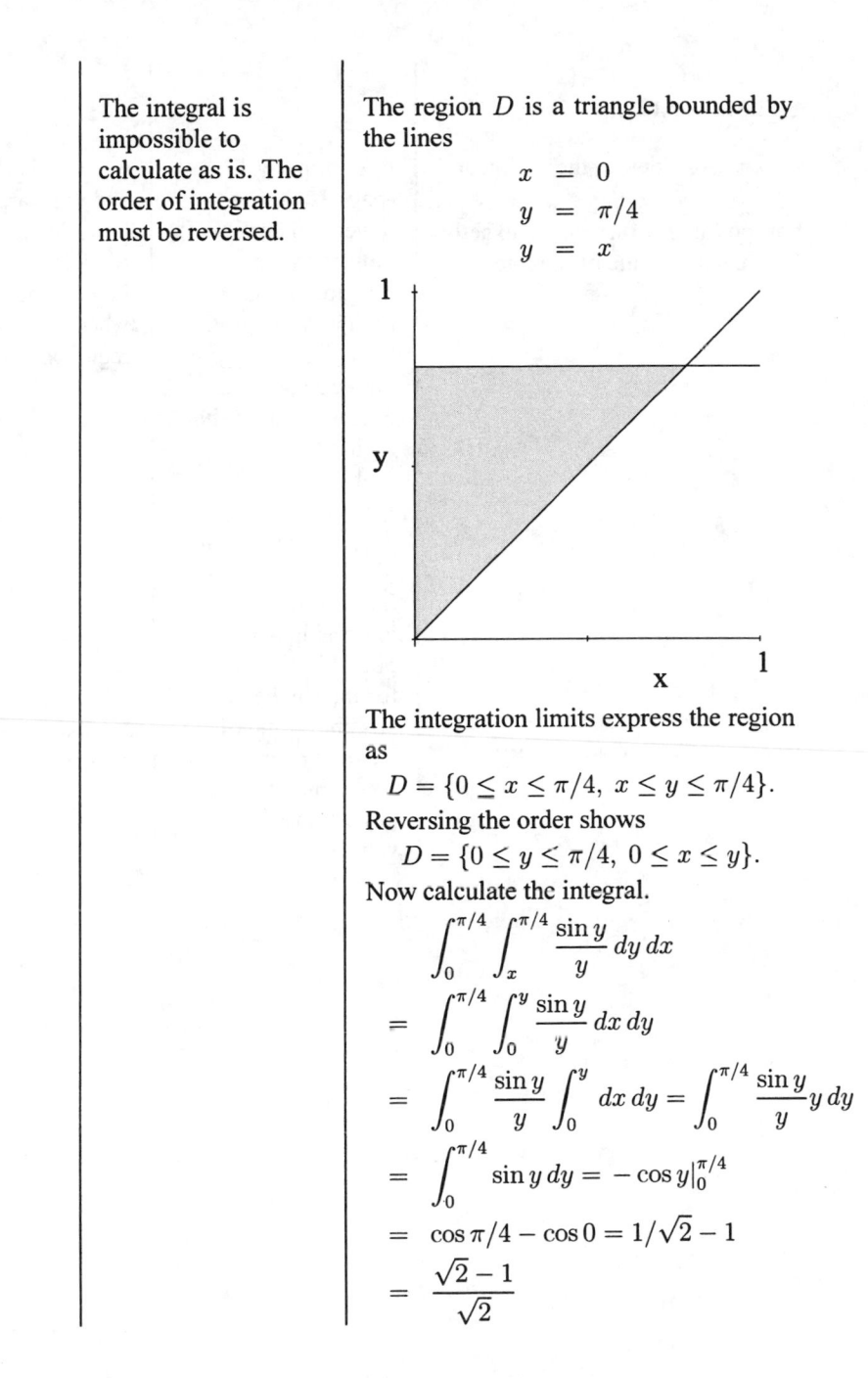

The integration limits express the region as

$$D = \{0 \le x \le \pi/4, \ x \le y \le \pi/4\}.$$

Reversing the order shows

$$D = \{0 \le y \le \pi/4, \ 0 \le x \le y\}.$$

Now calculate the integral.

$$\int_0^{\pi/4} \int_x^{\pi/4} \frac{\sin y}{y} \, dy \, dx$$

$$= \int_0^{\pi/4} \int_0^y \frac{\sin y}{y} \, dx \, dy$$

$$= \int_0^{\pi/4} \frac{\sin y}{y} \int_0^y dx \, dy = \int_0^{\pi/4} \frac{\sin y}{y} y \, dy$$

$$= \int_0^{\pi/4} \sin y \, dy = -\cos y \big|_0^{\pi/4}$$

$$= \cos \pi/4 - \cos 0 = 1/\sqrt{2} - 1$$

$$= \frac{\sqrt{2} - 1}{\sqrt{2}}$$

687

SkillMaster 12.6.

· Consider the integral $\int_0^1 \int_0^1 e^{x^2-y^2} \, dx \, dy$ which is not possible to evaluate exactly. Estimate the value of the integral.

Use Property 11 (page 861). Find lower and upper bounds to the integral on the region of integration. If $m \le e^{x^2-y^2} \le M$ on the region and if A is the area of the region then

$$mA \le \int_0^1 \int_0^1 e^{x^2-y^2} \, dx \, dy \le MA.$$

Since e^u is increasing a lower bound is obtained for the smallest possible value of u and an upper bound is obtained for the largest possible value of u.

$x^2 - y^2$ is smallest when x^2 is smallest and y^2 is largest. On the rectangle $[0, 1] \times [0, 1]$ this occurs when $x = 0$ and $y = 1$. Similarly, $x^2 - y^2$ is largest when x^2 is largest and y^2 is smallest. This occurs when $x = 1$ and $y = 0$. The area of the region is $A = 1$.

$$e^{x^2-y^2} \ge e^{0-1}$$
$$= e^{-1}$$
$$= m$$

$$e^{x^2-y^2} \le e^{1-0}$$
$$= e^1$$
$$= M$$

So

$$e^{-1} = m$$
$$\le \int_0^1 \int_0^1 e^{x^2-y^2} \, dx \, dy$$
$$\le e = M$$

Section 12.4 – Double Integrals in Polar Coordinates

Key Concepts:

- Polar coordinate Forms of Double Integrals

Skills to Master:

- Evaluate double integrals over polar regions using polar coordinates.

Discussion:

Sometimes a region D in the $xy-$ plane is more easily described by using polar coordinates than by using rectangular coordinates. This sections discusses how to evaluate double integrals using polar coordinates. If you need to review the basic facts about polar coordinates, now is the time to do so.

Key Concept: Polar coordinate Forms of Double Integrals

A polar rectangle R is a region in the xy plane given by
$$R = \{(r, \theta) \mid a \le r \le b, \alpha \le \theta \le \beta\}$$
where (r, θ) are polar coordinates.

If $f(x, y)$ is continuous on a polar rectangle R, and if $\beta - \alpha \le 2\pi$, then we you can evaluate the double integral of f over R by

$$\iint\limits_{R} f(x, y) dA = \int_{\alpha}^{\beta} \int_{a}^{b} f(r \cos \theta, r \sin \theta) r \, dr \, d\theta.$$

Make sure that you remember the additional factor of r in $r \, dr \, d\theta$. Study *Figures 3*

page 864.

and 4 and the discussion that goes with these figures to gain a better understanding of why double integrals over polar rectangles can be evaluated this way.

SkillMaster 12.7: Evaluate double integrals over polar regions using polar coordinates.

page 866-867.

To evaluate double integrals using polar coordinates, you need to first express the region that you are integrating over in polar form. Then use the formula from the previous Key Concept to evaluate the integral. Pay careful attention to the figures in this section to get a better geometric understanding of this process. Study *Examples 1-3* to see how the process is carried out.

Worked Examples

For each of the following examples, first try to find the solution without looking at the middle or right columns. Cover the middle and right columns with a piece of paper. If you need a hint, uncover the middle column. If you need to see the worked solution, uncover the right column.

Example	Tip	Solution

SkillMaster 12.7.

· Evaluate the double integral

$$\iint_R \frac{\ln(\sqrt{x^2 + y^2})}{x^2 + y^2} \, dA$$

where R is the region in the first quadrant between the unit circle $x^2 + y^2 = 1$ and the circle of radius e, $x^2 + y^2 = e^2$.

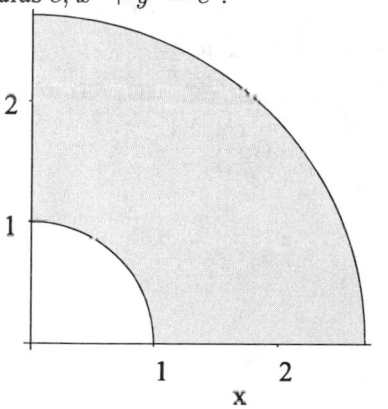

The region is shown. This is a polar rectangle, $R = \{0 \leq \theta \leq \pi/2, 1 \leq r \leq e\}$. When converting to polar coordinates, don't forget to add the extra factor r which appears in the differential $r \, dr \, d\theta$.

$$\iint_R \frac{\ln(\sqrt{x^2 + y^2})}{x^2 + y^2} \, dA$$

$$= \int_0^{\pi/2} \int_1^e \frac{\ln(r)}{r^2} r \, dr \, d\theta$$

$$= \left(\int_0^{\pi/2} d\theta \right) \left(\int_1^e \frac{\ln(r)}{r} \, dr \right)$$

$$= \frac{\pi}{2} \int_1^e \frac{\ln(r)}{r} \, dr$$

Use the substitution

$$u = \ln r$$
$$du = \frac{1}{r} \, dr$$
$$r = 1 \Rightarrow u = 0$$
$$r = e \Rightarrow u = 1.$$

The integral equals

$$\frac{\pi}{2} \int_0^1 u \, du$$

$$= \frac{\pi u^2}{4} \Big|_0^1$$

$$= \frac{\pi}{4}$$

· Evaluate the double integral

$$\int\int_R 6y\,dA$$

where R is the region in the first quadrant bounded above by the circle $(x-1)^2 + y^2 = 1$ and bounded below by the line $y = x$.

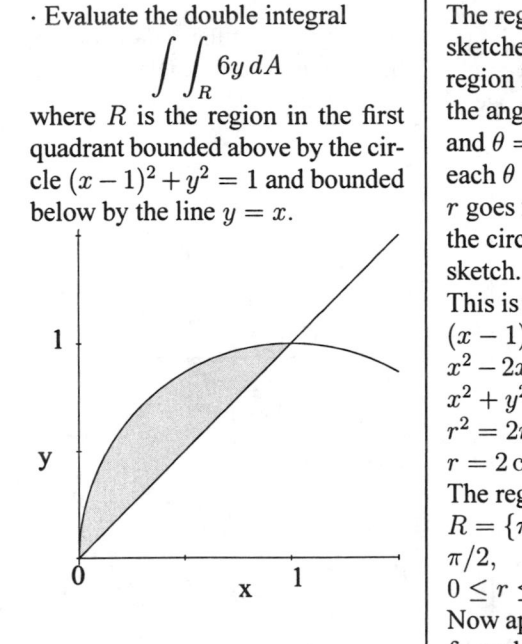

The region is sketched above. The region lies between the angles $\theta = \pi/4$ and $\theta = \pi/2$. For each θ in this range, r goes from $r = 0$ to the circle in the sketch.

This is equation is
$$(x-1)^2 + y^2 = 1$$
$$x^2 - 2x + 1 + y^2 = 1$$
$$x^2 + y^2 = 2x$$
$$r^2 = 2r\cos\theta$$
$$r = 2\cos\theta.$$

The region is
$$R = \{\pi/4 \le \theta \le \pi/2,$$
$$0 \le r \le 2\cos\theta\}.$$
Now apply the formula for integrating in polar coordinates.

$$\int\int_R 6y\,dA$$

$$= \int_{\pi/4}^{\pi/2} \int_0^{2\cos\theta} 6r\sin\theta\, r\,dr\,d\theta$$

$$= \int_{\pi/4}^{\pi/2} \sin\theta \int_0^{2\cos\theta} 6r^2\,dr\,d\theta$$

$$= \int_{\pi/4}^{\pi/2} \sin\theta \left(2r^3\big|_0^{2\cos\theta}\right)\,d\theta$$

$$= 16\int_{\pi/4}^{\pi/2} \sin\theta\cos^3\theta\,d\theta$$

Now make the substitution
$$u = \cos\theta$$
$$du = -\sin\theta\,d\theta$$
$$\theta = \pi/4 \Rightarrow u = 1/\sqrt{2}$$
$$\theta = \pi/2 \Rightarrow u = 0.$$

The integral is equal to

$$16\int_0^{1/\sqrt{2}} u^3\,du$$

$$= 16\frac{u^4}{4}\bigg|_0^{1/\sqrt{2}}$$

$$= 4(1/\sqrt{2})^4$$

$$= 1.$$

· Use polar coordinates to find the volume of the solid between the paraboloid
$$z = 3x^2 + 3y^2$$
and the circular cone
$$z = 6\sqrt{x^2 + y^2}.$$

First calculate the intersection of the two surfaces to find the region of integration. First convert to polar coordinates.
$$z = 3r^2$$
$$z = 6r$$
Then set these equal to find the intersection curve.
$$3r^2 = 6r$$
$$r^2 - 2r = 0$$
$$r(r - 2) = 0$$
This gives $r = 2$ as the intersection curve, that is a circle of radius two centered at the origin. The z value is $z = 12$. The region of integration is the disk of radius 2, $R = \{0 \le \theta \le 2\pi,\ 0 \le r \le 2\}$. The volume lies below the cone and above the paraboloid.
The part of the volume in the first octant is shown at the right.

The volume is
$$\int\int_R (6r - 3r^2)\, dA$$
$$= \int_0^{2\pi} \int_0^2 (6r - 3r^2) r\, dr\, d\theta$$
$$= 3\int_0^{2\pi} \int_0^2 (2r^2 - r^3)\, dr\, d\theta$$
$$= 3\left(\int_0^{2\pi} d\theta\right)\left(\int_0^2 (2r^2 - r^3)\, dr\right)$$
$$= 3(2\pi)\left(\frac{2r^3}{3} - \frac{r^4}{4}\right)\Big|_0^2$$
$$= 6\pi\left(\frac{16}{3} - \frac{16}{4}\right)$$
$$= 8\pi$$

· Evaluate the double integral

$$\int_0^1 \int_0^{\sqrt{1-x^2}} \frac{dy\,dx}{\sqrt{(x^2+y^2)}\tan^{-1}(y/x)}.$$

This integral is difficult to solve as it appears, so convert to polar coordinates. First, find the region of integration. In rectangular coordinates it is
$R = \{0 \le x \le 1,$
$0 \le y \le \sqrt{1-x^2}\}$
= the set of points in the first coordinate with
$x^2 + y^2 \le 1.$
This is the quarter unit circle.
$R = \{0 \le \theta \le$
$\pi/2,\ 0 \le r \le 1\}.$
In the switch to polar coordinates remember that
$\tan^{-1}(y/x) = \theta.$

The integral equals

$$\int_0^{\pi/2} \int_0^1 \frac{r\,dr\,d\theta}{\sqrt{r^2\theta}}$$

$$= \int_0^{\pi/2} \int_0^1 \frac{dr\,d\theta}{\sqrt{\theta}}$$

$$= \left(\int_0^{\pi/2} \frac{d\theta}{\sqrt{\theta}}\right)\left(\int_0^1 dr\right)$$

$$= \left(\int_0^{\pi/2} \theta^{-1/2}\,d\theta\right)(1)$$

$$= 2\theta^{1/2}\Big|_0^{\pi/2}$$

$$= 2\sqrt{\pi/2}$$

$$= \sqrt{2\pi}.$$

Section 12.5 – Applications of Double Integrals

Key Concepts:

- Density, Mass, and Centers of Mass
- Moments of Inertia
- Probability and Expected Values

Skills to Master:

- Use double integrals to compute mass.

- Compute moments of inertia and centers of mass.

- Compute probabilities and expected values.

Discussion:

This section introduces more applications of double integrals. You have already seen how to use double integrals to find volumes. You will now see how to use these integrals to find mass and centers of mass and moments of inertia. You will also learn about the use of double integrals in probability. Make sure that you work most of the problems for each application

Key Concept: Density and Mass

If you are given a thin plate or lamina lying over a region D in the xy plane, the density at a point (x, y), $\rho(x, y)$, is given by

$$\lim \frac{\Delta m}{\Delta A}$$

where Δm represents the mass of a small rectangular region of the plate containing (x, y), ΔA is the area of the region, and the limit is taken as the area of the

695

region goes to 0. If you know the density, the mass is given by

$$m = \iint_D \rho(x,y)dA.$$

page 870.

Study Figures 1 and 2 to gain a better understanding of density and mass.

Key Concept: Moments of Inertia and Centers of Mass

Consider a thin plate or lamina lying over a region D in the xy plane as in the previous Key Concept.

The moment of the plate about the $x-$ axis is defined to be

$$M_x = \iint_D y\rho(x,y)dA$$

and the moment of the plate about the $y-$ axis is defined to be

$$M_y = \iint_D x\rho(x,y)dA$$

The center of mass of the plate is at the point $(\overline{x}, \overline{y})$ given by

$$\overline{x} = \frac{1}{m}\iint_D x\rho(x,y)dA = \frac{M_y}{m} \text{ and }$$

$$\overline{y} = \frac{1}{m}\iint_D y\rho(x,y)dA = \frac{M_x}{m}.$$

page 871-872.

Study *Figures 4 and 5* to see illustrations of the center of mass.

The moments of inertia of the plate about the $x-$ axis and $y-$ axis respectively are defined to be

$$I_x = \iint_D y^2\rho(x,y)dA \text{ and } I_y = \iint_D y\rho(x,y)dA.$$

The moment of inertia about the origin, or polar moment of inertia is defined to be

$$I_o = \iint_D (x^2 + y^2)\rho(x,y)dA.$$

page 873.

Study *Example 4* to see computations of moments of inertia.

Key Concept: Probability and Expected Values

Review the material on probability density functions from Section 6.7 if you need to. If X and Y are continuous random variables, the *joint density function* of X and Y is a function $f(x, y)$ such that the probability that (X, Y) lies in a region D in the plane,

$$P\left((X,Y) \in D\right) = \iint_D f(x,y)dA.$$

The X mean and Y mean, or expected values of X and Y are defined to be

$$\mu_1 = \iint_{R^2} xf(x,y)dA \text{ and } \mu_2 = \iint_{R^2} yf(x,y)dA$$

where $\iint_{R^2} xf(x,y)dA$ is defined to be the limit of the double integral over ex-

panding squares that fill up the plane and where $\iint_{R^2} yf(x,y)dA$ is defined simi-

page 876-877.

larly. Study *Examples 6 and 7* for computations involving these ideas.

SkillMaster 12.8: Use double integrals to compute mass.

page 871.

To use double integrals to compute mass, first make sure that you understand the region of the plate or lamina and density that is involved. Then take the double integral of the density over the region. Study the first part of *Example 2* to see a computation involving mass.

SkillMaster 12.9: Compute moments of inertia and centers of mass.

To compute moments of inertia and centers of mass, first make sure that you understand the region associated with the thin plate or lamina under consideration. Then use the double integral formulas for

$$M_x, M_y, m, I_x, I_y, \text{ and } I_o$$

page 871-873.

presented in the Key Concept above. Study *Examples 2, 3 and 4* for details of these computations.

SkillMaster 12.10: Compute probabilities and expected values.

If X and Y are continuous random variables with a joint density function $f(x,y)$, use the formula

$$P\left((X,Y) \in D\right) = \iint\limits_{D} f(x,y)dA.$$

to compute the probability that $(X,Y) \in D$.

page 875-877.

Use the formulas for the expected values μ_1 and μ_2 of X and Y presented in the Key Concept above to compute expected values. Study *Examples 5, 6 and 7* to see computations about probability using double integrals.

Worked Examples

For each of the following examples, first try to find the solution without looking at the middle or right columns. Cover the middle and right columns with a piece of paper. If you need a hint, uncover the middle column. If you need to see the worked solution, uncover the right column.

Example	Tip	Solution

SkillMaster 12.8.

· A tissue sample is modeled geometrically so that it occupies the region bounded by the line $y = x + 2$ and by the parabola $y = x^2$. If the density of the tissue, modeled in this fashion, is equal to the square of the distance from the y-axis, find the mass of the tissue sample.

First sketch the region. Note that the curves intersect when
$x + 2 = x^2$
$x^2 - x - 2 = 0$
$(x - 2)(x + 1) = 0,$
$x = 2, \ x = -1.$

This is a type I region,
$D = \{-1 \le x \le 2,$
$x^2 \le y \le x + 2\}.$
The density function
$\rho(x, y) = x^2.$

The mass is the value of the integral,

$$m$$
$$= \int \int_D \rho(x, y) \, dA$$
$$= \int_{-1}^{2} \int_{x^2}^{x+2} x^2 \, dy \, dx$$
$$= \int_{-1}^{2} x^2 \left(\int_{x^2}^{x+2} dy \right) dx$$
$$= \int_{-1}^{2} x^2 (x + 2 - x^2) \, dx$$
$$= \int_{-1}^{2} x^3 + 2x^2 - x^4 \, dx$$
$$= \left. \frac{x^4}{4} + \frac{2x^3}{3} - \frac{x^5}{5} \right|_{-1}^{2}$$
$$= \frac{63}{20}.$$

· Find the center of mass of the tissue from the previous problem.

The center of mass is $(\overline{x}, \overline{y})$ where
$\overline{x} =$
$\frac{1}{m} \int \int_D x\rho(x, y)\, dA,$
$\overline{y} =$
$\frac{1}{m} \int \int_D y\rho(x, y)\, dA.$

$$\overline{x} = \frac{20}{63} \int \int_D x\rho(x, y)\, dA$$

$$= \frac{20}{63} \int_{-1}^{2} \int_{x^2}^{x+2} x^3 \, dy \, dx$$

$$= \frac{20}{63} \int_{-1}^{2} x^3 \int_{x^2}^{x+2} dy \, dx$$

$$= \frac{20}{63} \int_{-1}^{2} x^3 (x + 2 - x^2) \, dx$$

$$= \frac{20}{63} \int_{-1}^{2} (x^4 + 2x^3 - x^5) \, dx$$

$$= \frac{20}{63} \left(\frac{x^5}{5} + \frac{x^4}{2} - \frac{x^6}{6} \right) \Big|_{-1}^{2}$$

$$= \frac{20}{63} \left(\frac{32}{5} + 8 - \frac{32}{3} \right.$$
$$\left. + \frac{1}{5} - \frac{1}{2} + \frac{1}{6} \right) = \frac{8}{7} \approx 1.14$$

Similarly compute $\overline{y}$.

$$\overline{y} = \frac{20}{63} \int \int_D y\rho(x, y)\, dA$$

$$= \frac{20}{63} \int_{-1}^{2} \int_{x^2}^{x+2} yx^2 \, dy \, dx$$

$$= \frac{20}{63} \int_{-1}^{2} x^2 \left(\int_{x^2}^{x+2} y \, dy \right) dx$$

$$= \frac{20}{63} \int_{-1}^{2} x^2 (1/2) \left((x+2)^2 - x^4 \right) dx$$

$$= \frac{10}{63} \int_{-1}^{2} x^4 + 4x^3 + 4x^2 - x^6 \, dx$$

$$= \frac{10}{63} \left(\frac{x^5}{5} + x^4 + \frac{4x^3}{3} - \frac{x^7}{7} \right) \Big|_{-1}^{2}$$

$$= \frac{10}{63} \left(\frac{531}{35} \right) = \frac{118}{49} \approx 2.41$$

The center of mass is
$$\left(\frac{8}{7}, \frac{118}{49} \right).$$

· Electric charge is distributed over the annular plate $1 \leq x^2 + y^2 \leq 3$. The charge density (100 Coulombs/m^2) is given by $\sigma(x,y) = K/(x^2 + y^2)$. If the total charge is 100 Coulombs, what is the constant K?

The total charge is $100 = \int\int_D \sigma(x,y)\, dA$. Integrate and solve for K.

$$100$$
$$= \int\int_D \sigma(x,y)\, dA$$
$$= \int\int_D \frac{K}{x^2 + y^2}\, dA$$
$$= \int_0^{2\pi} \int_1^{\sqrt{3}} \frac{K}{r^2}(r\, dr\, d\theta)$$
$$= \left(\int_0^{2\pi} d\theta\right)\left(\int_1^{\sqrt{3}} \frac{K}{r}\, dr\right)$$
$$= 2\pi K \ln r|_1^{\sqrt{3}}$$
$$= 2\pi K \ln\sqrt{3}$$
$$= \pi K \ln 3$$

So
$$K = \frac{100}{\pi \ln 3}.$$

· For electrical charge as in the previous problem, how much charge is concentrated on the part of the plate shown.
The lines form angles of $\pi/4$ with the y axis.

The region is a polar rectangle
$R = \{\pi/4 \leq \theta \leq 3\pi/4,\ 1 \leq r \leq \sqrt{3}\}. : .7854 : 2.3562$
The total charge is $\int\int_R \sigma(x,y)\, dA$.

$$\int_{\pi/4}^{3\pi/4} \int_1^3 \frac{K}{r}\, dr\, d\theta$$
$$= \left(\int_{\pi/4}^{3\pi/4} d\theta\right)\left(\int_1^3 \frac{K}{r}\, dr\right)$$
$$= (\pi/2)\, K \ln 3$$
$$= \frac{1}{4}((2\pi)K \ln 3)$$
$$= \frac{1}{4}(\text{total charge})$$
$$= \frac{100}{4}$$
$$= 25.$$

701

SkillMaster 12.9.

· A lamina occupies a region $D = \{0 \le x \le 1, \; x^3 \le y \le x^{1/2}\}$ with a mass density $= 5x$ and total mass $= 1$. Find the center of mass.

The center of mass is $(\overline{x}, \overline{y})$ where

$\overline{x} = \frac{1}{m} \int \int_D x\rho(x,y) \, dA,$

$\overline{y} = \frac{1}{m} \int \int_D y\rho(x,y) \, dA.$

Now $m = 1$ and $\rho(x,y) = 5x$.

$$\begin{aligned}
\overline{x} &= \int_0^1 \int_{x^3}^{x^{1/2}} x(5x) \, dy \, dx \\
&= \int_0^1 (5x^2) \int_{x^3}^{x^{1/2}} dy \, dx \\
&= \int_0^1 (5x^2)(x^{1/2} - x^3) \, dx \\
&= 5 \int_0^1 (x^{5/2} - x^5) \, dx \\
&= 5\left(\frac{2}{7} - \frac{1}{6}\right) \\
&= \frac{25}{42} \approx 0.60
\end{aligned}$$

$$\begin{aligned}
\overline{y} &= \int_0^1 \int_{x^3}^{x^{1/2}} y(5x) \, dy \, dx \\
&= \int_0^1 (5x) \int_{x^3}^{x^{1/2}} y \, dy \, dx \\
&= \int_0^1 (5x) \left(\frac{y^2}{2}\Big|_{x^3}^{x^{1/2}}\right) dx \\
&= (5/2) \int_0^1 x(x - x^6) \, dx \\
&= (5/2) \int_0^1 (x^2 - x^7) \, dx \\
&= 5\left(\frac{1}{3} - \frac{1}{8}\right) \\
&= \frac{25}{48} \approx 0.52
\end{aligned}$$

The center of mass is
$$\left(\frac{25}{42}, \frac{25}{48}\right).$$

· Find the moments of inertial $I_x, I_y,$ and $I_0.$ for the lamina from the previous problem.

$I_x =$
$\int \int_D x^2 \rho(x,y)\, dA,$
$I_y =$
$\int \int_D y^2 \rho(x,y)\, dA,$
and $I_0 = I_x + I_y.$

$$I_x = \int \int_D x^2 \rho(x,y)\, dA$$

$$= \int_0^1 \int_{x^3}^{x^{1/2}} x^2(5x)\, dy\, dx$$

$$= 5 \int_0^1 x^3 \int_{x^3}^{x^{1/2}} dy\, dx$$

$$= 5 \int_0^1 x^3 (x^{1/2} - x^3)\, dx$$

$$= 5 \int_0^1 (x^{7/2} - x^6)\, dx$$

$$= 5\left(\frac{2}{9} - \frac{1}{7}\right) = \frac{25}{63} \approx 0.40.$$

Similarly

$$I_y = \int \int_D y^2 \rho(x,y)\, dA$$

$$= \int_0^1 \int_{x^3}^{x^{1/2}} y^2(5x)\, dy\, dx$$

$$= 5 \int_0^1 x \int_{x^3}^{x^{1/2}} y^2\, dy\, dx$$

$$= 5 \int_0^1 x \left(\frac{y^3}{3}\Big|_{x^3}^{x^{1/2}}\right) dx$$

$$= (5/3) \int_0^1 x(x^{3/2} - x^9)\, dx$$

$$- (5/3) \int_0^1 (x^{5/2} - x^{10})\, dx$$

$$= 5\left(\frac{2}{7} - \frac{1}{11}\right) = \frac{25}{77} \approx 0.32.$$

Finally

$$I_0 = I_x + I_y$$

$$= \frac{25}{63} + \frac{25}{77} = \frac{500}{693} \approx 0.72.$$

SkillMaster 12.10.

· The joint probability function for random variables X and Y is $f(x,y) = C(\frac{x}{y} + \frac{y}{x})$ for (x,y) in $[1,2] \times [2,8]$ and zero otherwise. Find the value of the constant C.

A joint probability must satisfy $\int\int f(x,y)\,dA = 1$. Use this to evaluate the constant.

$$1$$
$$= \int\int f(x,y)\,dA$$
$$= \int_1^2 \int_2^8 C(\frac{x}{y} + \frac{y}{x})\,dy\,dx$$

$$1/C$$
$$= \int_1^2 \int_2^8 (\frac{x}{y} + \frac{y}{x})\,dy\,dx$$
$$= \int_1^2 \int_2^8 \frac{x}{y}\,dy\,dx$$
$$+ \int_1^2 \int_2^8 \frac{y}{x}\,dy\,dx$$
$$= \left(\int_1^2 x\,dx\right)\left(\int_2^8 \frac{1}{y}\,dy\right)$$
$$+ \left(\int_1^2 \frac{1}{x}\,dx\right)\left(\int_2^8 y\,dy\right)$$
$$= (\frac{2^2}{2} - \frac{1}{2})(\ln 8 - \ln 2)$$
$$+ (\ln 2 - \ln 1)(\frac{8^2}{2} - \frac{2^2}{2})$$
$$= \frac{3}{2}\ln(8/2) + 30\ln 2$$
$$= 3\ln 2 + 30\ln 2$$
$$= 33\ln 2$$

Thus
$$C = \frac{1}{33\ln 2}.$$

· For the random variables from the previous problem, what is the probability that

$$X \leq 4/3 \text{ and } Y \geq 6?$$

This is the same as the probability that (X, Y) is in $[1, 4/3] \times [6, 8]$

$$= C \cdot$$
$$\int_1^{4/3} \int_6^8 f(x, y) \, dy \, dx$$

The probability is

$$= C \int_1^{4/3} \int_6^8 (\frac{x}{y} + \frac{y}{x}) \, dy \, dx$$

$$= C \left(\int_1^{4/3} x \, dx \right) \left(\int_6^8 \frac{1}{y} \, dy \right)$$

$$+ C \left(\int_1^{4/3} \frac{1}{x} \, dx \right) \left(\int_6^8 y \, dy \right)$$

$$= C(\frac{(4/3)^2}{2} - \frac{1}{2})(\ln 8 - \ln 6)$$

$$+ C(\ln(4/3) - \ln 1)(\frac{8^2}{2} - \frac{6^2}{2})$$

$$= C(\frac{7}{18}) \ln(8/6) + C(14) \ln(4/3)$$

$$= C(\frac{259}{18}) \ln(4/3)$$

$$= \frac{1}{33 \ln 2} \frac{259}{18} \ln(4/3)$$

$$= \frac{259}{594} \frac{\ln(4/3)}{\ln 2} \approx 0.18.$$

Section 12.6 – Surface Area

Key Concepts:

- Surface Area of Smooth Parametric Surfaces
- Surface Area of a Surface given as a Graph

Skills to Master:

- Compute the surface area of a parametric surface.
- Compute the surface area of a surface given as the graph of a function.

Discussion:

This section shows you how to compute the surface area of various surfaces using double integrals. Make sure that you study the geometric reasoning leading up to the formulas for surface area. The figures in this section are particularly helpful in developing this geometric understanding.

Key Concept: Surface Area of Smooth Parametric Surfaces

---

page 734.

Recall from *Section 10.5* that a parametric surface S is defined by a vector valued function of two variables

$$\mathbf{r}(u, v) = x(u, v)\mathbf{i} + y(u, v)\mathbf{j} + z(u, v)\mathbf{k}.$$

Each of the x, y, and z components of points on the surface are given by functions of the two variables u and v. The surface is smooth if the normal vector to the surface

$$\mathbf{r}_u \times \mathbf{r}_v$$

is never the $\mathbf{0}$ vector. Here $\mathbf{r}_u$ and $\mathbf{r}_v$ are the partial derivatives of $\mathbf{r}$ with respect to u and v respectively.

If S is a smooth surface given by $\mathbf{r}(u, v)$ for (u, v) in a region D, and if the surface is traced out just once as (u, v) range throughout D, then the surface area of S is given by

$$A(S) = \iint_D |\mathbf{r}_u \times \mathbf{r}_v| \, dA.$$

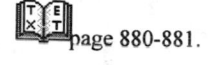page 880-881.

Study *Figures 2 and 3* and the explanation that goes with these figures to see how this formula is derived.

Key Concept: Surface Area of a Surface given as a Graph

If a surface S is given by an equation $z = f(x, y)$ the formula for the surface area of a parametric surface can be used to derive a formula for the surface area of S. The formula is

$$A(S) = \iint_D \sqrt{1 + \left(\frac{\partial z}{\partial x}\right)^2 + \left(\frac{\partial z}{\partial y}\right)^2} \, dA.$$

page 883.

Make sure that you understand how this formula is derived. Study *Figure 4* to see an example of such a surface.

SkillMaster 12.11: Compute the surface area of a parametric surface.

To compute the surface area of smooth parametric surfaces, first compute the partial derivatives $\mathbf{r}_u$ and $\mathbf{r}_v$. Next compute $|\mathbf{r}_u \times \mathbf{r}_v|$. Finally, use your results to compute

$$A(S) = \iint_D |\mathbf{r}_u \times \mathbf{r}_v| \, dA.$$

page 881.

It helps to be able to visualize what the surface looks like. Study *Example 1* to see a detailed computation of surface area for a parametric surface.

SkillMaster 12.12: Compute the surface area of a surface given as the graph of a function.

To compute the surface area of a surface given as the graph of a function $z = f(x, y)$, first compute the partial derivatives

$$\frac{\partial z}{\partial x} \text{ and } \frac{\partial z}{\partial y}.$$

Then use the results of these computations in the formula

$$A(S) = \iint_D \sqrt{1 + \left(\frac{\partial z}{\partial x}\right)^2 + \left(\frac{\partial z}{\partial y}\right)^2}\, dA.$$

page 882.

Study *Example 2* to see a detailed computation of surface area for this type of surface.

Worked Examples

For each of the following examples, first try to find the solution without looking at the middle or right columns. Cover the middle and right columns with a piece of paper. If you need a hint, uncover the middle column. If you need to see the worked solution, uncover the right column.

Example	Tip	Solution

SkillMaster 12.11.

· Find the surface area of that portion of the parametric surface
$$\mathbf{r}(u,v) = \left\langle \frac{e^{2u} - v^2}{2}, ve^u, \frac{e^{2u} + v^2}{2} \right\rangle$$
corresponding to the parameter rectangle
$\{0 \leq u \leq 1, \ 0 \leq v \leq 1\}.$

The surface area is
$$\iint dS$$
$$= \iint |\mathbf{r}_u \times \mathbf{r}_v| \, du \, dv$$

First calculate the tangent vectors, then the cross product to get the normal vector.

$\mathbf{r}_u$
$$= \left\langle e^{2u}, ve^u, e^{2u} \right\rangle$$

$\mathbf{r}_v$
$$= \left\langle -v, e^u, v \right\rangle$$

$\mathbf{r}_u \times \mathbf{r}_v$
$$= \begin{vmatrix} \mathbf{i} & \mathbf{j} & \mathbf{k} \\ e^{2u} & ve^u & e^{2u} \\ -v & e^u & v \end{vmatrix}$$

$$= e^u \begin{vmatrix} \mathbf{i} & \mathbf{j} & \mathbf{k} \\ e^u & v & e^u \\ -v & e^u & v \end{vmatrix}$$

$$e^u \left\langle v^2 - e^{2u}, -2ve^u, v^2 + e^{2u} \right\rangle$$

Now find the magnitude.

$$|\mathbf{r}_u \times \mathbf{r}_v|$$
$$= e^u((v^2 - e^{2u})^2 + (2ve^u)^2 + (v^2 + e^{2u})^2)^{1/2}$$
$$= e^u(v^4 - 2v^2e^{2u} + e^{4u} + 4v^2e^{2u} + v^4 + 2v^2e^{2u} + e^{4u})^{1/2}$$
$$= e^u(2v^4 + 4v^2e^{2u} + e^{4u})^{1/2}$$
$$= e^u\sqrt{2}(v^2 + e^{2u})$$

Evaluate the integral.

The surface area is

$$\iint dS$$

$$= \int_0^1 \int_0^1 \sqrt{2}e^u(v^2 + e^{2u}) \, dv \, du$$

$$= \int_0^1 \int_0^1 \sqrt{2}e^u v^2 \, dv \, du$$

$$+ \int_0^1 \int_0^1 \sqrt{2}e^{3u}) \, dv \, du$$

$$= \sqrt{2} \left(\int_0^1 e^u \, du \right) \left(\int_0^1 v^2 \, dv \right)$$

$$+\sqrt{2} \left(\int_0^1 e^{3u} \, du \right) \left(\int_0^1 dv \right)$$

$$= \sqrt{2}(e-1)(1/3)$$
$$+\sqrt{2}(e^3 - 1)(1/3)$$
$$= (\sqrt{2}/3)(e^3 + e - 2).$$

SkillMaster 12.12.

· Find the surface area of the portion of the hyperbolic paraboloid $z = f(x,y) = xy$ for (x,y) in the disk of radius 2 centered at the origin.

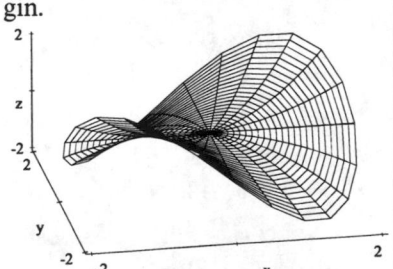

This surface is given as the graph of a function so it may be evaluated using
$$\iint dS$$
where dS is
$$\sqrt{1 + f_x^2 + f_y^2} \, dx \, dy.$$
The partial derivatives are
$fx = y$ and $f_y = x$.

The surface area is
$\iint_D \sqrt{1 + y^2 + x^2} \, dx \, dy.$

This is clearly a job for polar coordinates since the region is a circle and the integrand is a function of $x^2 + y^2$.

$$\int\int_D \sqrt{1 + y^2 + x^2}\, dx\, dy$$

$$= \int_0^{2\pi} \int_0^2 \sqrt{1 + r^2}\, r\, dr\, d\theta$$

$$= 2\pi \int_0^2 r\sqrt{1 + r^2}\, dr$$

Use the substitution

$$u = 1 + r^2$$
$$du = 2r\, dr$$
$$(1/2)\, du = r\, dr$$
$$r = 0 \Rightarrow u = 1$$
$$r = 2 \Rightarrow u = 5$$

The surface area equals

$$2\pi \int_1^5 \sqrt{u}(1/2)\, du$$

$$= \pi \int_1^5 u^{1/2}\, du$$

$$= \pi \left(\frac{2}{3} u^{3/2} \right) \Big|_1^5$$

$$= \frac{2\pi}{3}(5\sqrt{5} - 1).$$

· Consider a sphere of radius a that is intersected by two parallel planes of distance h apart. Show that the surface area of the sphere between these two planes is $2\pi ah$. This is a surprising fact because the surface area does not depend on where the planes hit the sphere, it only depends on how far apart the planes are.

Rotate the picture until the planes are parallel to the $y, z-$axis. The surface area in question may be described as the area of a surface of revolution. The curve is $\sqrt{a^2 - x^2}$. Set the intersection of the first plane with the x axis to be b and the intersection with the second plane to be $b + h$.

The formula for surfaces area is
$$2\pi \int_b^{b+h} f(x)\sqrt{1 + [f'(x)]^2}\, dx.$$

$$f(x) = \sqrt{a^2 - x^2}$$
$$f'(x) = \frac{-x}{\sqrt{a^2 - x^2}}$$
$$1 + [f'(x)]^2 = 1 + \frac{x^2}{a^2 - x^2}$$
$$= \frac{a^2}{a^2 - x^2}$$

The surface area is
$$2\pi \int_b^{b+h} \sqrt{a^2 - x^2}\frac{a}{\sqrt{a^2 - x^2}}\, dx$$
$$= 2\pi a \int_b^{b+h} dx$$
$$= 2\pi ah.$$

· Suppose D is a region in the $x, y-$plane and there is a surface over D is given by a graph $z = f(x, y)$. Show that the surface area S is greater than or equal to the area of D. Thus a soap bubble with a circular wire frame will lie in the plane of the wire frame since this minimizes the surface area.

Use the formula for the surface area of a surface given as a graph of a function. Use the fact that the square of a number is always nonnegative and use the properties of integrals.

The surface area is
$$\iint_D \sqrt{1 + (f_x)^2 + (f_y)^2}\, dx\, dy$$
$$\geq \iint_D \sqrt{1}\, dx\, dy$$
$$= \iint_D dx\, dy$$
$$= \text{area of } D.$$

712

Section 12.7 – Triple Integrals

Key Concepts:

- Triple Integrals
- Fubini's Theorem for Triple Integrals
- Applications of Triple Integrals

Skills to Master:

- Compute triple integrals
- Apply triple integrals to problems of volume, density, and mass.

Discussion:

This section generalizes the results about double integrals from the earlier sections in this chapter to triple integrals. Triple integrals are defined over three dimensional solid regions rather than over two dimensional regions. Since the ideas and definitions are similar to what you have already seen, the material is presented rather quickly. If you are have difficulty with any of the topics presented, look back at the corresponding material on double integrals.

Key Concept: Triple Integrals

Triple integrals are first defined for functions $f(x, y, z)$ defined on a rectangular box

$$\{B = (x, y, z) \mid a \le x \le b, c \le y \le d, r \le z \le s\}.$$

The box is divided into sub-boxes B_{ijk} and triple Riemann sums are defined as

713

sums of the form

$$\sum_{i=1}^{l}\sum_{j=1}^{m}\sum_{k=1}^{n} f(x_{ijk}^*, y_{ijk}^*, z_{ijk}^*)\Delta V$$

page 884.

where $(x_{ijk}^*, y_{ijk}^*, z_{ijk}^*)$ is in B_{ijk} and where ΔV is the volume of B_{ijk}. See Figure 1 for a geometric picture of this. The triple integral of f over the box B is then defined to be

$$\iiint_B f(x, y, z)dV = \lim_{l,m,n\to\infty} \sum_{i=1}^{l}\sum_{j=1}^{m}\sum_{k=1}^{n} f(x_{ijk}^*, y_{ijk}^*, z_{ijk}^*)\Delta V$$

provided this limit exists. For continuous functions, the limit always exists.

Triple integrals of $f(x, y, z)$ defined on bounded regions E in R^3 are defined by enclosing the region E in a box B and be defining a new function $F(x, y, z)$ that agrees with f on E and is 0 elsewhere. Then the triple integral is defined by

$$\iiint_E f(x, y, z)dV = \iiint_B F(x, y, z)dV.$$

Key Concept: Fubini's Theorem for Triple Integrals

Fubini's Theorem for triple integrals states that if f is continuous on a rectangular box $B = [a, b] \times [c, d] \times [r, s]$, then

$$\iiint_B f(x, y, z)dV = \int_r^s \left(\int_c^d \left(\int_a^b f(x, y, z)dx \right) dy \right) dz$$

where the inner integral is evaluated as a single integral by holding y and z fixed and the middle integral is evaluated as a single integral by holding z fixed. In fact, the order of integration can be done in and of the five additional ways below and the result remains the same.

$$\int_c^d \int_r^s \int_a^b dxdzdy \qquad \int_r^s \int_a^b \int_c^d dydxdz \qquad \int_a^b \int_r^s \int_c^d dydzdx$$

$$\int_c^d \int_a^b \int_r^s dzdxdy \qquad \int_a^b \int_c^d \int_r^s dzdydx$$

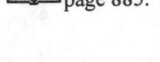
page 885.

See *Example 1* for a computation involving Fubini's Theorem.

A solid region E is of type I if $\{E = (x, y, z) \mid (x, y) \in D$ and $\phi_1(x, y) \le z \le \phi_2(x, y)\}$. That is, E lies between the graphs of ϕ_1 and ϕ_2.

page 886-887.

See *Figures 2, 3 and 4* for pictures of Type I solid regions. For $f(x, y, z)$ defined

on this type of region,

$$\iiint_E f(x,y,z)dV = \iint_D \left[\int_{\phi_1(y,z)}^{\phi_2(y,z)} f(x,y,z)dz \right] dA$$

where the inner integral is evaluated by holding y and z fixed and where the outer integral is evaluated as a double integral by one of the techniques that you already know.

page 887-888.

Solid regions of Types II and III are similarly defined as are integrals over these regions. See *Figures 7 and 8* and *Example 3* for details on triple integrals on these types of regions.

Key Concept: Applications of Triple Integrals

Just as with double integrals, we can consider various physical properties of a solid object filling a region E in three dimensional space. If the object has density $\rho(x,y,z)$ at (x,y,z) the mass is

$$m = \iiint_E \rho(x,y,z)dV.$$

The moments about the three coordinate planes are

$$M_{yz} = \iiint_E x\rho(x,y,z)dV \qquad M_{xz} = \iiint_E y\rho(x,y,z)dV$$

$$M_{xy} = \iiint_E z\rho(x,y,z)dV$$

The center of mass is at the point $(\overline{x}, \overline{y}, \overline{z})$ where

$$\overline{x} = \frac{M_{yz}}{m} \qquad \overline{y} = \frac{M_{xz}}{m} \qquad \overline{z} = \frac{M_{xy}}{m}.$$

The moments of inertia about the three coordinate axes are defined by

$$I_x = \iiint_E (y^2 + z^2)\rho(x,y,z)dV \qquad I_y = \iiint_E (x^2 + z^2)\rho(x,y,z)dV$$

$$I_z = \iiint_E (x^2 + y^2)\rho(x,y,z)dV.$$

page 891.

See *Figure 14 and Example 5* for a computation involving these quantities.

See the *discussion in the text* for applications of triple integrals to electric charge and probability.

page 890.

SkillMaster 12.13: Compute triple integrals

page 885.

To compute triple integrals of $f(x, y, z)$ over a region E, first make sure that you understand what the region looks like. If the region is a rectangular box, apply Fubini's theorem to evaluate the integral. You may need to change the order of integration to make the integral easier to evaluate. See *Example 1* for details on a triple integral evaluated over a rectangular box.

page 887-888.

If the region is not a rectangular box, see if you can express the region as a solid region of type I, II, or III. Then use the appropriate form of the triple integral to evaluate the integral. See *Examples 2 and 3* for details on triple integrals over these types of regions.

SkillMaster 12.14: Apply triple integrals to problems of volume, density, and mass.

page 891.

To find mass, center of mass or moments of inertia using triple integrals, you need to understand what the solid object under consideration looks like and what is being asked. After you understand the problem, use the appropriate formula from the Key Concept above to set up the integral. Then evaluate the integral by one of the techniques that you have learned. See *Example 5* for details on this type of problem.

Worked Examples

For each of the following examples, first try to find the solution without looking at the middle or right columns. Cover the middle and right columns with a piece of paper. If you need a hint, uncover the middle column. If you need to see the worked solution, uncover the right column.

Example	Tip	Solution

SkillMaster 12.13.

· Find the constant $k > 0$ so that
$$\int\int\int_E x^2 y^k \sin z \, dV = 1,$$
where
$$E = [-1, 1] \times [0, 1] \times [0, \pi].$$

Tip: The integral is over a product of rectangles and the integrand is separable so that this may be reduced to a product of one-variable integrals.

Solution:

$$\int\int\int_E x^2 y^k \sin z \, dV$$

$$= \int_{-1}^{1} \int_0^1 \int_0^\pi x^2 y^k \sin z \, dz \, dy \, dx$$

$$= \int_{-1}^{1} x^2 \, dx \int_0^1 y^k \, dy \int_0^\pi \sin z \, dz$$

$$= 2 \left(\int_0^1 x^2 \, dx \right) \left(\frac{1}{k+1} \right) \left(-\cos z \big|_0^\pi \right)$$

$$= 2 \left(\frac{1}{3} \right) \left(\frac{1}{k+1} \right) \left(-(-1-1) \right)$$

$$= \frac{4}{3(k+1)} = 1$$

So

$$k + 1 = \frac{4}{3}$$

$$k = \frac{4}{3} - 1$$

$$= \frac{1}{3}.$$

· Evaluate the integral
$$\int\int\int_{E}\sqrt{1-x}\,dV$$
where E is the solid tetrahedron in the first quadrant with vertices $(0,0,0)$, $(1,0,0)$, $(0,2,0)$, and $(0,0,3)$.

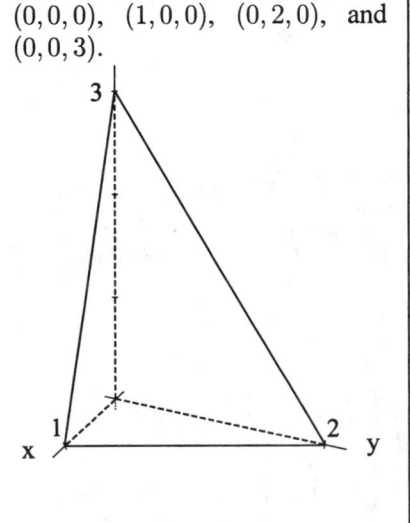

The upper plane that bounds this region passes through points $(1,0,0)$, $(0,2,0)$, and $(0,0,3)$. This plane has the equation $x + y/2 + z/3 = 1$. Thus this region lies over the region D in the $x,y-$plane. The region D may be written as
$$D = \{0 \le x \le 1, \\ 0 \le y \le 2(1-x)\}.$$

$$\int\int\int_{E}\sqrt{1-x}\,dV$$
$$= \int\int_{D}\int_{0}^{3(1-x-y/2)}\sqrt{1-x}\,dz\,dA$$
$$= \int\int_{D}\sqrt{1-x}\left(\int_{0}^{3(1-x-y/2)}dz\right)dA$$
$$= \int\int_{D}\sqrt{1-x}\,3(1-x-y/2)\,dA$$
$$= 3\int_{0}^{1}\sqrt{1-x}\int_{0}^{2(1-x)}(1-x-y/2)\,dy\,dx$$
$$= 3\int_{0}^{1}\sqrt{1-x}[(1-x)^2 - \\ (2(1-x))^2/4]\,dx$$
$$= 3\int_{0}^{1}\sqrt{1-x}\left[(1-x)^2\right]\,dx$$
$$= 3\int_{0}^{1}(1-x)^{5/2}\,dx$$
$$= 3\frac{-(1-x)^{7/2}}{7/2}\Big|_{0}^{1}$$
$$= \frac{6}{7}[-0-(-1)]$$
$$= \frac{6}{7}$$

· Evaluate the integral
$$\int \int \int_E x \, dV$$
where E is the solid tetrahedron described in the previous worked out example.

Notice that the calculation in the preceding example will work until
$$\int \int \int_E x \, dV$$
$$=$$
$$3 \int_0^1 x \left[(1-x)^2\right] dx$$

$$\int \int \int_E x \, dV$$
$$= 3 \int_0^1 x \left[(1-x)^2\right] dx$$
$$= 3 \int_0^1 x \left[1 - 2x + x^2\right] dx$$
$$= 3 \int_0^1 x - 2x^2 + x^3 \, dx$$
$$= 3 \left[\frac{1}{2} - \frac{2}{3} + \frac{1}{4}\right]$$
$$= 3(\frac{1}{12})$$
$$= \frac{1}{4}$$

SkillMaster 12.14.

· Find the average value of the function
$$f(x, y, x) = e^{ax + by + cz}$$
over the solid cube
$$E = [0, 2] \times [0, 2] \times [0, 2].$$

The average value of a function over a region is the integral of that function over the region divided by the volume of the region. In this case the region is a cube with volume $2^3 = 8$. Also use the properties of exponents to write the integrand as a product and separate the variables.

The average value is

$$\frac{1}{8} \int \int \int_E e^{ax + by + cz} \, dV$$
$$= \frac{1}{8} \int_0^2 \int_0^2 \int_0^2 e^{ax + by + cz} \, dz \, dy \, dx$$
$$= \frac{1}{8} \int_0^2 \int_0^2 \int_0^2 e^{ax} e^{by} e^{cz} \, dz \, dy \, dx$$
$$= \frac{1}{8} \int_0^2 e^{ax} \, dx \int_0^2 e^{by} \, dy \int_0^2 e^{cz} \, dz$$
$$= \frac{1}{8} \left(\frac{e^{ax}}{a}\Big|_0^2\right) \left(\frac{e^{by}}{b}\Big|_0^2\right) \left(\frac{e^{cz}}{c}\Big|_0^2\right)$$
$$= \frac{1}{8} \left(\frac{e^{2a} - 1}{a}\right) \left(\frac{e^{2b} - 1}{b}\right) \left(\frac{e^{2c} - 1}{c}\right).$$

· Find the mass of the solid E, which is bounded by the two cylinders

$$z = y^2,$$
$$z = 8 - y^2$$

and by the planes

$$x = 0,$$
$$x = 3/4,$$

if the density of E is

$$f(x, y, x) = x^2 z.$$

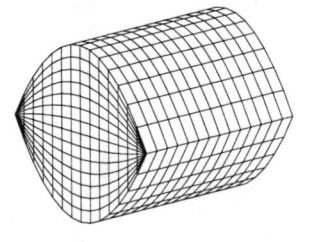

This cylinder is parallel to the x−axis. The cross-sectional region D in the y, z−plane is bounded by the curves $z = y^2$ and $z = 8 - y^2$. Thus the mass is given by the integral $\int \int \int_E x^2 z \, dV$
$=$
$\int \int_D \int_0^{3/4} x^2 z \, dx \, dA.$

$$\int \int \int_E x^2 z \, dV$$

$$= \int \int_D \int_0^{3/4} x^2 z \, dx \, dA$$

$$= \int \int_D z \left(\int_0^{3/4} x^2 \, dx \right) dA$$

$$= \int \int_D z \left. \frac{x^3}{3} \right|_0^{3/4} dA$$

$$= \frac{9}{64} \int \int_D z \, dA$$

$$= \frac{9}{64} \int_{-2}^2 \int_{y^2}^{8-y^2} z \, dz \, dy$$

$$= \frac{9}{64} \int_{-2}^2 \left. \frac{z^2}{2} \right|_{y^2}^{8-y^2} dy$$

$$= \frac{9}{128} \int_{-2}^2 (8 - y^2)^2 - y^4 \, dy$$

$$= \frac{9}{64} \int_0^2 (8 - y^2)^2 - y^4 \, dy$$

$$= \frac{9}{64} \int_0^2 64 - 16y^2 + y^4 - y^4 \, dy$$

$$= \frac{9}{64} \int_0^2 64 - 16y^2 \, dy$$

$$= \frac{9}{4} \int_0^2 4 - y^2 \, dy$$

$$= \frac{9}{4} \left. \left(4y - \frac{y^3}{3} \right) \right|_0^2$$

$$= \frac{9}{4} \left(4(2) - \frac{2^3}{3} \right)$$

$$= \frac{9}{4} (8 - \frac{8}{3})$$

$$= 12$$

· Find the center of mass of the tetrahedron E located in the first quadrant, with vertices $(0,0,0)$, $(1,0,0,)$, $(0,1,0)$, and $(0,0,1)$. The density of the mass is constant.

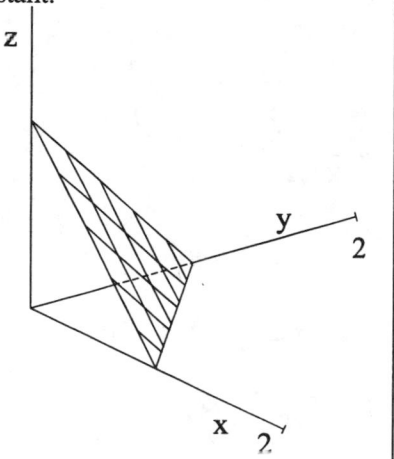

First find the volume of the region. The region of integration is

$E =$
$\{0 \le x \le 1,$
$0 \le y \le 1 - x,$
$0 \le z \le 1 - x - y\}$

The volume is the integral

$$\iiint_E dV$$

$$= \int_0^1 \int_0^{1-x} \int_0^{1-x-y} dz\, dy\, dx$$

$$= \int_0^1 \int_0^{1-x} (1 - x - y)\, dy\, dx$$

$$= \int_0^1 \left. (1-x)y - \frac{y^2}{2} \right|_0^{1-x} dx$$

$$= \int_0^1 (1-x)^2 - \frac{(1-x)^2}{2}\, dx$$

$$= \frac{1}{2} \int_0^1 (1-x)^2\, dx$$

$$= \frac{1}{2} \left. \left(\frac{-(1-x)^3}{3} \right) \right|_0^1$$

$$= \frac{1}{2} \left(\frac{0 - (-1)}{3} \right)$$

$$= \frac{1}{6}$$

Actually the cube of side length 1 may be dissected into 6 congruent copies of the tetrahedron. To find the $x-$coordinate of the center of mass compute $\frac{1}{vol(E)} \int \int \int_E x\, dV.$ Hint: notice that the volume integral computed above would give the same computation here until the step where the dx integration is carried out. Here there will be an extra factor of x.

$$\int \int \int_E x\, dV$$

$$= \int_0^1 x \int_0^{1-x} \int_0^{1-x-y} dz\, dy\, dx$$

$$= \frac{1}{2} \int_0^1 x(1-x)^2\, dx$$

$$= \frac{1}{2} \int_0^1 x - 2x^2 + x^3\, dx$$

$$= \frac{1}{2} \left(\frac{x^2}{2} - \frac{2x^3}{3} + \frac{x^4}{4} \right)\Big|_0^1$$

$$= \frac{1}{2}\left(\frac{1}{2} - \frac{2}{3} + \frac{1}{4} \right)$$

$$= \frac{1}{24}$$

The $x-$coordinate of the center of mass is

$$\overline{x} = (1/24)/(1/6)$$
$$= 1/4.$$

By symmetry this must also be the value of the y and z coordinates.

$$\overline{y} = 1/4$$
$$\overline{z} = 1/4$$

so then center of mass is

$$\left(\frac{1}{4}, \frac{1}{4}, \frac{1}{4} \right).$$

Section 12.8 – Triple Integrals in Cylindrical and Spherical Coordinates

Key Concepts:

- Triple Integrals in Cylindrical Coordinates
- Triple Integrals in Spherical Coordinates

Skills to Master:

- Compute triple integrals using cylindrical coordinates.
- Compute triple integrals using spherical coordinates.

Discussion:

page 692.

In this section, you will learn to compute triple integrals over regions that are best expressed in cylindrical or spherical coordinates. Now is the time to review *Section 9.7* if you need more practice on these types of coordinates. Just as certain double integrals were best evaluated using polar coordinates, certain triple integrals will be best evaluated using cylindrical or spherical coordinates. The picture below reviews cylindrical and spherical coordinates.

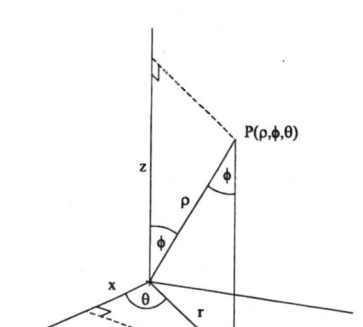

Key Concept: Triple Integrals in Cylindrical Coordinates

If $f(x, y, z)$ is continuous on a Type I solid region E of the form
$$\{E = (x, y, z) \mid (x, y) \in D \text{ and } \phi_1(x, y) \leq z \leq \phi_2(x, y)\}$$

where D can be given in polar coordinates
$$D = \{(r, \theta) \mid \alpha \leq \theta \leq \beta, h_1(\theta) \leq r \leq h_2(\theta)\},$$
then the triple integral of f can be evaluated using cylindrical coordinates:
$$\iiint_E f(x, y, z)dV = \int_\alpha^\beta \int_{h_1(\theta)}^{h_2(\theta)} \int_{\phi_1(r\cos\theta, r\sin\theta)}^{\phi_2(r\cos\theta, r\sin\theta)} f(r\cos\theta, r\sin\theta, z)r\,dz\,dr\,d\theta.$$

page 895-896.

Note that dV is replaced by $r\,dz\,dr\,d\theta$ just as dA was replaced by $r\,dr\,d\theta$ for double integrals with polar coordinates. See *Figures 2 and 3* for a geometric picture of this situation.

Key Concept: Triple Integrals in Spherical Coordinates

If $f(x, y, z)$ is continuous on a spherical wedge E of the form
$$E = \{(\rho, \theta, \phi) \mid a \leq \rho \leq b, \alpha \leq \theta \leq \beta, c \leq \phi \leq d\}$$
the triple integral of f over E can often be evaluated using spherical coordinates. In this case,
$$\iiint_E f(x, y, z)dV = \int_c^d \int_\alpha^\beta \int_a^b f(\rho\sin\phi\cos\theta, \rho\sin\phi\sin\theta, \rho\cos\phi)\rho^2\sin\phi\,d\rho\,d\theta\,d\phi.$$

page 896-897.

Study *Figures 7 and 8* to gain a geometric understanding of where this formula comes from. Note that dV is replaced by $\rho^2\sin\phi\,d\rho\,d\theta\,d\phi$.

SkillMaster 12.15: Compute triple integrals using cylindrical coordinates.

To compute triple integrals using cylindrical coordinates, first check the region can be expressed as in the Key Concept above. Change the triple integral to cylindrical coordinates and carry out the integration using the techniques that you already know. See *Examples 1 and 2* for details on how to do this.

page 895.

SkillMaster 12.16: Compute triple integrals using spherical coordinates.

To compute triple integrals using spherical coordinates, first check that the region can be expressed in spherical coordinates as in the Key Concept above. Then change the integral to spherical coordinates and carry out the integration. Remember to replace dV by $\rho^2 \sin \phi \, d\rho \, d\theta \, d\phi$. See *Examples 3 and 4* for details on how to do this.

page 8997-898.

Worked Examples

For each of the following examples, first try to find the solution without looking at the middle or right columns. Cover the middle and right columns with a piece of paper. If you need a hint, uncover the middle column. If you need to see the worked solution, uncover the right column.

Example	Tip	Solution

SkillMaster 12.15.

· Compute the triple integral

$$\int\int\int_E z^2\, dV$$

where E is the solid region enclosed by the two paraboloids
$$z = \pm(1 - x^2 - y^2).$$

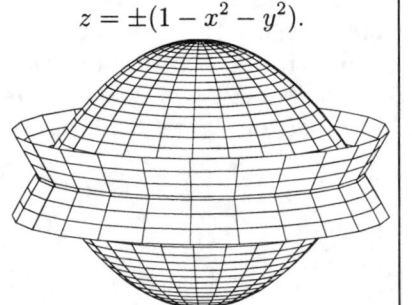

First compute the intersection curve of the two paraboloids to find the region of integrations.

$$z = 1 - x^2 - y^2$$
$$z = -(1 - x^2 - y^2)$$
Equate these two expressions,
$$1 - x^2 - y^2$$
$$= -(1 - x^2 - y^2)$$

$$2 = 2x^2 + 2y^2$$
$$1 = x^2 + y^2.$$
This is the unit circle in the $x, y-$plane. Notice that for (x, y) on the unit circle the $z-$coordinate is 0.

The region of integration is	The integral is
$0 \le \theta \le 2\pi$, $0 \le r \le 1$, $-(1 - x^2 - y^2) \le z \le (1 - x^2 - y^2)$, or $-(1 - r^2) \le z \le (1 - r^2)$.	$\displaystyle \int \int \int_E z^2 \, dV$

$$= \int_0^{2\pi} \int_0^1 \int_{-(1-r^2)}^{(1-r^2)} z^2 r \, dz \, dr \, d\theta$$

$$= 2 \int_0^{2\pi} \int_0^1 r \int_0^{1-r^2} z^2 \, dz \, dr \, d\theta$$

$$= 2 \int_0^{2\pi} \int_0^1 r \left. \frac{z^3}{3} \right|_0^{1-r^2} dr \, d\theta$$

$$= \frac{2}{3} \int_0^{2\pi} \int_0^1 r(1 - r^2)^3 \, dr \, d\theta$$

$$= \frac{2}{3} \int_0^{2\pi} \left. -(1 - r^2)^4/8 \right|_0^1 d\theta$$

$$= \frac{1}{12} \int_0^{2\pi} 0 - (-1) \, d\theta$$

$$= \frac{1}{12} \int_0^{2\pi} d\theta$$

$$= \frac{\pi}{6}.$$

· Compute

$$\int\int\int_E \frac{\sqrt{z}}{x^2+y^2}\,dV$$

where E is the solid region above the $xy-$plane, below the cone

$$z = \sqrt{x^2+y^2}$$

and inside the cylinder

$$x^2+y^2 = 9.$$

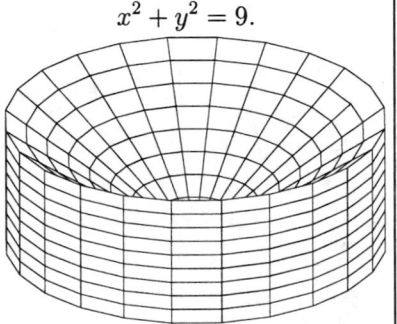

Convert this problem to cylindrical coordinates. The region of integration is

$$\{0 \le r \le 3, 0 \le z \le r, 0 \le \theta \le 2\pi\}.$$

$$\int\int\int_E \frac{\sqrt{z}}{x^2+y^2}\,dV$$

$$= \int\int\int_E \frac{\sqrt{z}}{r^2}\,dV$$

$$= \int_0^{2\pi}\int_0^3\int_0^r \frac{\sqrt{z}}{r^2}r\,dz\,dr\,d\theta$$

$$= \int_0^{2\pi}\int_0^3 \frac{1}{r}\int_0^r \sqrt{z}\,dz\,dr\,d\theta$$

$$= \int_0^{2\pi}\int_0^3 \frac{1}{r}\frac{2}{3}z^{3/2}\Big|_0^r\,dr\,d\theta$$

$$= \int_0^{2\pi}\int_0^3 \frac{1}{r}\frac{2}{3}r^{3/2}\,dr\,d\theta$$

$$= \frac{2}{3}\int_0^{2\pi}\int_0^3 r^{1/2}\,dr\,d\theta$$

$$= \frac{2}{3}\int_0^{2\pi} \frac{2}{3}r^{3/2}\Big|_0^3\,d\theta$$

$$= \frac{4}{9}\int_0^{2\pi} 3^{3/2}\,d\theta$$

$$= \frac{4}{\sqrt{3}}\int_0^{2\pi} d\theta$$

$$= \frac{8\pi}{\sqrt{3}}$$

· Find the volume of the infinite Gaussian ravioli which is defined to be the solid between the two surfaces

$$z = \pm e^{-(x^2+y^2)/2}.$$

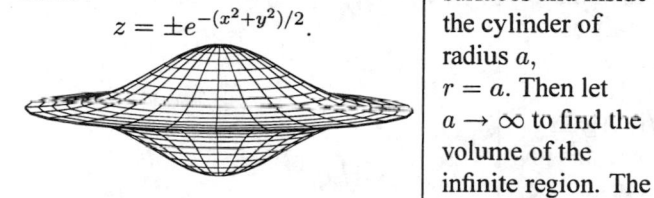

Find the volume of the solid region between the two surfaces and inside the cylinder of radius a, $r = a$. Then let $a \to \infty$ to find the volume of the infinite region. The bounding surfaces are easier to express in cylindrical coordinates

$$z = \pm e^{-r^2/2}$$

To find the volume take the limit as $a \to \infty$.

The region inside the cylinder of radius a is

$$\{0 \le \theta \le 2\pi, 0 \le r \le a, -e^{-r^2/2} \le z \le e^{-r^2/2}\}$$

The volume is

$$\int_0^{2\pi} \int_0^a \int_{-e^{-r^2/2}}^{e^{-r^2/2}} r \, dz \, dr \, d\theta$$

$$= 2 \int_0^{2\pi} \int_0^a \int_0^{e^{-r^2/2}} r \, dz \, dr \, d\theta$$

$$= 2 \int_0^{2\pi} \int_0^a e^{-r^2/2} r \, dr \, d\theta$$

$$= 2 \int_0^{2\pi} -e^{-r^2/2} \Big|_0^a \, dr \, d\theta$$

$$= 2 \int_0^{2\pi} 1 - e^{-a^2/2} \, d\theta$$

$$= 2 \left(1 - e^{-a^2/2}\right) \int_0^{2\pi} d\theta$$

$$= 4\pi \left(1 - e^{-a^2/2}\right).$$

$$\lim_{a \to \infty} 4\pi \left(1 - e^{-a^2/2}\right)$$
$$= \qquad 4\pi.$$

· Find the center of mass of the half infinite Gaussian ravioli which is the solid region above the xy−plane and below the surface

$$z = e^{-(x^2+y^2)/2}$$

Because of the symmetry of the region the x and y coordinates of the center of mass are both 0. The z coordinate is $\int \int \int z \, dV$. Again first compute this for the region inside the cylinder of radius a and then take the limit as $a \to \infty$.

$$\int \int \int z \, dV$$

$$= \int_0^{2\pi} \int_0^a \int_0^{e^{-r^2/2}} zr \, dz \, dr \, d\theta$$

$$= \int_0^{2\pi} \int_0^a r \frac{z^2}{2} \Big|_0^{e^{-r^2/2}} \, dr \, d\theta$$

$$= \frac{1}{2} \int_0^{2\pi} \int_0^a re^{-r^2} \, dr \, d\theta$$

$$= \frac{1}{2} \int_0^{2\pi} -e^{-r^2}/2 \Big|_0^a \, d\theta$$

$$= \frac{1}{4}(1 - e^{-a^2}) \int_0^{2\pi} d\theta$$

$$= \frac{\pi}{2}(1 - e^{-a^2})$$

The limit as $a \to \infty$ is

$$\lim_{a \to \infty} \frac{\pi}{2}(1 - e^{-a^2})$$

$$= \frac{\pi}{2}.$$

The z coordinate of the center of mass is this value divided by the volume,

$$\overline{z}$$

$$= (\frac{\pi}{2})/(4\pi)$$

$$= \frac{1}{8}.$$

The center of mass is

$$(\overline{x}, \overline{y}, \overline{z})$$

$$= (0, 0, \frac{1}{8}).$$

SkillMaster 12.16.

· Find the constant $K > 0$ so that
$$\int\int\int_E \frac{K}{1 + x^2 + y^2 + z^2} dV = 1$$
where E is the solid unit ball
$$E = \{x^2 + y^2 + z^2 \leq 1\}.$$

This is a natural for spherical coordinates since everything is a function of $\rho^2 = x^2 + y^2 + z^2$.

Evaluate the integral and set it equal to 1.
$$\int\int\int_E \frac{K}{1 + x^2 + y^2 + z^2} dV$$
$$= \int_0^{2\pi} \int_0^\pi \int_0^1 \frac{K}{1 + \rho^2}\rho^2 \sin\varphi \, d\rho \, d\varphi \, d\theta$$
$$= K\int_0^{2\pi} d\theta \int_0^\pi \sin\varphi \, d\varphi \int_0^1 \frac{\rho^2}{1 + \rho^2} d\rho$$
$$= K(2\pi)(-\cos\varphi|_0^\pi)\int_0^1 1 - \frac{1}{1 + \rho^2} d\rho$$
$$= 2K\pi(-(-1) + 1)\left(\rho - \tan^{-1}(\rho)|_0^1\right)$$
$$= 4\pi K(1 - \pi/4 + 0)$$
$$= K\pi(4 - \pi)$$
Setting this expression equal to 1 allows the value of K to be found.
$$1 = K\pi(4 - \pi)$$
$$K = \frac{1}{\pi(4 - \pi)} \approx 0.3708.$$

· Find the total mass of the solid between the two spheres
$$x^2 + y^2 + z^2 = a^2$$
and
$$x^2 + y^2 + z^2 = b^2$$
for $0 < a < b$, where the mass density function is
$$f(x, y, z) = (x^2 + y^2 + z^2)^{-3/2}.$$

The region of integration is the solid $a \leq \rho \leq b$ and the density function is $(\rho^2)^{-3/2} = \rho^{-3}$.

The total mass is
$$\int_0^{2\pi} \int_0^\pi \int_a^b \rho^{-3}\rho^2 \sin\varphi \, d\rho \, d\varphi \, d\theta$$
$$= \int_0^{2\pi} d\theta \int_0^\pi \sin\varphi \, d\varphi \int_a^b \rho^{-1} d\rho$$
$$= (2\pi)(-\cos\theta|_0^\pi)(\ln\rho|_a^b)$$
$$= 4\pi(\ln b - \ln a)$$
$$= 4\pi \ln(b/a).$$

· Find the volume of the snowcone, that is the solid region inside the sphere of radius a and within the cone that makes an angle α with the z−axis.

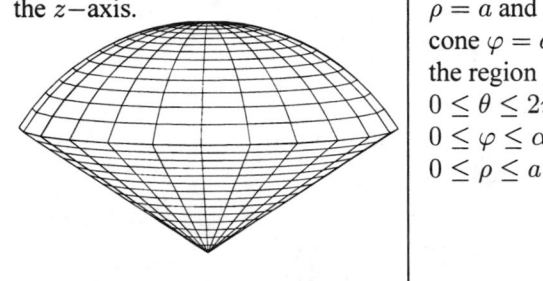

In spherical coordinates the region is the solid contained with $\rho = a$ and above the cone $\varphi = \alpha$, that is the region
$0 \leq \theta \leq 2\pi$,
$0 \leq \varphi \leq \alpha$, and
$0 \leq \rho \leq a$.

The volume is

$$\int_0^{2\pi} \int_0^{\alpha} \int_0^a dV$$

$$= \int_0^{2\pi} \int_0^{\alpha} \int_0^a \rho^2 \sin \varphi \, d\rho \, d\varphi \, d\theta$$

$$= \int_0^{2\pi} d\theta \int_0^{\alpha} \sin \varphi \, d\varphi \int_0^a \rho^2 \, d\rho$$

$$= (2\pi)(-\cos \varphi|_0^{\alpha})(\frac{1}{3}\rho^3|_0^a)$$

$$= (2\pi)(-\cos \alpha + 1)\frac{a^3}{3}$$

$$= \frac{2\pi a^3(1 - \cos \alpha)}{3}$$

Section 12.9 – Change of Variables in Multiple Integrals

Key Concepts:

- The Jacobian of a Transformation
- Change of Variables in Double Integrals
- Change of Variables in Triple Integrals

Skills to Master:

- Find the image of a set under a transformation.
- Compute the Jacobian of a transformations.
- Use change of variables to compute multiple integrals.

Discussion:

Using polar coordinates for double integrals and using cylindrical or spherical coordinates for triple integrals are examples of using a change of variables to simplify an integral. In this section, you will learn the general form of using a change of variables to simplify double and triple integrals.

Key Concept: The Jacobian of a Transformation

A transformation from a region in the uv–plane to the xy–plane, $T(u,v) = (x,y)$ is given by equations of the form

$$x = g(u,v), y = h(u,v) \text{ or}$$
$$x = x(u,v), y = y(u,v).$$

The transformation is C^1 if the first order partial derivatives of g and h are con-

733

tinuous.

The Jacobian of the transformation T is denoted
$$\frac{\partial(x,y)}{\partial(u,v)}$$
and is given by
$$\frac{\partial(x,y)}{\partial(u,v)} = \begin{vmatrix} \frac{\partial x}{\partial u} & \frac{\partial x}{\partial v} \\ \frac{\partial y}{\partial u} & \frac{\partial y}{\partial v} \end{vmatrix} = \frac{\partial x}{\partial u}\frac{\partial y}{\partial v} - \frac{\partial x}{\partial v}\frac{\partial y}{\partial u}.$$

page 904. The Jacobian is a measure of how the transformation T changes area. See *Figures 3 and 4* for a geometric picture of this.

A transformation from a region in $uvw-$space to $xyz-$space, $S(u,v,w) = (x,y,z)$ is given by equations of the form
$$x = g(u,v,w), y = h(u,v,w), z = k(u,v,w)$$
The transformation is C^1 if the first order partial derivatives of g and h and k are continuous.

The Jacobian of the transformation S is denoted
$$\frac{\partial(x,y,z)}{\partial(u,v,w)}$$
and is given by
$$\frac{\partial(x,y,z)}{\partial(u,v,w)} = \begin{vmatrix} \frac{\partial x}{\partial u} & \frac{\partial x}{\partial v} & \frac{\partial x}{\partial w} \\ \frac{\partial y}{\partial u} & \frac{\partial y}{\partial v} & \frac{\partial y}{\partial w} \\ \frac{\partial z}{\partial u} & \frac{\partial z}{\partial v} & \frac{\partial z}{\partial w} \end{vmatrix}.$$
To evaluate this Jacobian, you evaluate the 3×3 determinant.

Key Concept: Change of Variables in Double Integrals

Suppose that T is a one-to-one C^1 transformation whose Jacobian is nonzero, and that T maps a region S in the $uv-$plane to a region R in the $xy-$plane. If f is continuous on R and of R and S are Type I or II plane regions, then
$$\iint_R f(x,y)dA = \iint_S f(x(u,v), y(u,v)) \left| \frac{\partial(x,y)}{\partial(u,v)} \right| dudv.$$

You should think of this as replacing $dA = dxdy$ by $\left| \frac{\partial(x,y)}{\partial(u,v)} \right| dudv$. The integral on the left is in x and y coordinates and the integral on the right is in u and v

page 906-907.

coordinates. See *Examples 2 and 3* for details on how to carry out this kind of change of variables.

Key Concept: Change of Variables in Triple Integrals

Suppose that T is a one-to-one C^1 transformation whose Jacobian is nonzero, and that T maps a region S in the $uvw-$space to a region R in $xyz-$space. If f is continuous on R and of R and S are Type I, II, or III solid regions, then

$$\iiint_R f(x,y,z)dV = \iiint_S f(x(u,v,w),y(u,v,w),z(u,v,w))\left|\frac{\partial(x,y,z)}{\partial(u,v,w)}\right| dudvdw.$$

You should think of this as replacing $dV = dxdydz$ by $\left|\frac{\partial(x,y,z)}{\partial(u,v,w)}\right| dudvdw$. The integral on the left is in x, y and z coordinates and the integral on the right is in u, v and w coordinates. See *Example 4* for details on how to carry out this kind of change of variables.

page 908.

SkillMaster 12.17: Find the image of a set under a transformation.

page 903.

To find the image of a set under a transformation, find the images of the boundaries of the set. See *Example 1* for details on how to do this. The worked examples for this SkillMaster provide additional practice.

SkillMaster 12.18: Compute Jacobians of transformations.

page 906.

To compute the Jacobians of transformations, compute the partial derivatives that you need to. Then substitute these partial derivatives into the appropriate determinant form the Key Concept above. The discussion in the text that goes with *Figure 7* shows how to do this.

SkillMaster 12.19: Use change of variables to compute multiple integrals.

To use the change of variables formulas to compute multiple integrals, first make sure that you understand what the transformation from (u, v) to (x, y), or form (u, v, w) to (x, y, z) coordinates is. Then make sure that you understand how the transformation changes the region under consideration. Next, compute the appropriate Jacobian. Finally, use the appropriate change of variables formula from the Key Concepts above and evaluate the resulting integrals. Study the examples in this section to see details on how to do this.

Worked Examples

For each of the following examples, first try to find the solution without looking at the middle or right columns. Cover the middle and right columns with a piece of paper. If you need a hint, uncover the middle column. If you need to see the worked solution, uncover the right column.

Example	Tip	Solution

SkillMaster 12.17.

· Fix an angle α and define a transformation T by
$$x = (\cos \alpha) u - (\sin \alpha)v + 2$$
$$y = (\sin \alpha)u + (\cos \alpha)v + 3.$$
Find the image of the unit square $0 \le u \le 1, \ 0 \le v \le 1$ in the xy−plane.

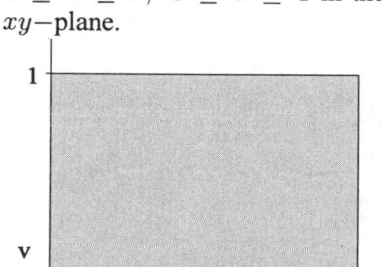

First find the images of the corners, then the sides. Notice that the transformation is a rigid motion. Specifically, this transformation rotates (u, v) by an angle α and then translates the result by adding $(2, 3)$.

The line segment on the u−axis may be parameterized by $\{(t, 0) : 0 \le t \le 1\}$. Plugging this into the equations for T one gets the straight line segment
$$x = (\cos \alpha)t + 2 \quad y = (\sin \alpha)t + 3$$
$$0 \le t \le 1.$$
Similarly, the line segment on the v−axis goes to
$$x = -(\sin \alpha)t + 2 \quad y = (\cos \alpha)t + 3$$
$$0 \le t \le 1.$$
The line segment parallel to the u−axis may be parameterized by $\{(t, 1) : 0 \le t \le 1\}$ giving the parametric line segment in the x, y−plane
$$x = (\cos \alpha)t - \sin \alpha + 2$$
$$y = (\sin \alpha)t + \cos \alpha + 3$$
$$0 \le t \le 1.$$
The final result is a square in the xy−plane whose lower left corner is at $(2, 3)$ and rotated by an angle α.

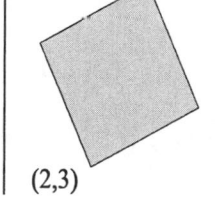

(2,3)

· Consider the transformation
$$x = u^2 + v^2$$
$$y = u.$$
What is the image of the quarter unit circle in the first quadrant of the xy−plane? This is the region in the first quadrant within the unit circle
$$u^2 + v^2 = 1.$$

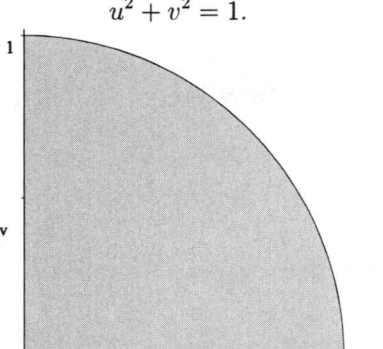

First make parametric representations of the sides of this region. The u−axis side may be parameterized by $u = t, v = 0$ for $0 \leq t \leq 1$.
Similarly the y−axis side may be parameterized by $u = 0, v = t$ for $0 \leq t \leq 1$.
The quarter circle may be parameterized by
$u = \cos\theta$,
$v = \sin\theta$,
for $0 \leq \theta \leq \pi/2$.

Transform these parametric representations in order. First the u−axis segment is transformed to
$$
\begin{aligned}
x &= u^2 + v^2 \\
&= t^2 + 0 \\
&= t^2 \\
y &= u \\
&= t.
\end{aligned}
$$
This traces out the graph of $y = \sqrt{x}$, $0 \leq x \leq 1$.
Second, the v−axis segment is transformed to
$$
\begin{aligned}
x &= u^2 + v^2 \\
&= 0 + t^2 \\
&= t^2 \\
y &= u \\
&= 0.
\end{aligned}
$$
This traces out the unit interval on the x−axis from $x = 0$ to $x = 1$.
Finally, the quarter circle transforms to
$$
\begin{aligned}
x &= u^2 + v^2 \\
&= (\cos\theta)^2 + (\sin\theta)^2 \\
&= 1 \\
y &= u \\
&= \cos\theta.
\end{aligned}
$$
This traces out the straight line segment from $(1, 1)$ to $(1, 0)$.

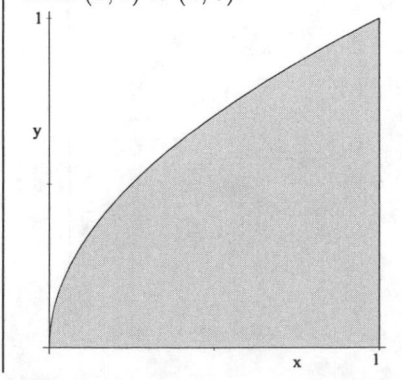

SkillMaster 12.18.

· Compute the Jacobian of the transformation in the first worked out example above,

$$x = (\cos\alpha)\,u - (\sin\alpha)v + 2$$
$$y = (\sin\alpha)u + (\cos\alpha)v + 3.$$

The Jacobian is the determinant of partial derivatives.

$$\frac{\partial(x,y)}{\partial(u,v)} =$$

$$\begin{vmatrix} \partial x/\partial u & \partial x/\partial v \\ \partial y/\partial u & \partial y/\partial v \end{vmatrix}$$

$$= \frac{\partial x}{\partial u}\frac{\partial y}{\partial v} - \frac{\partial x}{\partial v}\frac{\partial y}{\partial u}.$$

$$\frac{\partial(x,y)}{\partial(u,v)}$$

$$= \begin{vmatrix} \cos\alpha & -\sin\alpha \\ \sin\alpha & \cos\alpha \end{vmatrix}$$

$$= (\cos\alpha)^2 + (\sin\alpha)^2$$

$$= 1.$$

· Compute the Jacobian of the transformation in the second worked out example above,

$$x = u^2 + v^2$$
$$y = u.$$

Solve this using the same method as the previous worked out example.

$$\frac{\partial(x,y)}{\partial(u,v)}$$

$$= \begin{vmatrix} 2u & 2v \\ 1 & 0 \end{vmatrix}$$

$$= -2v.$$

SkillMaster 12.19.

· Use the transformation in the previous worked out examples to compute the integral

$$\int\int_D \sqrt{x - y^2}\,dA$$

where D is the region

$$\{0 \le x \le 1,\ 0 \le y \le \sqrt{x}\}.$$

The region D is transformed into the region E which is the quarter disk contained inside the unit circle and in the first quadrant.

$$\int\int_D \sqrt{x - y^2}\,dA$$

$$= \int\int_E \sqrt{(u^2 + v^2) - u^2}\,\frac{\partial(x,y)}{\partial(u,v)}\,dudv$$

$$= \int\int_E \sqrt{u^2}(-2v)\,dudv$$

$$= -2\int\int_E uv\,dudv$$

$$= -2\int_0^1 u \int_0^{\sqrt{1-u^2}} v\,dv\,du$$

$$= -2\int_0^1 u \left(\frac{v^2}{2}\Big|_0^{\sqrt{1-u^2}}\right) du$$

$$= -2\int_0^1 u \frac{(\sqrt{1-u^2})^2}{2}\,du$$

$$= -\int_0^1 u(1 - u^2)\,du = \int_0^1 -u + u^3\,du$$

$$= -\frac{1}{2} + \frac{1}{4} = -\frac{1}{4}.$$

· Fine the center of mass of the tetrahedron E with corners at $(0,0,0)$, $(a,0,0)$, $(0,b,0)$ and $(0,0,c)$ by using the transformation

$$x = au$$
$$y = bv$$
$$z = cw$$

to transform the tetrahedron E in uvw coordinates into the tetrahedron D with vertices $(0,0,0)$, $(1,0,0)$, $(0,1,0)$, and $(0,0,1)$. Recall that the center of mass for this tetrahedron has already been computed in the worked out example.

To find the center of mass first compute the volume of the region
$$\text{vol}(E) = \int \int \int_E dV.$$
Then to find the $x-$coordinate of the center of mass compute
$$\bar{x} =$$
$$\frac{1}{\text{vol}(E)} \int \int \int_E x \, dV.$$
The other coordinates may be found by symmetry.

The transformation sends E to D. The Jacobian needs to be computed.

$$\frac{\partial(x,y,z)}{\partial(u,v,w)}$$

$$= \begin{vmatrix} \partial x/\partial u & \partial x/\partial v & \partial x/\partial w \\ \partial y/\partial u & \partial y/\partial v & \partial y/\partial w \\ \partial z/\partial u & \partial z/\partial v & \partial z/\partial w \end{vmatrix}$$

$$= \begin{vmatrix} a & 0 & 0 \\ 0 & b & 0 \\ 0 & 0 & c \end{vmatrix} = abc.$$

$$\text{vol}(E)$$

$$= abc \int \int \int_D dV = \frac{abc}{6}$$

Recall that this integral was already computed in 12.7.
Similarly the integral

$$\int \int \int_E x \, dV$$

$$= abc \int \int \int_D au \, dV$$

$$= a^2bc \int \int \int_D u \, dV = a^2bc \frac{1}{24}.$$

This was also calculated in 12.7. The $x-$coordinate of the center of mass is

$$\bar{x}$$

$$= \frac{1}{\text{vol}(E)} \int \int \int_E x \, dV$$

$$= \frac{6}{abc} \frac{a^2bc}{24} = \frac{a}{4}.$$

By symmetry the other coordinates have a similar form so that the center of mass is

$$\left(\frac{a}{4}, \frac{b}{4}, \frac{c}{4}\right).$$

Notice that this is the average of the four vertices of the tetrahedron.

SkillMasters for Chapter 12

SkillMaster 12.1:	Use double Riemann sums to approximate double integrals.
SkillMaster 12.2:	Evaluate double integrals by computing volumes.
SkillMaster 12.3:	Use the properties of double integrals to estimate or evaluate double integrals.
SkillMaster 12.4:	Evaluate double integrals using iterated integrals.
SkillMaster 12.5:	Evaluate Integrals over General Regions.
SkillMaster 12.6:	Estimate double integrals using the properties of double integrals.
SkillMaster 12.7:	Evaluate double integrals over polar regions using polar coordinates.
SkillMaster 12.8:	Use double integrals to compute mass.
SkillMaster 12.9:	Compute moments of inertia and centers of mass.
SkillMaster 12.10:	Compute probabilities and expected values.
SkillMaster 12.11:	Compute the surface area of a parametric surface
SkillMaster 12.12:	Compute the surface area of a surface given as the graph of a function.
SkillMaster 12.13:	Compute triple integrals
SkillMaster 12.14:	Apply triple integrals to problems of volume, density, and mass.
SkillMaster 12.15:	Compute triple integrals using cylindrical coordinates
SkillMaster 12.16:	Compute triple integrals using spherical coordinates.
SkillMaster 12.17:	Find the image of a set under a transformation.
SkillMaster 12.18:	Compute Jacobians of transformations.
SkillMaster 12.19:	Use change of variables to compute multiple integrals.

Chapter 13 - Vector Calculus

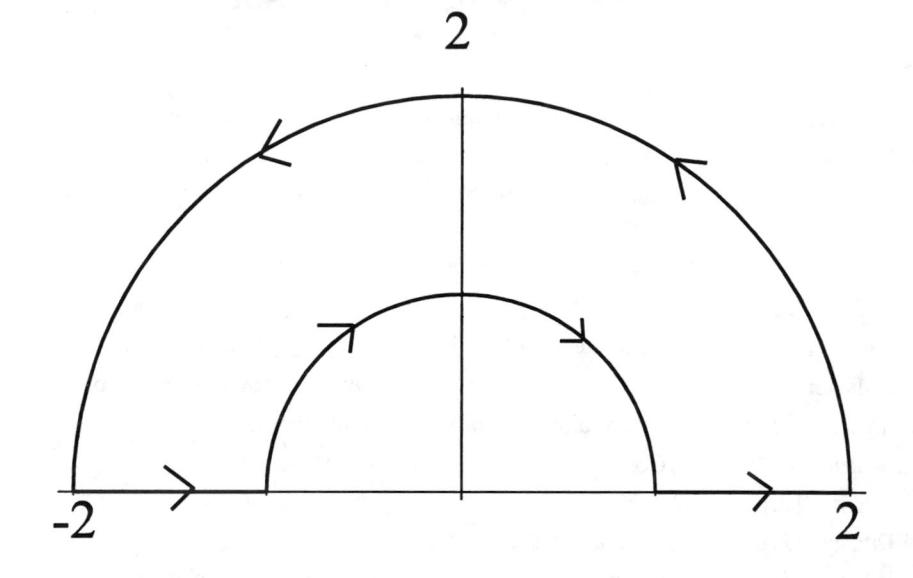

Section 13.1 – Vector Fields

Key Concepts:

- Vector Fields in R^2 and R^3
- Conservative Vector Fields and Potential Functions

Skills to Master:

- Sketch and interpret vector fields.
- Compute gradient fields.

Discussion:

This section introduces vector fields in R^2 and R^3. Vector fields can best be thought of as functions that assign to each point in a region under consideration in R^2 or R^3 a vector. You are already familiar with real valued functions from R to R. These kinds of functions are what you studied in single variable calculus. Vector valued functions defined from R to R^2 or R^3 were studied in *Chapter 10* in relation to plane and space curves. Functions of several variables from R^2 or R^3 to R were studied in *Chapter 11*. The following table should help you with these various types of functions.

page 702.

page 746.

Domain	Range	Example	Terminology
R	R	$f(x) = x^2 \sin x$	real valued function
R	R^2	$\mathbf{F}(x) = \langle x, \sin(x^2) \rangle$	vector valued function
R	R^3	$\mathbf{F}(x) = \langle x, \sin x, -\cos x \rangle$	vector valued function
R^2	R	$g(x,y) = 4 - x^2 - y^2$	function of several variables
R^3	R	$g(xyz) = x^2 + y^2 - z^2$	function of several variables
R^2	R^2	$\mathbf{F}(x,y) = \langle -y^2, xy \rangle$	vector field
R^3	R^3	$\mathbf{F}(x,y,z) = \langle -y^2, xyz, -x \rangle$	vector field

Key Concept: Vector Fields in R^2 and R^3

A vector field on a plane region D in R^2 is a function $\mathbf{F}$ that assigns to each point (x, y) in D a two dimensional vector $\mathbf{F}(x, y)$. A vector field on a space region E in R^3 is a function $\mathbf{F}$ that assigns to each point (x, y, z) in E a three dimensional vector $\mathbf{F}(x, y, z)$. Velocity fields, force fields, gradient vector fields, and gravitational fields are examples that often occur. Study *Figures 2, 3, and 4* to see geometric pictures of some vector fields.

page 919.

Key Concept: Conservative Vector Fields and Potential   Functions

A vector field $\mathbf{F}$ is called a conservative vector field if it is the gradient of a scalar function f. That is,

$$\mathbf{F} = \nabla f.$$

The function f is called a potential function for $\mathbf{F}$. Study *Figure 14* to see the relation between the contour lines of f and the vector field $\mathbf{F}$.

page 922.

SkillMaster 13.1: Sketch and interpret vector fields.

To sketch a vector field, pick a number of representative points in the region under consideration. Sketch the vector associated with each of these points. *Examples 1 through 5* show how to do this. If the problem you are working on comes from an actual physical situation, try to interpret the magnitude and direction of the vectors in the vector field in terms of this situation.

page 919-921.

SkillMaster 13.2: Compute gradient fields.

Given a scalar function f of two or three variables, you can compute the gradient

page 922.

vector field associated with f by taking the gradient of f. You already know how to do this from Chapter 11. Study *Example 6 and the following discussion* to get a better understanding of how to compute gradient fields.

Worked Examples

For each of the following examples, first try to find the solution without looking at the middle or right columns. Cover the middle and right columns with a piece of paper. If you need a hint, uncover the middle column. If you need to see the worked solution, uncover the right column.

Example	Tip	Solution
SkillMaster 13.1.		
· Sketch the vector field $(y - 2)\mathbf{i}$.	First notice that each vector in the vector field is parallel to the x−axis. The vectors along the line $y = 2$ are zero. For y larger than 2 the vectors are in the positive x direction and for y smaller than 2 the vectors are in the negative x direction. The magnitude increases as the y values become farther from the line $y = 2$.	The vector field is shown here. The vectors are not drawn at their actual length. 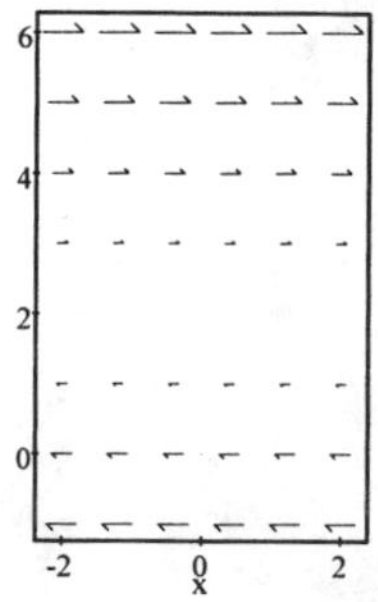
· Sketch the vector field $< -x/2, -y/2>$.	Sketch a few vectors in the field, such as $< -1/2, -1/2 >$ at $(1, 1)$ and others. Notice that the vectors appear to be pointing to the origin.	The vectors are not drawn at their actual length. 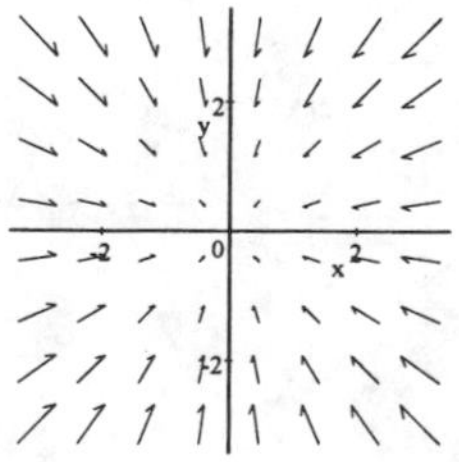

· Sketch the vector field $< y, -x >$.

Sketch a few vectors in the field. Notice that they appear to be going clockwise around the origin. The magnitude of these vectors grows as the point is farther from the origin. Check these observations by first taking the scalar product of the vector with the vector from the origin to the point and then by calculating the magnitude of the vector.

The dot product of the vector and the vector from its beginning point to the origin is

$$\langle y, -x \rangle \cdot \langle x, y \rangle = yx - xy = 0.$$

Thus the vector in the field is orthogonal to the vector the origin to the point.. The magnitude of this vector is

$$|\langle y, -x \rangle| = \sqrt{y^2 + x^2} = r,$$

the distance from the point to the origin. Here is a diagram of the vector field. The vectors are not drawn at their actual length.

· Sketch the vector field $< y - x/2, -x - y/2 >$

This is the vector sum of the two vector fields in the preceding worked examples,
$\langle y, -x \rangle +$
$\langle -x/2, -y/2 \rangle$.
Sketch a few vectors, then fill in using this analysis.

The vector field is sketched here. The vectors are not drawn at their actual length.

747

· Match the vector fields with the graphs. Give reasons for your answers.
The vectors are not drawn at their actual length.

(a) $x^2 \mathbf{i} + y\mathbf{j}$

(b)
$$< \frac{y}{1 + \sqrt{x^2 + y^2}}, \frac{-x}{1 + \sqrt{x^2 + y^2}} >$$

(c) $< 2xy, x^2 - y^2 >$

I.

II.

III.

Evaluate the vector fields at a point in each quadrant.

The vector filed (a) corresponds to II.

This vector field points upward when y is positive and downward when y is negative.

The vector field (b) corresponds to I.

This vector field points to the right when y is positive and to the left when y is negative.

The vector field (c) corresponds to III.

This vector field points to the right when xy is positive and to the left when xy is negative.

SkillMaster 13.2.

· Find the gradient field of the potential function
$f(x,y) = \ln(\cos x) - \log(\cos(y))$

The gradient vector field is defined by
$$\nabla f = \frac{\partial f}{\partial x}\mathbf{i} + \frac{\partial f}{\partial y}\mathbf{j}$$

The gradient field is
$$\nabla f = \frac{\partial f}{\partial x}\mathbf{i} + \frac{\partial f}{\partial y}\mathbf{j} = \frac{-\sin x}{\cos x}\mathbf{i} - \frac{-\sin y}{\cos y}\mathbf{j}$$
$$= -\tan x\mathbf{i} + \tan y\mathbf{j}.$$
It is shown here. The vectors are not drawn at their actual length.

· Find the gradient field of the potential function
$f(x,y,z) = xyz$

Use the definition of gradient vector field.

$$\nabla f = \frac{\partial f}{\partial x}\mathbf{i} + \frac{\partial f}{\partial y}\mathbf{j} + \frac{\partial f}{\partial z}\mathbf{k}.$$
$$= yz\mathbf{i} + xz\mathbf{j} + xy\mathbf{k}$$

· Below is the contour plot for a function and a plot of the gradient field for the same function. How are these related and why?

The vectors in the gradient field appear to be orthogonal to the contours. Use the interpretation of the gradient field to explain the reason this is true.

The gradient points in the direction that is the steepest uphill on the surface. The tangent line to level curves is in the direction that does not change (since the contours are curves with the same function values). These two directions must be orthogonal.

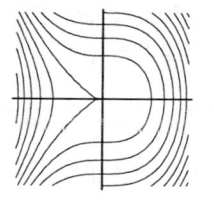

Section 13.2 – Line Integrals

Key Concepts:

- Line Integral of a Function Along a Curve
- Line Integrals in Space
- Line Integrals of Vector Fields

Skills to Master:

- Compute line integrals.
- Compute line integrals of vector fields.

Discussion:

This section introduces the concept of line integrals. These integrals are set up in a way that is similar to other kinds of integrals that you have encountered. In physical applications, these kinds of integrals arise in situations where you are trying to find the work done in moving an object through a force field along a curve. Study the examples carefully. You will need a good understanding of the various types of line integrals for future sections in this chapter.

Key Concept: Line Integral of a Function Along a Curve

If a scalar function f is defined on a smooth curve C in the plane, the line integral of f along C is defined to be

$$\int_C f(x, y)ds = \lim_{n \to \infty} \sum_{i=1}^{n} f(x_i^*, y_i^*)\Delta s_i$$

provided this limit exists. Notice the similarity of this definition to definitions of other types of integrals. In this definition, the curve C is divided into n subarcs, (x_i^*, y_i^*) is a point in the ith subarc, and Δs_i is the length of the ith subarc. This line integral is called the line integral of f with respect to arc length. See *Figure 1* for a geometric picture of this.

page 924.

If f is continuous, then the line integral always exists and is equal to

$$\int_C f(x, y)ds = \int_a^b f(x(t), y(t)) \sqrt{\left(\frac{dx}{dt}\right)^2 + \left(\frac{dy}{dt}\right)^2} dt$$

where $\mathbf{r}(t) = \langle x(t), y(t) \rangle$ is a parameterization of C that traces out C exactly once as t increases from a to b.

Key Concept: Line Integrals in Space

If a scalar function f is defined on a space curve C, the line integral of f along C is defined to be

$$\int_C f(x, y, z)ds = \lim_{n \to \infty} \sum_{i=1}^{n} f(x_i^*, y_i^*, z_i^*)\Delta s_i$$

provided this limit exists. In this definition, the curve C is again divided into n subarcs, (x_i^*, y_i^*) is a point in the ith subarc, and Δs_i is the length of the ith subarc. As before, if f is continuous on some region containing C, the integral always exists and is equal to

$$\int_C f(x, y, z)ds = \int_a^b f(x(t), y(t), z(t)) \sqrt{\left(\frac{dx}{dt}\right)^2 + \left(\frac{dy}{dt}\right)^2 + \left(\frac{dz}{dt}\right)^2} dt$$

page 930.

where $\mathbf{r}(t) = \langle x(t), y(t), z(t) \rangle$ is a parameterization of C that traces out C exactly once as t increases from a to b. Study *Examples 5 and 6* to see how this works.

Key Concept: Line Integrals of Vector Fields

If $\mathbf{F}$ is a continuous vector field defined on a smooth curve C given by a vector function $\mathbf{r}(t)$, $a \leq t \leq b$, then the line integral of $\mathbf{F}$ along C is defined by

$$\int_C \mathbf{F} \cdot d\mathbf{r} = \int_C \mathbf{F} \cdot \mathbf{T} ds = \int_a^b \mathbf{F}(\mathbf{r}(t)) \cdot \mathbf{r}'(t)dt.$$

Note that this line integral is the line integral of the scalar function $\mathbf{F} \cdot \mathbf{T}$ as defined

page 932.

above. Here, $\mathbf{T}$ is the unit tangent vector to the curve. Study *Figure 11* to see a motivation for this definition.

SkillMaster 13.3: Compute line integrals.

To compute line integrals, use the fact that

$$\int_C f(x,y)ds = \int_a^b f(x(t),y(t))\sqrt{\left(\frac{dx}{dt}\right)^2 + \left(\frac{dy}{dt}\right)^2}\, dt \text{ and}$$

$$\int_C f(x,y,z)ds = \int_a^b f(x(t),y(t),z(t))\sqrt{\left(\frac{dx}{dt}\right)^2 + \left(\frac{dy}{dt}\right)^2 + \left(\frac{dz}{dt}\right)^2}\, dt.$$

page 925-926.

Study *Examples 1, 2, and 3* to see how specific computations of line integrals.

Another kind of line integral is the line integral of f along C with respect to x or y. This kind of line integral is given by

$$\int_C f(x,y)dx = \int_a^b f(x(t),y(t))x'(t)dt \text{ and}$$

$$\int_C f(x,y)dy = \int_a^b f(x(t),y(t))y'(t)dt.$$

The definitions for line integrals of f along space curves with respect to x, y, or z are similar. Study *Example 4* to see a computation of this type of line integral.

page 925-926.

SkillMaster 13.4: Compute line integrals of vector fields.

To compute line integrals of vector fields, use the definition

$$\int_C \mathbf{F} \cdot d\mathbf{r} = \int_C \mathbf{F} \cdot \mathbf{T}ds = \int_a^b \mathbf{F}(\mathbf{r}(t)) \cdot \mathbf{r}'(t)dt.$$

page 933.

Study *Examples 7 and 8* to see how this is done. Note that the line integral of a vector field can also be expressed as

$$\int_C \mathbf{F} \cdot d\mathbf{r} = \int_C Pdx + Qdy + Rdz \text{ where } \mathbf{F} = P\mathbf{i} + Q\mathbf{j} + R\mathbf{k}.$$

Worked Examples

For each of the following examples, first try to find the solution without looking at the middle or right columns. Cover the middle and right columns with a piece of paper. If you need a hint, uncover the middle column. If you need to see the worked solution, uncover the right column.

Example	Tip	Solution

SkillMaster 13.3.

· Evaluate the line integral along the path C, the first quarter of the unit circle oriented in the counter-clockwise direction.

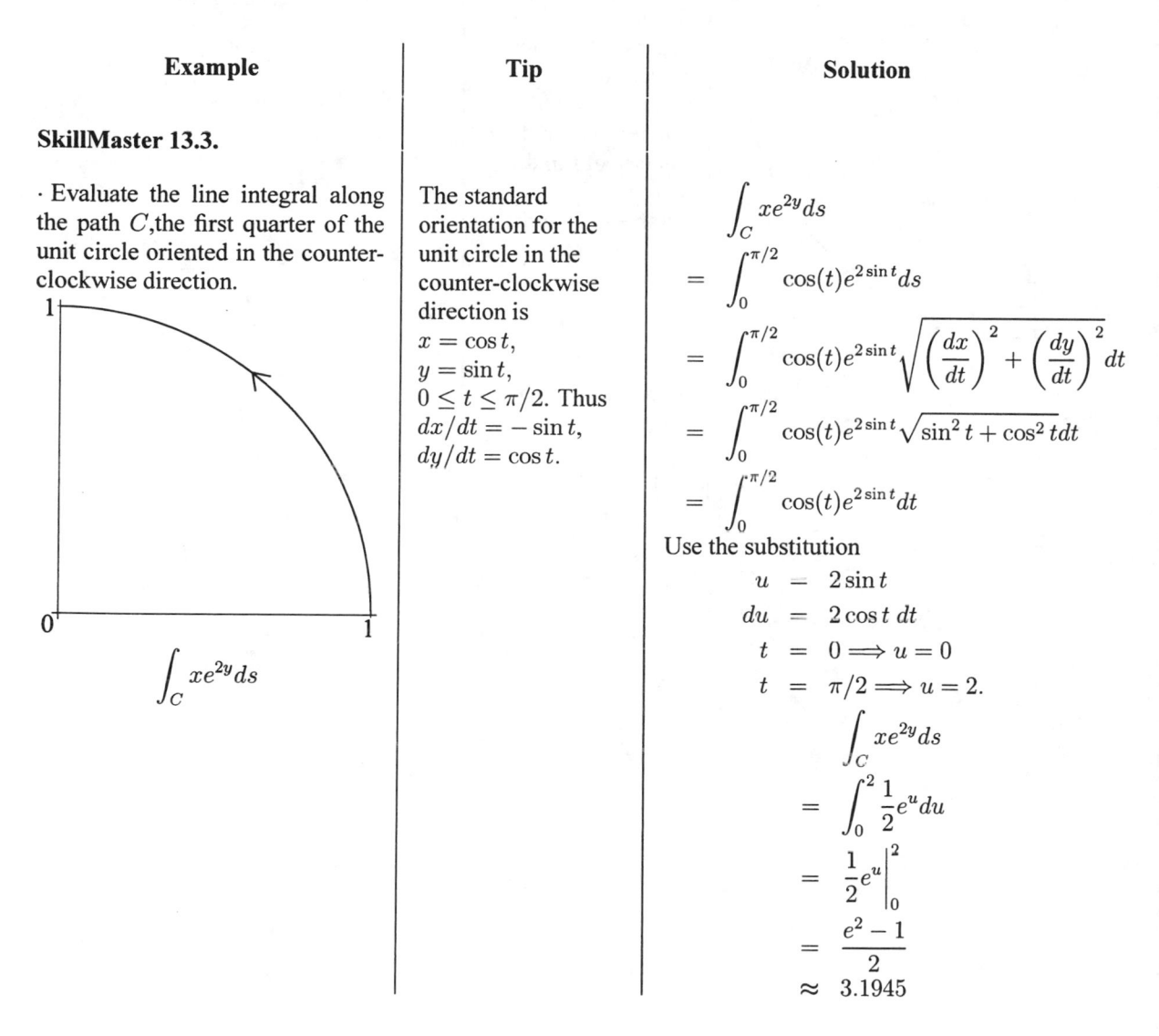

$$\int_C xe^{2y}ds$$

Tip: The standard orientation for the unit circle in the counter-clockwise direction is
$x = \cos t$,
$y = \sin t$,
$0 \le t \le \pi/2$. Thus
$dx/dt = -\sin t$,
$dy/dt = \cos t$.

Solution:

$$\int_C xe^{2y}ds$$

$$= \int_0^{\pi/2} \cos(t)e^{2\sin t}ds$$

$$= \int_0^{\pi/2} \cos(t)e^{2\sin t}\sqrt{\left(\frac{dx}{dt}\right)^2 + \left(\frac{dy}{dt}\right)^2}\,dt$$

$$= \int_0^{\pi/2} \cos(t)e^{2\sin t}\sqrt{\sin^2 t + \cos^2 t}\,dt$$

$$= \int_0^{\pi/2} \cos(t)e^{2\sin t}dt$$

Use the substitution

$$\begin{aligned} u &= 2\sin t \\ du &= 2\cos t\ dt \\ t &= 0 \Longrightarrow u = 0 \\ t &= \pi/2 \Longrightarrow u = 2. \end{aligned}$$

$$\int_C xe^{2y}ds$$

$$= \int_0^2 \frac{1}{2}e^u du$$

$$= \frac{1}{2}e^u\Big|_0^2$$

$$= \frac{e^2 - 1}{2}$$

$$\approx 3.1945$$

· Evaluate the line integral along the path C, the quarter of the unit circle in the fourth quadrant, oriented in the clockwise direction.

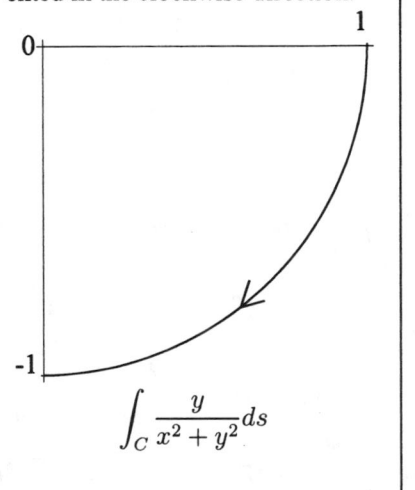

$$\int_C \frac{y}{x^2+y^2}ds$$

This goes in the opposite direction of the standard parameterization of the unit circle and begins at the same place. To parameterize this quarter circle put $-t$ in place of t in the standard parameterization:
$x = \cos(-t)$
$= \cos t,$
$y = \sin(-t)$
$= -\sin t,$
$0 \le t \le \pi/2.$
So $dx/dt = -\sin t,$
$dy/dt = -\cos t,$
$\sqrt{\left(\frac{dx}{dt}\right)^2 + \left(\frac{dy}{dt}\right)^2}$
$= 1.$

$$\int_C \frac{y}{x^2+y^2}ds$$
$$= \int_0^{\pi/2} \frac{-\sin t}{\cos^2 t + \sin^2 t}dt$$
$$= \int_0^{\pi/2} -\sin t\,dt$$
$$= \cos t\big|_0^{\pi/2}$$
$$= 0 - 1$$
$$= -1.$$

· Calculate the line integral
$$\int_C \frac{\ln(x+y)}{z}\,ds$$
where C is the straight line path from $(2,-3,1)$ to $(4,-6,4)$.

The standard parameterization from a point P to a point Q is
$(Q-P)t + P,$
$0 \le t \le 1.$
In this case, the calculation is
$((4,-6,4) -$
$(2,-3,1))t$
$+(2,-3,1) =$
$(t+2, t-3, 3t+1).$

$$
\begin{aligned}
x &= t+2 \\
y &= t-3 \\
z &= 3t+1 \\
dx/dt &= 1 \\
dy/dt &= 1 \\
dz/dt &= 3
\end{aligned}
$$

$$\int_C \frac{\ln(2x+y)}{z}\,ds$$
$$= \int_0^1 \frac{\ln(2t+4+t-5)}{3t+1}\,ds$$
$$= \int_0^1 \frac{\ln(3t-1)}{3t+1}\sqrt{1+1+9}\,dt$$
$$= \sqrt{11}\int_0^1 \frac{\ln(3t+1)}{3t+1}\,dt$$

Use the substitution
$$
\begin{aligned}
u &= \ln(3t+1) \\
du &= \frac{3}{3t+1}\,dt \\
t &= 0 \Longrightarrow u = 0 \\
t &= 1 \Longrightarrow u = \ln 4.
\end{aligned}
$$
$$\int_C \frac{\ln(2x+y)}{z}\,ds$$
$$= \sqrt{11}/3 \int_0^{\ln 4} u\,du$$
$$= \frac{\sqrt{11}}{6}(\ln 4)^2.$$

· Find the integral
$$\oint_C \left(\frac{dy}{x} - \frac{dx}{y}\right)$$
where C is the counter-clockwise oriented ellipse
$$\frac{x^2}{a^2} + \frac{y^2}{b^2} = 1.$$

The standard parameterization for the ellipse is
$$
\begin{aligned}
x &= a\cos t \\
y &= b\sin t, \\
0 &\le t \le 2\pi. \\
dx/dt &= -a\sin t \\
dy/dt &= b\cos t.
\end{aligned}
$$

$$\oint_C \left(\frac{dy}{x} - \frac{dx}{y}\right)$$
$$= \int_0^{2\pi} \frac{b\cos t}{a\cos t} - \frac{-a\sin t}{b\sin t}\,dt$$
$$= \int_0^{2\pi} \frac{b}{a} + \frac{a}{b}\,dt$$
$$= \frac{a^2+b^2}{ab}\int_0^{2\pi} dt$$
$$= 2\pi\frac{a^2+b^2}{ab}$$

755

· A fence, modeled mathematically below, is to be painted with expensive enamel paint. If each can of this paint can cover 260 square feet, how many cans of paint should be purchased to cover both sides of this fence? The path of the fence is given parametrically by

$$x = 30\cos^3 t$$
$$y = 30\sin^3 t,$$
$$0 \le t \le \pi.$$

The height of the fence at the location (x, y) is

$$f(x,y) = 1 + \frac{1}{3}y.$$

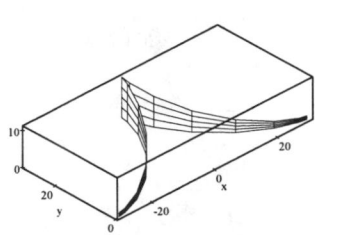

It is required to measure the area of the fence. This is one interpretation of the line integral.
$A = \int_C f(x,y)ds,$
$dx/dt =$
$-90\cos^2 t \sin t,$
$dy/dt =$
$90\sin^2 t \cos t.$
$ds/dt =$
$[(-90\cos^2 \sin t)^2$
$+(90\sin^2 t \cos t)^2]^{1/2}$
$= 90\cos t \sin t \sqrt{\cos^2 t + \sin^2 t}$
$= 90\cos t \sin t.$

By symmetry, the area of the surface is twice the area of the fence from $t = 0$ to $t = \pi/2$.

$$
\begin{aligned}
A &= 2\int_0^{\pi/2}(1 + 10\sin^3 t)ds \\
&= 2\int_0^{\pi/2}(1 + 10\sin^3 t)(90\cos t \sin t)dt \\
&= 180\int_0^{\pi/2}\cos t \sin t\,dt \\
&\quad +1800\int_0^{\pi/2}\sin^4 t \cot t\,dt \\
&= 180\left.\frac{\sin^2 t}{2}\right|_0^{\pi/2} \\
&\quad +1800\left.\frac{\sin^5 t}{5}\right|_0^{\pi/2} \\
&= 90 + 360 \\
&= 450 \text{ ft}^2.
\end{aligned}
$$

This is the area of only one side of the fence. The total area of the fence is $2(450) = 900$ ft^2. Each can of paint covers 260 ft^2 So the ratio is $900/260 = 3.461$. It is necessary to buy 4 cans of paint.

SkillMaster 13.4.

· A wire has the shape of that part of the circular helix

$$\mathbf{r}(t) = <3\cos t, 3\sin t, 4t>,$$

where $0 \le t \le \pi$.

Find the work done by the force

$$\mathbf{F} = <2x, 3y, z>$$

moving a bead on the wire from $(3, 0, 0)$ to $(-3, 0, 4\pi)$.

The force at time t is

$$\mathbf{F}(t) =$$
$$\langle 2x(t), 3y(t), z(t) \rangle$$
$$=$$
$$\langle 6\cos t, 9\sin t, 4t \rangle .$$

The work done is
$W = \int_C \mathbf{F} \cdot d\mathbf{r}.$
$d\mathbf{r}/dt =$
$\langle -3\sin t, 3\cos t, 4 \rangle.$
The dot product is
$6\cos(t)(-3\sin t) +$
$9\sin(t)(3\cos t)$
$+4t(4)$
$= 9\sin t \cos t + 16t.$

$$W = \int_C \mathbf{F} \cdot d\mathbf{r}$$

$$= \int_0^\pi 9\sin t \cos t + 16t \, dt$$

$$= \left(\frac{9}{2}\sin^2 t + 8t^2 \right)\Big|_0^\pi$$

$$= 8\pi^2 \approx 78.96$$

757

Section 13.3 – The Fundamental Theorem for Line Integrals

Key Concepts:

- Line Integrals of Conservative Vector Fields
- Path Independence of Line Integrals for Conservative Vector Fields

Skills to Master:

- Determine if vector fields are conservative.
- Find the potential function for a conservative vector field.
- Evaluate line integrals of conservative vector fields.

Discussion:

This section continues the discussion of line integrals. For line integrals of vector fields that are conservative, an easy way of computing the line integral is presented. There is a parallel between the material presented in this section and the material from single variable calculus on the Fundamental Theorem of Calculus. Keep this parallel in mind as you study this section.

Key Concept: Line Integrals of Conservative Vector Fields

If C is a smooth curve given by $\mathbf{r}(t)$, $a \leq t \leq b$, and if f is a scalar function whose gradient vector ∇f is continuous on C, then

$$\int_C \nabla f \cdot d\mathbf{r} = f(\mathbf{r}(b)) - f(\mathbf{r}(a)).$$

Note that can also be written as

$$\int_C \mathbf{F} \cdot d\mathbf{r} = f(\mathbf{r}(b)) - f(\mathbf{r}(a))$$

where $\mathbf{F} = \nabla f$. Thus if a vector field $\mathbf{F}$ is conservative, its line integral can easily be evaluated by evaluating f at the endpoints of C. Make sure that you understand why this works by reading the *explanation* in the text.

page 937.

Key Concept: Path Independence of Line Integrals for Conservative Vector Fields

---

If $\mathbf{F}$ is conservative, the previous key concept shows that

$$\int_C \mathbf{F} \cdot d\mathbf{r}$$

depends only on the endpoints of C and not on the particular way that C is traced out from one endpoint to the other. In general, if $\mathbf{F}$ is a continuous vector field with domain D, $\int_C \mathbf{F} \cdot d\mathbf{r}$ is independent of path if

$$\int_{C_1} \mathbf{F} \cdot d\mathbf{r} = \int_{C_2} \mathbf{F} \cdot d\mathbf{r}$$

for any two paths C_1 and C_2 in D that have the same initial and terminal points. Study *Figures 2 and 3* to understand the following result:

page 938.

$$\int_C \mathbf{F} \cdot d\mathbf{r} \text{ is independent of path in } D$$

$$\text{if and only if } \int_C \mathbf{F} \cdot d\mathbf{r} = 0 \text{ for every closed path } C \text{ in } D.$$

The converse of this result is also true. If $\mathbf{F}$ is a vector field that is continuous on an open connected region D, and if $\int_C \mathbf{F} \cdot d\mathbf{r}$ is independent of path in D, then $\mathbf{F}$ is a conservative vector field on D. Study *Figure 4 and the explanation in the text* to better understand this result.

page 939.

SkillMaster 13.5: Determine if vector fields are conservative.

If a vector field $\mathbf{F}(x, y) = P(x, y)\mathbf{i} + Q(x, y)\mathbf{j}$ is a conservative vector field, where

P and Q have continuous partial derivatives on a region D, then throughout D

$$\frac{\partial P}{\partial y} = \frac{\partial Q}{\partial x}.$$

This gives a way of determining if $\mathbf{F}(x, y)$ is conservative. If these partial derivatives are not equal, the vector field is not conservative. If they are equal and the region D is simply connected (every simple closed curve in D encloses only points of D), then $\mathbf{F}$ is conservative. Study *Figures 6 and 7* to gain a geometric understanding of the terminology used in these results. Study *Examples 2 and 3* to see how to apply these results.

page 940.

page 941.

SkillMaster 13.6: Find the potential function for a conservative vector field.

To find the potential function f for a conservative vector field $\mathbf{F} = P\mathbf{i} + Q\mathbf{j}$, you need to find a scalar function so that

$$\nabla f = \mathbf{F}.$$

If f exists,

$$\frac{\partial f}{\partial x} = P \text{ and } \frac{\partial f}{\partial y} = Q.$$

Integrate the first of these equations with respect to x, then differentiate the result with respect to y and compare the result with the second equation.

page 941.

Study *Example 4* to see how to do this.

SkillMaster 13.7: Evaluate line integrals of conservative vector fields.

To evaluate line integrals of conservative vector fields, first find a potential function for the vector field. then use the Fundamental Theorem for line integrals to evaluate the line integral. Study *Example 4* to see how to do this.

page 941.

Worked Examples

For each of the following examples, first try to find the solution without looking at the middle or right columns. Cover the middle and right columns with a piece of paper. If you need a hint, uncover the middle column. If you need to see the worked solution, uncover the right column.

Example	Tip	Solution
SkillMaster 13.5. · Which of the following vector fields are conservative? (a) $\quad \mathbf{F} =< x^2 - y^2, 2xy >$.	Recall that a $\mathbf{F} =< P, Q >$ is conservative means that $\mathbf{F} = \nabla f$ for some potential function, $z = f(x, y)$. This means that $P = f_x$ and $Q = f_y$. Clairaut's Theorem implies in this case that $P_y = f_{xy} = Q_x$ (as long as these partials derivatives are continuous). A test for conservative is to check whether $P_y = Q_x$.	$$P = x^2 - y^2$$ $$Q = 2xy$$ $$P_y = -2y$$ $$Q_x = 2y.$$ Since $P_y \neq Q_x$ the vector field is not conservative.
· (b) $\quad \mathbf{F} =< x^2 + y^2, 2xy >$.	Theorem 6. says that if $P_y = Q_x$ for an open simply connected domain then the force field is conservative.	$$P = x^2 + y^2$$ $$Q = 2xy$$ $$P_y = 2y$$ $$Q_x = 2y.$$ Since $P_y = Q_x$ on all of $\mathbb{R}^2$ the vector field is conservative.

761

· (c)

$$\mathbf{F} = < ye^{xy} + 2x, xe^{xy} - 2y >$$

Check to see if
$P_y = Q_x$.

$$P = ye^{xy} + 2x$$
$$Q = xe^{xy} - 2y$$

$$P_y = e^{xy} + xye^{xy}$$
$$Q_x = e^{xy} + xye^{xy}$$

So the vector field is conservative.

SkillMaster 13.6.

· Find the potential functions in the Worked Problems just above which are conservative.

(b)

$$\mathbf{F} = < x^2 + y^2, 2xy >$$

The potential function f satisfies
$$f_x = x^2 + y^2$$
$$f_y = 2xy.$$
Integrate the first equality with respect to x, then differentiate with respect to y and compare.

$$f = \frac{1}{3}x^3 + xy^2 + \varphi.$$

$$f_y$$
$$= 2xy + \varphi'(y)$$
$$= 2xy.$$

Thus
$$\varphi'(y) = 0,$$
$$\varphi(y) = K.$$

The potential function is
$$f = \frac{1}{3}x^3 + xy^2 + K.$$

· (c)

$$\mathbf{F} = < ye^{xy} + 2x, xe^{xy} - 2y >$$

Again integrate P with respect to x then differentiate with respect to y.

$$f_x = ye^{xy} + 2x$$
$$f = e^{xy} + x^2 + \varphi(y)$$

$$f_y = xe^{xy} + \varphi'(y)$$
$$= xe^{xy} - 2y$$

So

$$\varphi'(y) = -2y$$
$$\varphi(y) = -y^2 + K.$$
$$f = e^{xy} + x^2 - y^2 + K.$$

SkillMaster 13.7.

· Evaluate the line integral for the conservative vector field
$$\mathbf{F} = y^2\mathbf{i} + (2xy - e^y)\mathbf{j}$$
over the curve
$$\mathbf{r(t)} = \cos(t)\mathbf{i} + \sin(t)\mathbf{j}$$
for $0 \le t \le \pi/2$.

First check if the vector field is conservative on a simply connected domain that includes the path of the curve. If so the integral is the difference of the potential function evaluated at the endpoints of the curve.

$$P_y = 2y = Q_x,$$
so the vector field is conservative. The potential function satisfies
$$
\begin{aligned}
f_x &= y^2 \\
f &= xy^2 + \varphi(y) \\
f_y &= 2xy + \varphi'(y) = 2xy - e^y, \\
f &= xy^2 - e^y.
\end{aligned}
$$
The integral is
$$
\begin{aligned}
\int_C \mathbf{F}\cdot d\mathbf{r} &= \int_C \nabla f \cdot d\mathbf{r} \\
&= f(\mathbf{r}(\pi)) - f(\mathbf{r}(0)) \\
&= f(1,0) - f(0,1) \\
&= -e - (-1) = 1 - e.
\end{aligned}
$$

· Find the work done by the force field
$$\mathbf{F} = \left\langle \frac{-y}{x^2+y^2}, \frac{x}{x^2+y^2} \right\rangle$$
in moving a particle from $(1,0)$ to $(1,1)$ along a parabolic curve with vertex $(1,1)$ and concave downward.

Find a potential function by direct comparison.

$$
\begin{aligned}
f_x &= \frac{-y}{x^2+y^2} \\
f &= \tan^{-1}\left(\frac{y}{x}\right) + \varphi(y) \\
f_y &= \frac{x}{x^2+y^2} \\
f &= \tan^{-1}\left(\frac{y}{x}\right) + \psi(x)
\end{aligned}
$$
By direct comparison,
$$f = \tan^{-1}\left(\frac{y}{x}\right).$$
The integral depends only on the end points of the path, that is the work done is
$$
\begin{aligned}
W &= \int_C \mathbf{F}\cdot d\mathbf{r} = \int_C \nabla f \cdot d\mathbf{r} \\
&= f(1,1) - f(1,0) \\
&= \tan^{-1}(1/1) - \tan^{-1}(0/1) \\
&= \pi/4 - 0 = \pi/4.
\end{aligned}
$$

· Find the work done by the force field

$$\mathbf{F} = e^x \sin y \cos z \mathbf{i}$$
$$+ e^x \cos y \cos z \mathbf{j}$$
$$- e^x \sin y \sin z \mathbf{k}$$

in moving a particle along the ellipse which is the intersection of the plane $x + z = 1$ and the cylinder $x^2 + 2y^2 = 1$ in the counterclockwise direction.

First try to find a potential function by integrating f_x with respect to x and then differentiating with respect to y..

$$f_x = e^x \sin y \cos z$$
$$f = e^x \sin y \cos z + \varphi(y, z)$$
$$f_y = e^x \cos y \cos z + \varphi_y(y, z)$$
$$= e^x \cos y \cos z$$

Thus $\varphi(y, z) = \psi(z)$ is a function of z alone.

$$f = e^x \sin y \cos z + \psi(z)$$
$$f_z = -e^x \sin y \sin z + \psi'(z)$$
$$= -e^x \sin y \sin z$$

$$f = e^x \sin y \cos z + K$$

Since there is a potential function which is defined everywhere the line integral over any closed curve is 0.

$$W = \oint \nabla f \cdot d\mathbf{r}$$
$$= 0.$$

Section 13.4 – Green's Theorem

Key Concepts:

- Green's Theorem Relates Line Integrals and Double Integrals

Skills to Master:

- Use Green's Theorem to evaluate line integrals over a closed curve.
- Use Green's Theorem to evaluate double integrals and to find area.

Discussion:

This section discusses Green's Theorem, which gives a relationship between the line integral around a simple closed curve C and a double integral over the plane region bounded by the curve C. Make sure that you understand the material on line integrals from the previous sections before you start to work on this section. If you need to review double integrals, now is the time to do so.

Key Concept: Green's Theorem Relates Line Integrals and Double Integrals

A simple closed curve C in the plane is positively oriented if it is oriented in a counterclockwise manner. Let C be such a curve and let D be the plane region bounded by C. If P and Q have continuous partial derivatives on an open region that contains D, then

$$\int_C P\,dx + Q\,dy = \iint_D \left(\frac{\partial Q}{\partial x} - \frac{\partial P}{\partial y} \right) dA.$$

This result is know as Green's Theorem. Study *Figure 3 and the accompanying explanation* to see why Green's Theorem holds in a special case. Note that if $\mathbf{F}(x,y) = P(x,y)\mathbf{i} + Q(x,y)\mathbf{j}$, the left hand side of this equation can be written

page 947.

765

as

$$\int_C \mathbf{F} \cdot d\mathbf{r}.$$

SkillMaster 13.8: Use Green's Theorem to evaluate line integrals over a closed curve.

To use Green's Theorem to evaluate line integrals over a closed curve C,

$$\int_C P\,dx + Q\,dy,$$

evaluate instead the double integral over the region D enclosed by C,

$$\iint_D \left(\frac{\partial Q}{\partial x} - \frac{\partial P}{\partial y} \right) dA.$$

page 948, 950.

Examples 1 and 2 and 4 show how to do this.

SkillMaster 13.9: Use Green's Theorem to evaluate double integrals and to find area.

To use Green's Theorem to evaluate double integrals of the form

$$\iint_D \left(\frac{\partial Q}{\partial x} - \frac{\partial P}{\partial y} \right) dA,$$

evaluate instead the line integral

$$\int_C P\,dx + Q\,dy.$$

To use Green's theorem to find the area of a region bounded by a simple closed curve C, use

$$A = \oint_C x\,dy = - \oint_C y\,dx = (1/2) \oint_C (x\,dy - y\,dx).$$

page 949.

See *Example 3* for details on how to do this.

Worked Examples

For each of the following examples, first try to find the solution without looking at the middle or right columns. Cover the middle and right columns with a piece of paper. If you need a hint, uncover the middle column. If you need to see the worked solution, uncover the right column.

Example	Tip	Solution

SkillMaster 13.8.

· Use Green's Theorem to evaluate the line integral

$$\oint \tan^{-1}(y)dx - \frac{y^2x}{1+y^2}dy$$

where C is the path which travels counter-clockwise along the unit square with vertices $(0,0)$, $(1,0)$, $(1,1)$, and $(0,1)$.

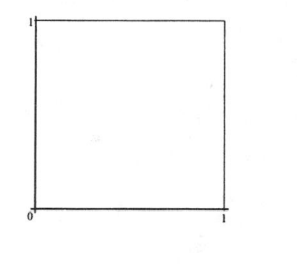

Green's Theorem states that given a closed positively oriented piecewise smooth curve C and given P and Q with continuous partial derivatives on an open region that contains C and its interior D, then
$\oint_C Pdx + Qdy$
$= \int \int (Q_x - P_y)dA.$
In this case the conclusions of Green's Theorem are true.

$$Q_x = \frac{-y^2}{1+y^2}$$

$$P_y = \frac{1}{1+y^2}$$

$$\oint_C \tan^{-1}(y)dx - \frac{y^2x}{1+y^2}dy$$
$$= \int\int_D \left(\frac{-y^2}{1+y^2} - \frac{1}{1+y^2}\right)dA$$
$$= -\int\int_D 1dA$$
$$= -\text{area}(D)$$
$$= -1^2$$
$$= -1.$$

· Use Green's Theorem to evaluate the line integral

$$\oint_C \left(xy + \ln(1 + x^2) \right) dx + xdy$$

where C is the triangle with vertices $(0,0)$, $(1,0)$, and $(1,2)$ oriented in a counterclockwise manner.

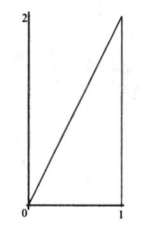

Again use Green's Theorem to reduce this line integral to an integral over the triangle D. The triangle is $D = \{0 \leq x \leq 1, 0 \leq y \leq 2x\}$.

$$Q_x = 1$$
$$P_y = x$$

$$\oint_C \left(xy + \ln(1 + x^2) \right) dx + xdy$$

$$= \int\int_D (1 - x) dA$$

$$= \int_0^1 \int_0^{2x} (1 - x) dy dx$$

$$= \int_0^1 (1 - x) \left[\int_0^{2x} dy \right] dx$$

$$= \int_0^1 (1 - x)(2x) dx$$

$$= \int_0^1 2x - 2x^2 dx$$

$$= x^2 - \frac{2}{3}x^3 \Big|_0^1$$

$$= 1 - 2/3$$

$$= 1/3.$$

· Use Green's Theorem to evaluate the line integral

$$\oint xy\,dx + (x+y)\,dy$$

where C is the path shown. The curved parts of the path are parts of circles centered at the origin.

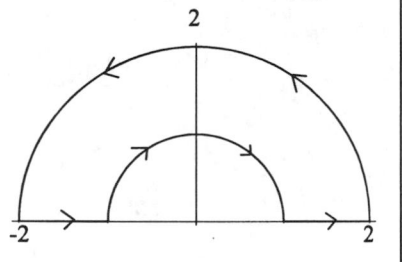

Since $P = xy$ and $Q = x + y$ have continuous partial derivatives everywhere Green's Theorem applies. The region is more conveniently described in polar coordinates.

$$D = \{1 \le r \le 2, 0 \le \theta \le \pi\}$$

$$Q_x = 1$$
$$P_y = x$$

$$\oint_C xy\,dx + (x+y)\,dy$$

$$= \int\int_D (1-x)\,dA$$

$$= \int_0^\pi \int_1^2 (1 - r\cos\theta)\,r\,dr\,d\theta$$

$$= \int_0^\pi \int_1^2 r\,dr\,d\theta - \int_0^\pi \int_1^2 r^2\cos\theta\,dr\,d\theta$$

$$= \int_0^\pi d\theta \int_1^2 r\,dr - \int_0^\pi \cos\theta\,d\theta \int_1^2 r^2\,dr$$

$$= \pi \int_1^2 r\,dr + (\sin\pi - \sin 0)\int_1^2 r^2\,dr$$

$$= \pi \int_1^2 r\,dr$$

$$= \pi \frac{r^2}{2}\Big|_1^2$$

$$= \pi\left(\frac{4-1}{2}\right)$$

$$= \frac{3\pi}{2}$$

· Calculate the work done by the force field
$$\mathbf{F} = < \sin x - 6x^2 y, 3xy^2 - x^3 >$$
in moving a particle from the origin $(0,0)$ along the x-axis to the point $(1,0)$ then counter-clockwise around the unit circle to the line $y = x$ then down this line back to the origin.

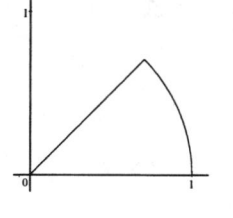

The components of the force are continuously differentiable everywhere so Green's Theorem applies.
The region in polar coordinates is
$$D = \{0 \le r \le 1, \\ 0 \le \theta \le \pi/4\}$$

$$Q_x = 3y^2 - 3x^2$$
$$P_y = -6x^2$$

The work done is

$$
\begin{aligned}
W &= \oint_C \mathbf{F} \cdot d\mathbf{r} \\
&= \int\int_D 3y^2 - 3x^2 - (-6x^2) dA \\
&= \int\int_D (3y^2 + 3x^2) dA \\
&= 3 \int\int_D x^2 + y^2 \, dA \\
&= 3 \int_0^{\pi/4} \int_0^1 r^2 r \, dr \, d\theta \\
&= 3 \int_0^{\pi/4} d\theta \int_0^1 r^3 dr \\
&= 3 \left(\frac{\pi}{4}\right)\left(\frac{1}{4}\right) \\
&= \frac{3\pi}{16}
\end{aligned}
$$

SkillMaster 13.9.

· Find the area of the region D bounded by the x-axis and one arch of the cycloid
$$x = a(t - \sin t)$$
$$y = a(1 - \cos t).$$

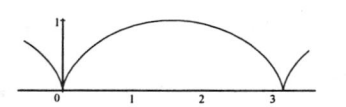

The boundary of the region is made up of two pieces. The line segment C_1 on the x-axis from $(0,0)$ to $(2\pi a, 0)$ and the arch C_2 on top which is traversed backwards, that is from $t = 2\pi$ to $t = 0$.
The area of the region by Formula 5 (using Green's Theorem) is
$$\tfrac{1}{2} \oint x\,dy - y\,dx$$
$$= \oint x\,dy$$
$$= - \oint y\,dx.$$
Choose the formula that is easiest to compute.

The curve C_1 is parameterized by
$$x = t$$
$$y = 0, 0 \le t \le 2\pi a$$

$$dx = dt$$
$$dy = 0.$$
The curve C_2 is parameterized by
$$x = a(t - \sin t)$$
$$y = a(1 - \cos t),$$
$$t \text{ begins at } 2\pi$$
$$t \text{ ends at } 0$$

$$dx = a(1 - \cos t)dt$$
$$dy = a \sin t\,dt$$
Choose the area formula based on dy (since $dy = 0$) for C_1.
$$\text{area}(D)$$
$$= \oint_{C_1} x\,dy + \oint_{C_2} x\,dy$$
$$= 0 + \oint_{C_2} x\,dy$$
$$= \int_{2\pi}^{0} a(t - \sin t)a \sin t\,dt$$
$$= -a^2 \int_{0}^{2\pi} t \sin t - \sin^2 t\,dt$$
$$= -a^2 \left(\sin t - t\cos t - \frac{1}{2}t + \frac{1}{4}\sin 2t\right)\Big|_{0}^{2\pi}$$
$$= -a^2(0 - 2\pi - \pi + 0)$$
$$\quad + a^2(0 - 0 - 0 + 0)$$
$$= 3\pi a^2$$

Section 13.5 – Curl and Divergence

Key Concepts:

- Curl of a Vector Field
- Divergence of a Vector Field
- Vector Form of Green's Theorem

Skills to Master:

- Compute the curl and divergence of a vector field.
- Use the curl and divergence to solve problems.

Discussion:

This section introduces the curl of a vector field $\mathbf{F} = P\mathbf{i} + Q\mathbf{j} + R\mathbf{k}$ defined in three dimensional space. The curl will be used to test if the vector field is conservative and will be used in a vector version of Green's Theorem. Another concept, that of the divergence of $\mathbf{F}$ is also introduced in this section. The divergence is also used in a different vector version of Green's Theorem. Make sure that you understand the material in the earlier sections of this chapter before reading through this section.

Key Concept: Curl of a Vector Field

If $\mathbf{F} = P\mathbf{i} + Q\mathbf{j} + R\mathbf{k}$ is a vector field defined in three dimensional space and if the partial derivatives of P, Q, and R all exist, then the curl of $\mathbf{F}$ is defined by

$$\text{curl } \mathbf{F} = \left(\frac{\partial R}{\partial y} - \frac{\partial Q}{\partial z} \right) \mathbf{i} + \left(\frac{\partial P}{\partial z} - \frac{\partial R}{\partial x} \right) \mathbf{j} + \left(\frac{\partial Q}{\partial x} - \frac{\partial P}{\partial y} \right) \mathbf{k}.$$

The third component should look familiar to you from earlier sections. Another way of writing this in a way that is easier to remember is

$$\text{curl } \mathbf{F} = \nabla \times \mathbf{F} = \begin{vmatrix} \mathbf{i} & \mathbf{j} & \mathbf{k} \\ \dfrac{\partial}{\partial x} & \dfrac{\partial}{\partial y} & \dfrac{\partial}{\partial z} \\ P & Q & R \end{vmatrix}$$

There is a close relationship between the concept of a conservative vector field and the concept of curl. If $\mathbf{F}$ is conservative, then

$$\text{curl } \mathbf{F} = 0.$$

A partial converse to this is the following. If $\mathbf{F}$ is a vector field on all of R^3 whose component functions have continuous partial derivatives, and if

$$\text{curl } \mathbf{F} = 0,$$

page 955.

then $\mathbf{F}$ is conservative. Study *Examples 2 and 3* to gain a better understanding of these results.

Key Concept: Divergence of a Vector Field

If $\mathbf{F} = P\mathbf{i} + Q\mathbf{j} + R\mathbf{k}$ is a vector field defined in three dimensional space and if the partial derivatives of P, Q, and R with respect to x, y, and z, respectively, exist, then the divergence of $\mathbf{F}$ is defined by

$$\text{div } \mathbf{F} = \nabla \cdot \mathbf{F} = \frac{\partial P}{\partial x} + \frac{\partial Q}{\partial y} + \frac{\partial R}{\partial z}.$$

The divergence of $\mathbf{F}$ measures the tendency of a fluid whose fluid flow is given by $\mathbf{F}$ to diverge from a point. A relationship between divergence and curl is given by

$$\text{div curl } \mathbf{F} = 0$$

page 957.

if $\mathbf{F}$ has components with continuous second order partial derivatives. Study *Example 5* to see an application of this.

Key Concept: Vector Form of Green's Theorem

The vector form of Green's Theorem is

$$\oint_C \mathbf{F} \cdot d\mathbf{r} = \iint_D (\text{curl } \mathbf{F}) \cdot \mathbf{k} \, dA.$$

A second vector form of Green's Theorem is

$$\oint_C \mathbf{F} \cdot \mathbf{n}\, ds = \iint_D (\operatorname{div} \mathbf{F})\, dA$$

page 959.

where $\mathbf{n}$ is the outward pointing unit normal to C. Study *Figure 2* to gain a better understanding of this. Make sure that you understand the derivation of both of these forms. The first says that the line integral of the tangential component of $\mathbf{F}$ is given by the double integral of the vertical component of the curl. The second says that the line integral of the normal component of $\mathbf{F}$ is given by the double integral of divergence. Both of these forms have physical applications.

SkillMaster 13.10: Compute the curl and divergence of a vector field.

---

page 954, 957.

To compute the curl and divergence of vector fields, compute the correct partial derivatives and use the definitions. Remember that the curl of a vector field is another vector field, whereas the divergence of a vector field is a scalar function. Study *Examples 1 and 4* to see details on the computation of curl and divergence.

SkillMaster 13.11: Use the curl and divergence to solve problems.

---

page 957.

You can use the concepts of curl and divergence to solve other types of problems. *Example 5* shows how to use these concepts to show that a certain vector field is not the curl of another vector field. The Worked Examples below give more applications of these concepts.

Worked Examples

For each of the following examples, first try to find the solution without looking at the middle or right columns. Cover the middle and right columns with a piece of paper. If you need a hint, uncover the middle column. If you need to see the worked solution, uncover the right column.

Example	Tip	Solution

SkillMaster 13.10.

· Compute the curl and divergence of the vector field
$$\mathbf{F} = \ln(y)\mathbf{i} + xyz\mathbf{j} + (y^2 + z^2)\mathbf{k}.$$

Tip: Recall the divergence is $div(\mathbf{F}) = \nabla \cdot \mathbf{F}$ and the curl is $curl(\mathbf{F}) = \nabla \times \mathbf{F}$.

Solution:

$$
\begin{aligned}
div(\mathbf{F}) \\
&= \nabla \cdot \mathbf{F} \\
&= (\frac{\partial}{\partial x}\mathbf{i} + \frac{\partial}{\partial y}\mathbf{j} + \frac{\partial}{\partial z}\mathbf{k}) \\
&\quad \cdot(\ln(y)\mathbf{i} + xyz\mathbf{j} + (y^2 + z^2)\mathbf{k}) \\
&= \frac{\partial(\ln y)}{\partial x} + \frac{\partial(xyz)}{\partial y} + \frac{\partial(y^2 + z^2)}{\partial z} \\
&= 0 + xz + 2z \\
&= z(x + 2).
\end{aligned}
$$

$$
\begin{aligned}
curl(\mathbf{F}) \\
&= \nabla \times \mathbf{F} \\
&= \begin{vmatrix} \mathbf{i} & \mathbf{j} & \mathbf{k} \\ \frac{\partial}{\partial x} & \frac{\partial}{\partial y} & \frac{\partial}{\partial z} \\ \ln y & xyz & y^2 + z^2 \end{vmatrix} \\
&= \frac{\partial(y^2 + z^2)}{\partial y} - \frac{\partial(xyz)}{\partial z}\mathbf{i} \\
&\quad -\frac{\partial(y^2 + z^2)}{\partial x} - \frac{\partial(\ln y)}{\partial z}\mathbf{j} \\
&\quad +\frac{\partial(xyz)}{\partial x} - \frac{\partial(\ln y)}{\partial y}\mathbf{k} \\
&= (2y - xy)\mathbf{i} + 0\mathbf{j} + (yz - 1/y)\mathbf{k} \\
&= (2y - xy)\mathbf{i} + (yz - 1/y)\mathbf{k}
\end{aligned}
$$

· Compute the curl and divergence of the vector field

$$\mathbf{F} = (y^2 + ze^{xz})\mathbf{i} + 2xy\mathbf{j} + xe^{xz}\mathbf{k}$$

The divergence is $div(\mathbf{F}) = \nabla \cdot \mathbf{F}$ and the curl is $curl(\mathbf{F}) = \nabla \times \mathbf{F}$.

$$div(\mathbf{F})$$
$$= \nabla \cdot \mathbf{F}$$
$$= (\frac{\partial}{\partial x}\mathbf{i} + \frac{\partial}{\partial y}\mathbf{j} + \frac{\partial}{\partial z}\mathbf{k})$$
$$\cdot((y^2 + ze^{xz})\mathbf{i} + 2xy\mathbf{j} + xe^{xz}\mathbf{k})$$
$$= \frac{\partial(y^2 + ze^{xz})}{\partial x} + \frac{\partial(2xy)}{\partial y}$$
$$+ \frac{\partial(xe^{xz})}{\partial z}$$
$$= z^2 e^{xz} + 2x + x^2 e^{xz}$$

$$curl(\mathbf{F})$$
$$= \nabla \times \mathbf{F}$$
$$= \begin{vmatrix} \mathbf{i} & \mathbf{j} & \mathbf{k} \\ \frac{\partial}{\partial x} & \frac{\partial}{\partial y} & \frac{\partial}{\partial z} \\ y^2 + ze^{xz} & 2xy & xe^{xz} \end{vmatrix}$$
$$= (\frac{\partial(xe^{xz})}{\partial y} - \frac{\partial(2xy)}{\partial z})\mathbf{i}$$
$$- (\frac{\partial(xe^{xz})}{\partial x} - \frac{\partial(y^2 + ze^{xz})}{\partial z})\mathbf{j}$$
$$+ (\frac{\partial(2xy)}{\partial x} - \frac{\partial(y^2 + ze^{xz})}{\partial y})\mathbf{k}$$
$$= 0\mathbf{i}$$
$$- (e^{xz} + xze^{xz} - e^{xz} - xze^{xz})\mathbf{j}$$
$$+ (2y - 2y)\mathbf{k}$$
$$= 0\mathbf{i} + 0\mathbf{j} + 0\mathbf{k}$$
$$= \mathbf{0}$$

SkillMaster 13.11.

· Show that the vector field $\mathbf{F} = 2xyz\mathbf{i} + (x^2z + y)\mathbf{j} + (x^2y + 3z^2)\mathbf{k}$ is conservative by finding a potential function.

Sometimes a potential function can be found by integrating $f_x = 2xyz$ with respect to x, $f_y = x^2z + y$ with respect to y, $f_z = x^2y + 3z^2$ with respect to x, and comparing the results to guess the potential function.

$$f_x = 2xyz$$
$$\implies f = x^2yz + \varphi(y, z)$$
$$f_y = x^2z + y$$
$$\implies x^2yz + \frac{1}{2}y^2 + \psi(x, z)$$
$$f_z = x^2y + 3z^2$$
$$\implies x^2yz + z^3 + \vartheta(x, y)$$

Putting these together (and including the terms from each equation) gives the guess

$$f = x^2yz + \frac{1}{2}y^2 + z^3.$$

This is a potential function for $\mathbf{F}$ as may be checked by taking the gradient.

· Is there a vector field $\mathbf{F}$ such that $\nabla \times \mathbf{F} = (e^{3x} + xz^2)\mathbf{i} + (x^2y - 3ye^{3x})\mathbf{j} + (zy^2 + 1)\mathbf{k}$?

Recall that every curl is incompressible, that is $\operatorname{div}(\operatorname{curl}(\mathbf{F})) = 0$ or $\nabla \cdot (\nabla \times \mathbf{F}) = 0$. If such a vector field existed then the relation $\nabla \cdot (\nabla \times \mathbf{F}) = 0$ would have to hold.

Check whether

$$0$$
$$= \nabla \cdot (\nabla \times \mathbf{F})$$
$$= \frac{\partial(e^{3x} + xz^2)}{\partial x}$$
$$+ \frac{\partial(x^2y - 3ye^{3x})}{\partial y}$$
$$+ \frac{\partial(zy^2 + 1)}{\partial z}$$
$$= 3e^{3x} + z^2 + x^2 - 3e^{3x} + y^2$$
$$= x^2 + y^2 + z^2.$$

This is a contradiction since

$$0 \neq x^2 + y^2 + z^2,$$

so no such vector field exists.

777

· Consider the linear vector field on $\mathbb{R}^3$ given by $\mathbf{F}$
$= (x + Ay + z)\mathbf{i}$
$+(x + By + Cz)\mathbf{j}$
$+(Dx+y+z)\mathbf{k}$. Find the values of A, B, C, D so that $\mathbf{F}$ is irrotational and incompressible.

First $div(\mathbf{F})$ must equal 0. Then $curl(\mathbf{F}) = \nabla \cdot \mathbf{F}$ must be the zero vector. Check these relations to see what A, B, C, D must be.

$$
\begin{aligned}
\text{div}(\mathbf{F}) \\
= \ & \nabla \cdot \mathbf{F} \\
= \ & \frac{\partial(x + Ay + z)}{\partial z} \\
& +\frac{\partial(x + By + Cz)}{\partial y} \\
& +\frac{\partial(Dx + y + z)}{\partial z} \\
= \ & 1 + B + 1 \\
= \ & 2 + B \\
= \ & 0,
\end{aligned}
$$

so
$$B = -2.$$

To check that $\mathbf{F}$ is irrotational (and using the fact that $B = -2$)

$$
\begin{aligned}
& \text{curl}(\mathbf{F}) \\
= \ & \nabla \cdot \mathbf{F} \\
= \ & \begin{vmatrix} \mathbf{i} & \mathbf{j} & \mathbf{k} \\ \partial x & \partial y & \partial z \\ x + Ay + z & x - 2y + Cz & Dx + y + z \end{vmatrix} \\
= \ & \left(\frac{\partial(Dx + y + z)}{\partial y} - \frac{\partial(x - 2y + Cz)}{\partial z} \right)\mathbf{i} \\
& - \left(\frac{\partial(Dx + y + z)}{\partial x} - \frac{\partial(x + Ay + z)}{\partial z} \right)\mathbf{j} \\
& + \left(\frac{\partial(x - 2y + Cz)}{\partial x} - \frac{\partial(x + Ay + z)}{\partial y} \right)\mathbf{k} \\
= \ & (1 - C)\mathbf{i} + (D - 1)\mathbf{j} + (1 - A)\mathbf{k} \\
= \ & \mathbf{0}
\end{aligned}
$$

Thus $A = C = D = 1$.

· Find the potential function for the vector field in the worked example above.

As before integrate $f_x = x + y + z$ with respect to x, $f_y = x - 2y + z$ with respect to y, and $f_z = x + y + z$ with respect to z. Compare the results and combine to find the potential function.

$$f_x = x + y + z$$
$$\implies f = \frac{1}{2}x^2 + xy + xz + \varphi(y, z)$$
$$f_y = x - 2y + z$$
$$\implies f = xy - y^2 + yz + \psi(x, z)$$
$$f_z = x + y + z$$
$$\implies f = xz + yz + \frac{1}{2}z^2 + \zeta(x, y).$$

Combining all these terms gives

$$f = \frac{1}{2}x^2 + xy + xz - y^2 + yz + \frac{1}{2}z^2 + K.$$

This is a potential function as may be verified by taking the partial derivatives.

Section 13.6 – Surface Integrals

Key Concepts:

- Surface Integrals of a Function over a Surface
- Orientation of a Surface
- Flux of a Vector Field across a Surface

Skills to Master:

- Evaluate surface integrals.
- Evaluate the flux across a surface.

Discussion:

This section introduces the concept of integrals of functions defined on surfaces. The definition of surface integrals follows the now familiar path of taking a limiting value of certain sums. Surface integrals are used extensively in the following two sections of this chapter, so make sure that you understand the material in this section. Now is the time to review the material on parametric surfaces from *Section 10.5* and the material on surface area from *Section 12.6*.

page 734.

page 879.

Key Concept: Surface Integrals of a Function over a Surface

If a surface S is given parametrically by
$$\mathbf{r}(u,v) = x(u,v)\mathbf{i} + y(u,v)\mathbf{j} + z(u,v)\mathbf{k}$$
for (u,v) in a rectangular region D, and if $f(x,y,z)$ is defined on S, the surface

integral of f over S is defined as follows.

$$\iint_S f(x,y,z)dS = \lim_{m,n\to\infty} f(P_{ij}^*)\Delta S_{ij}$$

page 962.

where the rectangle D is divided into subrectangles R_{ij}, where ΔS_{ij} is the surface area of $\mathbf{r}(R_{ij})$, and where P_{ij}^* is a point in $\mathbf{r}(R_{ij})$. See *Figure 1* to gain a geometric understanding of this definition.

If the f and the components of $\mathbf{r}_u$ and $\mathbf{r}_v$ are continuous and if $\mathbf{r}_u$ and $\mathbf{r}_v$ are nonzero and nonparallel in the interior of D it can be shown that

$$\iint_S f(x,y,z)dS = \iint_D f(\mathbf{r}(u,v)) \ |\mathbf{r}_u \times \mathbf{r}_v| \ dA.$$

Think of $|\mathbf{r}_u \times \mathbf{r}_v| \, dA$ as representing an element of surface area, dS. This formula holds even if the region D is not rectangular.

If the surface S is given as the graph of $z = g(x,y)$ over a region D in the xy plane, the formula for the surface integral becomes

$$\iint_S f(x,y,z)dS = \iint_D f(x,y,g(x,y))\sqrt{\left(\frac{\partial z}{\partial x}\right)^2 + \left(\frac{\partial z}{\partial y}\right)^2 + 1}dA.$$

page 964-965.

See *Figures 2 and 3* for examples of such surfaces.

Key Concept: Orientation of a Surface

A surface S in R^3 is orientable if it is two sided. Another way of saying this is that the surface is orientable if it is possible to choose a unit normal $\mathbf{n}$ at each point of the surface so that $\mathbf{n}$ varies continuously over S. See *Figures 4, 5 and 6* to gain a geometric understanding of these concepts. If S is a smooth orientable surface given in parametric form by $\mathbf{r}(u,v)$, the unit normal vector can be chosen to be

page 966.

$$\mathbf{n} = \frac{\mathbf{r}_u \times \mathbf{r}_v}{|\mathbf{r}_u \times \mathbf{r}_v|}.$$

For a closed surface that is the boundary of a solid region in space, the positive orientation for the surface is the one for which the unit normal vectors point outward. See *Figures 8 and 9* for a geometric picture of this concept.

page 967.

Key Concept: Flux of a Vector Field across a Surface

If $\mathbf{F}$ is a continuous vector field defined on an oriented surface S with unit normal vector $\mathbf{n}$, then the surface integral of $\mathbf{F}$ over S in the direction of $\mathbf{n}$ is defined to be

$$\iint_S \mathbf{F} \cdot d\mathbf{S} = \iint_S \mathbf{F} \cdot \mathbf{n} \, dS.$$

This is also called the flux of $\mathbf{F}$ across S in the direction of $\mathbf{n}$. If S is given by $\mathbf{r}(u, v)$ for (u, v) in a region D, this can be computed by

$$\iint_S \mathbf{F} \cdot d\mathbf{S} = \iint_D \mathbf{F} \cdot (\mathbf{r}_u \times \mathbf{r}_v) \, dA.$$

If S is given by a graph $z = g(x, y)$, then

$$\iint_S \mathbf{F} \cdot d\mathbf{S} = \iint_D \left(-P\frac{\partial g}{\partial x} - Q\frac{\partial g}{\partial y} + R \right) dA.$$

 page 968.

See *Example 4* for a computation of flux.

SkillMaster 13.12: Evaluate surface integrals.

 page 963-965.

To evaluate surface integrals of functions f on a surface S, first decide whether you are dealing with a parametric surface or a surface given as the graph of a function. Then use the appropriate form of the surface integral from the Key Concept above. Study *Examples 1, 2, and 3* to see computations of surface integrals.

SkillMaster 13.13: Evaluate the flux across a surface.

To evaluate the flux of a vector field $\mathbf{F}$ across an orientable surface S in the direction of a unit normal $\mathbf{n}$, use the formula

$$\iint_S \mathbf{F} \cdot d\mathbf{S} = \iint_D \mathbf{F} \cdot (\mathbf{r}_u \times \mathbf{r}_v) \, dA.$$

 page 969-971.

First compute $\mathbf{r}_u$ and $\mathbf{r}_v$, then take the cross product. Take the dot product of the result with $\mathbf{F}$. Finally, carry out the integration. Study *Examples 5 and 6* to see computations of flux.

Worked Examples

For each of the following examples, first try to find the solution without looking at the middle or right columns. Cover the middle and right columns with a piece of paper. If you need a hint, uncover the middle column. If you need to see the worked solution, uncover the right column.

Example	Tip	Solution

SkillMaster 13.12.

· Evaluate $\int\int_S f(x,y,z)dS$ where S is the surface of the paraboloid above the x-axis, $\{z = 4 - x^2 - y^2, z \geq 0\}$ and $f(x,y,z) = x^2 + 2y^2 + z - 4$.

Tip: First note the surface is given by a graph $z = g(x,y) = 4 - x^2 - y^2$ over the disk D in the x,y-plane with the origin as center and with radius 2. $D = \{x^2 + y^2 \leq 4, z = 0\}$.
Use the formula
$\int\int_S f(x,y,z)dS = \int\int_D f(x,y,g(x,y))\sqrt{(\frac{\partial g}{\partial x})^2 + (\frac{\partial g}{\partial y})^2 + 1}\,dA$.

Solution:
$$f(x,y,g(x,y))$$
$$= x^2 + 2y^2 + (4 - x^2 - y^2) - 4$$
$$= y^2$$
$$\int\int_S f(x,y,z)dS =$$
$$\int\int_D f(x,y,g(x,y))\sqrt{(\frac{\partial g}{\partial x})^2 + (\frac{\partial g}{\partial y})^2 + 1}\,dA$$
$$= \int\int_D y^2\sqrt{(2x)^2 + (2y)^2 + 1}\,dA$$
$$= \int\int_D y^2\sqrt{4x^2 + 4y^2 + 1}\,dA$$

Tip: Switch to polar coordinates.
$x^2 + y^2 = r^2$
$y = r\sin\theta$
$D = \{0 \leq r \leq 2, 0 \leq \theta \leq 2\pi\}$

Solution:
$$= \int_0^{2\pi}\int_0^2 r^2\sin^2\theta\sqrt{4r^2 + 1}\,r\,dr\,d\theta$$
$$= \int_0^{2\pi}\sin^2\theta d\theta \int_0^2 r^3\sqrt{4r^2 + 1}\,dr$$
$$= \frac{1}{2}\int_0^{2\pi} 1 - \cos 2\theta d\theta \int_0^2 r^3\sqrt{4r^2 + 1}\,dr$$
$$= \pi\int_0^2 r^3\sqrt{4r^2 + 1}\,dr$$

Use the substitution
$u^2 = 4r^2 + 1$
$r^2 = (u^2 - 1)/4$
$udu = 4rdr$
$r = 0 \Longrightarrow u = 1$
$r = 2 \Longrightarrow u = \sqrt{17}$

$$= \frac{\pi}{4}\int_1^{\sqrt{17}} \frac{1}{4}(u^2-1)\sqrt{u^2}du$$
$$= \frac{\pi}{16}\int_1^{\sqrt{17}} u^3 - udu$$
$$= \frac{\pi}{16}\left(\frac{u^4}{4} - \frac{u^2}{2}\right)\Big|_1^{\sqrt{17}}$$
$$= \frac{\pi}{16}[(\frac{17^2}{4} - \frac{17}{2}) - (\frac{1}{4} - \frac{1}{2})]$$
$$= \frac{\pi}{64}[17^2 - 34 - 1 + 2]$$
$$= 4\pi.$$

· Evaluate $\int\int_S f(x,y,z)dS$ where S is the outer surface of the cylinder $S = \{x^2 + y^2 = 4, 0 \le z \le 3\}$ and $f(x,y,z) = 6 - x^2 - y^2$.

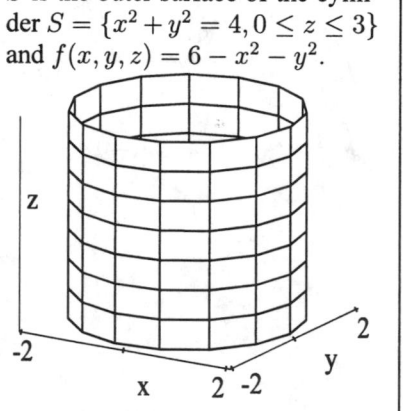

First evaluate the function and notice that it is constant on S. You do not have to do any integration to solve this problem.

On S the function $x^2 + y^2$ is the constant 4.

$$\int\int_S f(x,y,z)dS$$
$$= \int\int_S 6 - x^2 - y^2 dS$$
$$= \int\int_S 6 - 4 dS$$
$$= 2\int\int_S dS$$
$$= 2\,\text{area}(S)$$
$$= 2(2\pi)(\text{radius})(\text{height})$$
$$= 4\pi(2)(3)$$
$$= 24\pi.$$

· Evaluate $\int\int_S f(x,y,z)dS$ where S is parametrized by $\mathbf{r}(u,v) = v\mathbf{i} + uv\mathbf{j} + u\mathbf{k}$, $0 \le u, v \le 1$ and where $f(x,y,z) = xz$.

Use the formula
$\int\int_S f(x,y,z)dS =$
$\int_0^1 \int_0^1 f(\mathbf{r}(u,v))$
$|\mathbf{r}_u \times \mathbf{r}_v| \, dudv$

$$\mathbf{r}_u = v\mathbf{j} + \mathbf{k}$$
$$\mathbf{r}_v = \mathbf{i} + u\mathbf{j}$$

$$|\mathbf{r}_u \times \mathbf{r}_v|$$
$$= \begin{vmatrix} \mathbf{i} & \mathbf{j} & \mathbf{k} \\ 0 & v & 1 \\ 1 & u & 0 \end{vmatrix}$$
$$= |-u\mathbf{i} + \mathbf{j} - v\mathbf{k}|$$
$$= \sqrt{u^2 + v^2 + 1}$$
$$f(\mathbf{r}(u,v))$$
$$= uv$$

$$\int\int_S f(x,y,z)dS$$
$$= \int_0^1 \int_0^1 f(\mathbf{r}(u,v)) |\mathbf{r}_u \times \mathbf{r}_v| \, dudv$$
$$= \int_0^1 \int_0^1 uv\sqrt{u^2 + v^2 + 1} \, dudv$$
$$= \int_0^1 v \left(\frac{1}{3} \left(u^2 + v^2 + 1 \right)^{3/2} \right) \Big|_{u=0}^{u=1} dv$$
$$= \int_0^1 v\frac{1}{3} \left(\left(v^2 + 2 \right)^{3/2} - \left(v^2 + 1 \right) \right) dv$$
$$= \frac{1}{3} \int_0^1 v \left(v^2 + 2 \right)^{3/2} dv$$
$$\quad - \frac{1}{3} \int_0^1 v \left(v^2 + 1 \right)^{3/2} dv$$
$$= \frac{1}{15} \left(\left(v^2 + 2 \right)^{5/2} \Big|_0^1 \right)$$
$$\quad - \frac{1}{15} \left(\left(v^2 + 1 \right)^{5/2} \Big|_0^1 \right)$$
$$= \frac{1}{15} \left[3^{5/2} - 2(2)^{5/2} + 1 \right]$$
$$\approx 0.3516.$$

SkillMaster 13.13.

· Calculate the flux of the vector field $\mathbf{F} = <xy, \sin x, \cos y>$ upward across the surface S given by the function
$z = g(x, y)$
$= x^2 + y, \ 0 \leq x, y \leq \pi$.

This is the surface integral
$\int \int_S \mathbf{F} \cdot d\mathbf{S}$.

$$\int \int_S \mathbf{F} \cdot d\mathbf{S}$$

$$= \int_0^\pi \int_0^\pi \left(-xy\frac{\partial g}{\partial x} - \sin x\frac{\partial g}{\partial y} + \cos y\right)dx\,dy$$

$$= \int_0^\pi \int_0^\pi \left(-xy(2x) - \sin x(1) + \cos y\right)dx\,dy$$

$$= \int_0^\pi \int_0^\pi \left(-2x^2 y - \sin x + \cos y\right)dx\,dy$$

$$= \int_0^\pi \frac{-2\pi^3}{3}y + 2 + \pi\cos y\,dy$$

$$= \int_0^\pi \frac{-2\pi^3}{3}y + 2 + \pi\cos y\,dy$$

$$= \frac{-\pi^3}{3}\pi^2 + 2\pi + \pi(0)$$

$$= \frac{-\pi^5}{3} + 2\pi$$

· Evaluate the flux of the vector field

$$\mathbf{F}(x, y, z) = \frac{x}{z}\mathbf{i} - \frac{y}{z}\mathbf{j}$$

across the surface S parametrized by

$$\mathbf{r} = ve^u\mathbf{i} + ue^v\mathbf{j} + uv\mathbf{k}$$

for $2 \leq u, v \leq 5$.

The formula for the flux is $\int_2^5 \int_2^5 \mathbf{F}(\mathbf{r}) \cdot (\mathbf{r}_u \times \mathbf{r}_v)dA$.

First calculate the constituents of this integral.

$$\mathbf{F}(\mathbf{r}) = \frac{ve^u}{uv}\mathbf{i} - \frac{ue^v}{uv}\mathbf{j};$$
$$\mathbf{r}_u = ve^u\mathbf{i} + e^v\mathbf{j} + v\mathbf{k};$$
$$\mathbf{r}_v = e^u\mathbf{i} + ue^v\mathbf{j} + u\mathbf{k}.$$

$$\mathbf{r}_u \times \mathbf{r}_v$$
$$= \begin{vmatrix} \mathbf{i} & \mathbf{j} & \mathbf{k} \\ ve^u & e^v & v \\ e^u & ue^v & u \end{vmatrix}$$
$$= (ue^v - uve^v)\mathbf{i}$$
$$\quad -(ve^u - uve^u)\mathbf{j}$$
$$\quad +(uve^{u+v} - e^{u+v})\mathbf{k}$$

$$\mathbf{F}(\mathbf{r}) \cdot (\mathbf{r}_u \times \mathbf{r}_v)$$
$$= \frac{e^u}{u}(ue^v - uve^v)$$
$$\quad - \frac{e^v}{v}(ve^u - uve^u)$$
$$= e^{u+v} - ve^{u+v} - e^{u+v} + ue^{u+v}$$
$$= (u - v)e^{u+v}$$

The flux integral is

$$\int_2^5 \int_2^5 \mathbf{F}(\mathbf{r}) \cdot (\mathbf{r}_u \times \mathbf{r}_v)dA$$
$$= \int_2^5 \int_2^5 (u - v)e^{u+v}dudv$$
$$= \int_2^5 \int_2^5 ue^{u+v}dudv$$
$$\quad - \int_2^5 \int_2^5 ve^{u+v}dudv$$
$$= 0,$$

by symmetry.

· Evaluate the flux of the vector field
$$\mathbf{F}(x, y, z) = \frac{1}{x}\mathbf{i} - \frac{1}{y}\mathbf{j} + z\mathbf{k} \text{ across}$$
the surface S, the unit sphere.

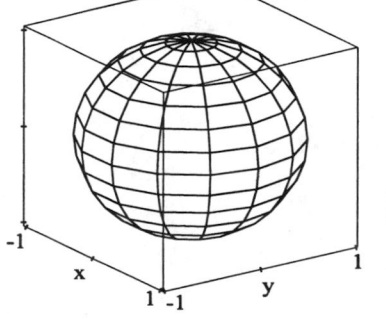

Parametrize the unit sphere is standard spherical coordinates.
$$\mathbf{r} = \sin\phi\cos\theta\mathbf{i}$$
$$+ \sin\phi\sin\theta\mathbf{j}$$
$$+ \cos\phi\mathbf{k}.$$
The cross product is (see pg 882)
$$\mathbf{r}_\phi \times \mathbf{r}_\theta =$$
$$\sin^2\phi\cos\theta\mathbf{i}$$
$$+ \sin^2\phi\sin\theta\mathbf{j}$$
$$+ \sin\phi\cos\theta\mathbf{k}.$$

$$\mathbf{F}(\mathbf{r})$$
$$= \frac{1}{\sin\phi\cos\theta}\mathbf{i}$$
$$+ \frac{1}{\sin\phi\cos\theta}\mathbf{j}$$
$$+ \cos\phi\mathbf{k}.$$
$$\mathbf{F}(\mathbf{r}) \cdot (\mathbf{r}_\phi \times \mathbf{r}_\theta)$$
$$= \sin\phi + \sin\phi + \sin\phi\cos\phi\cos\theta$$
$$= 2\sin\phi + \frac{1}{2}\sin 2\phi$$
$$\iint_S \mathbf{F}{\cdot}dS$$
$$= \iint_D \mathbf{F}(\mathbf{r}) \cdot (\mathbf{r}_\phi \times \mathbf{r}_\theta)dA$$
$$= \int_0^\pi \int_0^{2\pi} 2\sin\phi + \frac{1}{2}\sin 2\phi \, d\theta \, d\phi$$
$$= \int_0^\pi 2\sin\phi \, d\phi \int_0^{2\pi} d\theta$$
$$+ \frac{1}{2}\int_0^\pi \sin 2\phi \, d\phi \int_0^{2\pi} d\theta$$
$$= 4\pi \int_0^\pi \sin\phi \, d\phi$$
$$+ \pi(0)$$
$$= 4\pi \left(-\cos\phi\big|_0^\pi\right)$$
$$= 8\pi.$$

Section 13.7 – Stokes' Theorem

Key Concepts:

- Stokes' Theorem, Surface Integrals, and Line Integrals

Skills to Master:

- Use Stokes' Theorem to compute surface integrals.
- Use Stokes' Theorem to compute line integrals of a closed curve.

Discussion:

page 646.

Stokes' Theorem can be viewed as a higher dimensional analogue of Green's Theorem. Stokes' theorem relates the line integral of the tangential component of a vector field around the boundary of a surface in space to a certain surface integral over the surface. Make sure that you completely understand the previous material on line integrals, surface integrals and flux before reading this section. Pay careful attention to the physical interpretation of curl given at the end of this section.

Key Concept: Stokes' Theorem, Surface Integrals, and Line Integrals

---

page 973.

If S is an oriented piecewise smooth surface bounded by a simple closed, piecewise smooth curve C, a positive orientation of the boundary is the orientation such that if you walk around the boundary with your head pointing in the direction of n, then the surface is on your left. See *Figure 1* for a geometric picture of this. In this case, if F is a vector field with components that have continuous partial derivatives on a region of space containing S, Stokes' theorem states that

$$\int_C \mathbf{F} \cdot d\mathbf{r} = \iint_S \text{curl } \mathbf{F} \cdot d\mathbf{S}.$$

This is a generalization of one of the vector forms of Green's theorem. Recall that

$$\int_C \mathbf{F} \cdot d\mathbf{r} = \int_C \mathbf{F} \cdot \mathbf{T} ds.$$

That is, this integral is the integral of the tangential component of $\mathbf{F}$. Also recall that

$$\iint_S \text{curl } \mathbf{F} \cdot d\mathbf{S} = \iint_S \text{curl } \mathbf{F} \cdot \mathbf{n} \, dS.$$

That is, this integral is the integral of the flux of the curl of $\mathbf{F}$ across the surface. Note that this theorem relates an integral over a region with another integral over the boundary of this region just as Green's Theorem does. Study the explanation of the *special case of Stokes' Theorem* to gain a better understanding of the theorem.

page 974.

page 976.

If $\mathbf{F}$ represents velocity $\mathbf{v}$ of a fluid moving through space, *Figures 5* shows how to interpret

$$\int_C \mathbf{v} \cdot d\mathbf{r}$$

page 977.

as the tendency of the fluid to move around C. *Figure 6* and the accompanying discussion gives a physical interpretation for curl.

SkillMaster 13.14: Use Stokes' Theorem to compute surface integrals.

You can use Stokes' theorem to compute surface integrals of the curl of a vector field $\mathbf{F}$ by replacing the surface integral with a line integral. Since Stokes' theorem also implies that the surface integral over two surfaces with the same boundary will be the same, you can replace a given surface by a simpler surface with the same boundary before computing the integral. See *Example 2* for details on how this works.

page 975.

SkillMaster 13.15: Use Stokes' Theorem to compute line integrals of a closed curve.

page 975.

You can use Stokes' theorem to compute the line integral around a closed curve of the tangential component of a vector field $\mathbf{F}$ by replacing the integral by the surface integral of the flux of the curl of $\mathbf{F}$ across a surface with boundary C. See *Example 1* for details on how this works.

Worked Examples

For each of the following examples, first try to find the solution without looking at the middle or right columns. Cover the middle and right columns with a piece of paper. If you need a hint, uncover the middle column. If you need to see the worked solution, uncover the right column.

Example	Tip	Solution

SkillMaster 13.14.

· Compute $\int\int_S \text{curl}(\mathbf{F}) \cdot d\mathbf{S}$ where S is the surface defined by $S = \{z = (x^2 + y^2)^{1/3}, 0 \le z \le 2\}$ and $\mathbf{F} = y\mathbf{i} + y\mathbf{j} + z\mathbf{k}$.

Tip: Stokes' Theorem says $\int\int_S\text{curl}(\mathbf{F}) \cdot d\mathbf{S} = \int_C \mathbf{F} \cdot d\mathbf{r}$. In this case C is the circle of radius $\sqrt{8}$ with center on the $z-$axis and lying on the plane $z = 2$. We parametrize this equation by $\mathbf{r}(t) = \sqrt{8}\cos t\mathbf{i} + \sqrt{8}\sin t\mathbf{j} + 2\mathbf{k}$, $0 \le t \le 2\pi$.

Solution:

$$\mathbf{r}'(t) = -\sqrt{8}\sin t\mathbf{i} + \sqrt{8}\cos t\mathbf{j}.$$

$$\mathbf{F}(\mathbf{r}(t))$$
$$= \sqrt{8}\sin t\mathbf{i} + \sqrt{8}\sin t\mathbf{j}+2\mathbf{k}.$$

$$\mathbf{F}(\mathbf{r}(t)) \cdot \mathbf{r}'(t)$$
$$= -8\sin^2 t + 8\cos t\sin t$$

$$\int\int_S \text{curl}(\mathbf{F}) \cdot d\mathbf{S}$$

$$= \int_C \mathbf{F} \cdot d\mathbf{r}$$

$$= \int_C \mathbf{F}(\mathbf{r}(t)) \cdot \mathbf{r}'(t)dt$$

$$= \int_0^{2\pi} -8\sin^2 t + 8\cos t\sin t dt$$

$$= \left(-4t + 2\sin 2t + 4\sin^2 t\right|_0^{2\pi}$$

$$= -8\pi$$

· Compute $\int\int_S \text{curl}(\mathbf{F}) \cdot d\mathbf{S}$ where $S = \{x^2 + y^2 + z^2, 0 \le z\}$ is the upper half of the unit sphere and $\mathbf{F} = (e^x - 1/y)\mathbf{i} + e^y\mathbf{j} + e^{x+y+z}\mathbf{k}$.

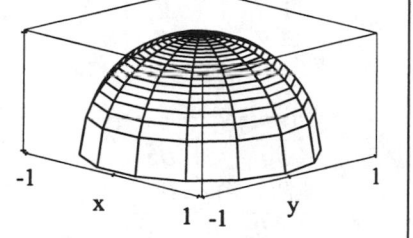

As before Stokes' Theorem says
$\int\int_S \text{curl}(\mathbf{F}) \cdot d\mathbf{S}$
$= \int_C \mathbf{F} \cdot d\mathbf{r}$.
Here C is the unit circle parametrized by
$\mathbf{r}(t) = \cos t + \sin t$,
$0 \le t \le 2\pi$.

$\mathbf{r}'(t) = -\sin t\mathbf{i} + \cos t\mathbf{j}$

$\mathbf{F}(\mathbf{r}(t))$
$= (e^{\cos t} - 1/\sin t)\mathbf{i}$
$+ e^{\sin t}\mathbf{j}$
$+ e^{\cos t + \sin t}\mathbf{k}$

$\mathbf{F}(\mathbf{r}(t)) \cdot \mathbf{r}'(t)$
$= -\sin t e^{\cos t} + 1 + \cos t e^{\sin t}$

$\int\int_S \text{curl}(\mathbf{F}) \cdot d\mathbf{S}$

$= \int_C \mathbf{F} \cdot d\mathbf{r}$

$= \int_C \mathbf{F}(\mathbf{r}(t)) \cdot \mathbf{r}'(t)dt$

$= \int_0^{2\pi} -\sin t e^{\cos t} + 1 + \cos t e^{\sin t}dt$

$= (e^{\cos t} + t + e^{\sin t}|_0^{2\pi}$

$= 2\pi$

793

SkillMaster 13.15.

· Compute $\int_C \mathbf{F} \cdot d\mathbf{r}$ where C is the boundary of the rectangular surface S with corner points $(0,0,1)$, $(0,1,1)$, $(1,0,2)$, and $(1,1,2)$ and $\mathbf{F} = xy^2\mathbf{i} + yz^2\mathbf{j} + zx^2\mathbf{k}$.

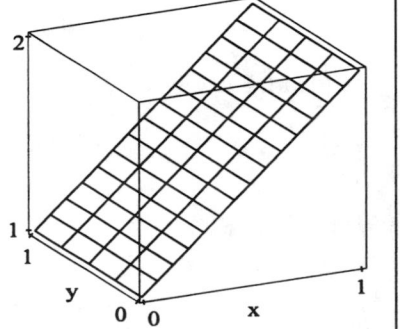

Stokes' Theorem says
$$\int_C \mathbf{F} \cdot d\mathbf{r} = \int \int_S \text{curl}(\mathbf{F}) \cdot d\mathbf{S}$$

The rectangular surface may be described by the function
$$g(x,y) = x + 1$$
with domain D the unit square,
$$D = \{0 \le x, y \le 1\}.$$

First compute the curl($\mathbf{F}$).

curl($\mathbf{F}$)

$$= \begin{vmatrix} \mathbf{i} & \mathbf{j} & \mathbf{k} \\ \frac{\partial}{\partial x} & \frac{\partial}{\partial y} & \frac{\partial}{\partial z} \\ xy^2 & yz^2 & zx^2 \end{vmatrix}$$

$$= (\frac{\partial(zx^2)}{\partial y} - \frac{\partial(yz^2)}{\partial z})\mathbf{i}$$
$$- (\frac{\partial(zx^2)}{\partial x} - \frac{\partial(xy^2)}{\partial z})\mathbf{j}$$
$$+ (\frac{\partial(yz^2)}{\partial x} - \frac{\partial(xy^2)}{\partial y})\mathbf{k}$$

$$= -2yz\mathbf{i} - 2xz\mathbf{j} - 2xy\mathbf{k}$$

$$\int_C \mathbf{F} \cdot d\mathbf{r}$$

$$= \int \int_S \text{curl}(\mathbf{F}) \cdot d\mathbf{S}$$

$$= \int \int_D \text{curl}(F) \cdot \left\langle -\frac{\partial g}{\partial x}, -\frac{\partial g}{\partial y}, 1 \right\rangle dA$$

$$= \int \int_D 2yz\frac{\partial g}{\partial x} + 2xz\frac{\partial g}{\partial y} - 2xy\, dx\, dy$$

$$= \int_0^1 \int_0^1 2xz + 0 - 2xy\, dx\, dy$$

$$= \int_0^1 \int_0^1 2xg(x,y) - 2xy\, dx\, dy$$

$$= \int_0^1 \int_0^1 2x(x+1) - 2xy\, dx\, dy$$

$$= \int_0^1 \int_0^1 2x^2 + 2x - 2xy\, dx\, dy$$

$$= \int_0^1 \frac{2}{3} + 1 - y\, dy = \frac{2}{3} + 1 - \frac{1}{2}$$

$$= \frac{7}{6}$$

Section 13.8 – The Divergence Theorem

Key Concepts:

- The Divergence Theorem and the Flux across a Closed Surface

Skills to Master:

- Use the Divergence Theorem to compute the flux across a closed surface.

Discussion:

Congratulations! You are nearly at the end of the text and have mastered a lot of material with many applications. Just as Stokes' theorem was a generalization of one of the vector forms of Green's Theorem, the Divergence Theorem is a generalization of the other vector form of Green's Theorem. Review these versions of Green's Theorem now if you need to.

Key Concept: The Divergence Theorem and the Flux across a Closed Surface

One of the vector forms of Green's Theorem from Section 13.5 was

$$\int_C \mathbf{F} \cdot \mathbf{n} \, ds = \iint_D div\mathbf{F}(x,y)dA.$$

A region E in space that is simultaneously a type 1, type 2 and type 3 region is called a simple solid region. The Divergence Theorem is true for more complicated regions, but is just stated for simple solid regions in this section. Let S be the boundary of such a region with positive outward orientation. Let $\mathbf{F}$ be a vector field whose component functions have continuous partial derivatives on an open region containing E. The the Divergence Theorem states that

$$\iint_S \mathbf{F} \cdot d\mathbf{S} = \iiint_E \text{div } \mathbf{F}(x,y,z) \, dV.$$

795

page 980.

That is, the flux of **F** through the surface S is equal to the triple integral of divergence. Study *Figure 1* and the accompanying explanation to gain a better understanding of why this theorem is true.

page 983.

The Divergence Theorem can be used to give a physical interpretation of the divergence of a fluid flow just as Stokes' theorem was used to give a physical interpretation of the curl of a fluid flow. See the *discussion* in the text.

SkillMaster 13.16: Use the Divergence Theorem to compute the flux across a closed surface.

---

page 981.

To use the Divergence Theorem to compute the flux across a closed surface, replace the surface integral that gives flux by the triple integral of divergence. See *Example 1* for details on how to so this.

Worked Examples

For each of the following examples, first try to find the solution without looking at the middle or right columns. Cover the middle and right columns with a piece of paper. If you need a hint, uncover the middle column. If you need to see the worked solution, uncover the right column.

Example	Tip	Solution

SkillMaster 13.16.

· Use the Divergence Theorem to calculate the flux of $\mathbf{F} = 2x^2y\mathbf{i} + 6y^2z\mathbf{j} + 2xz\mathbf{k}$ over the surface bounded by the unit cube $B = \{0 \le x, y, z \le 1\}$

The Divergence Theorem says the flux is the integral $\iint_S \mathbf{F} \cdot d\mathbf{S}$ $= \iiint_B \text{div } \mathbf{F} \, dV$ $= \int_0^1 \int_0^1 \int_0^1 \text{div } \mathbf{F} \, dxdydz$.

$$\text{div } \mathbf{F}$$
$$= \frac{\partial(2x^2y)}{\partial x} + \frac{\partial(6y^2z)}{\partial y} + \frac{\partial(2xz)}{\partial z}$$
$$= 4xy + 12yz + x$$

$$\iint_S \mathbf{F} \cdot d\mathbf{S}$$
$$= \iiint_B \text{div } \mathbf{F} dV$$
$$= \int_0^1 \int_0^1 \int_0^1 \text{div } \mathbf{F} dxdydz$$
$$= \int_0^1 \int_0^1 (4xy + 12yz + 2x)dxdydz$$
$$= \int_0^1 \int_0^1 2y + 12yz + 1 dydz$$
$$= \int_0^1 1 + 6z + 1 dz$$
$$= 2 + 3$$
$$= 5$$

· Evaluate $\int\int_S \mathbf{F} \cdot d\mathbf{S}$ where $\mathbf{F}(x,y,z) = (x+y)\mathbf{i}+(y+z)\mathbf{j}+(z+x)\mathbf{k}$ and S is the surface of a cone B with height 6 and radius 4.

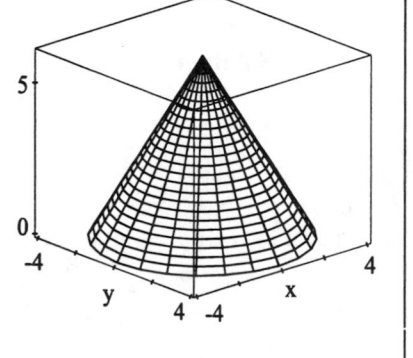

Use the Divergence Theorem.

div $\mathbf{F} =$
$\frac{\partial(x+y)}{\partial x} + \frac{\partial(y+z)}{\partial y} + \frac{\partial(z+x)}{\partial z}$
$= 1 + 1 + 1 = 3.$

$$\int\int_S \mathbf{F} \cdot d\mathbf{S}$$
$$= \int\int\int_B \text{div } \mathbf{F} dV$$
$$= \int\int\int_B 3 dV$$
$$= 3 \text{ vol}(B)$$
$$= 3(\frac{\pi}{3}4^2 6)$$
$$= 96\pi$$

· Evaluate $\int\int_S \mathbf{F} \cdot d\mathbf{S}$ where $\mathbf{F}(x,y,z) = x^2\mathbf{i}+y\mathbf{j}+z^2\mathbf{k}$ and S is the surface of the solid B bounded by the paraboloid $9 - x^2 - y^2$ and the xy−plane.

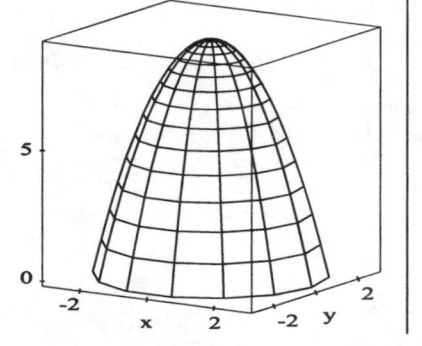

Use the Divergence Theorem.

div $\mathbf{F} =$
$\frac{\partial(x^2)}{\partial x} + \frac{\partial(y)}{\partial y} + \frac{\partial(z^2)}{\partial z}$
$= 2x + 1 + 2z.$

$$\int\int_S \mathbf{F} \cdot d\mathbf{S}$$
$$= \int\int\int_B \text{div } \mathbf{F} dV$$
$$= \int\int\int_B (2x + 1 + 2z) dV$$

Switch to cylindrical coordinates.	$\displaystyle = \int_0^{2\pi} \int_0^3 \int_0^{9-r^2} (2r\cos\theta + 1 + 2z) dz\,dr\,d\theta$

$$= \int_0^{2\pi} \int_0^3$$
$$2r\cos\theta(9-r^2) + (9-r^2) + (9-r^2)^2 dr\,d\theta$$

$$= \int_0^{2\pi} \cos\theta\, d\theta \int_0^3 2r(9-r^2)dr$$

$$+ \int_0^{2\pi} d\theta \int_0^3 \left((9-r^2) + (9-r^2)^2 \right) dr$$

$$= 0 + 2\pi \int_0^3 \left((9-r^2) + (9-r^2)^2 \right) dr$$

$$= 2\pi \int_0^3 90 - 19r^2 + r^4 dr$$

$$= 2\pi \left(90(3) - \frac{19}{3}3^3 + \frac{1}{5}3^5 \right)$$

$$= \frac{2\pi}{15} \left(15(90)(3) - 19(3^3)5 + 3^6 \right)$$

$$= 568.8\pi$$

$$\approx 1786.94.$$

SkillMasters for Chapter 13

SkillMaster 13.1:   Sketch and interpret vector fields.

SkillMaster 13.2:   Compute gradient fields.

SkillMaster 13.3:   Compute line integrals.

SkillMaster 13.4:   Compute line integrals of vector fields.

SkillMaster 13.5:   Determine if vector fields are conservative.

SkillMaster 13.6:   Find the potential function for a conservative vector field.

SkillMaster 13.7:   Evaluate line integrals of conservative vector fields.

SkillMaster 13.8:   Use Green's Theorem to evaluate line integrals over a closed curve.

SkillMaster 13.9:   Use Green's Theorem to evaluate double integrals and to find area.

SkillMaster 13.10:   Compute the curl and divergence of a vector field.

SkillMaster 13.11:   Use the curl and divergence to solve problems.

SkillMaster 13.12:   Evaluate surface integrals.

SkillMaster 13.13:   Evaluate the flux across a surface.

SkillMaster 13.14:   Use Stokes' Theorem to compute surface integrals.

SkillMaster 13.15:   Use Stokes' Theorem to compute line integrals of a closed curve.

SkillMaster 13.16:   Use the Divergence Theorem to compute the flux across a closed surface.